Advanced X-ray Imaging of Electrochemical Energy Materials and Devices

Jiajun Wang
Editor

Advanced X-ray Imaging of Electrochemical Energy Materials and Devices

Springer

Editor
Jiajun Wang
Chemical Engineering Institute
Harbin Institute of Technology
Harbin, China

ISBN 978-981-16-5327-8 ISBN 978-981-16-5328-5 (eBook)
https://doi.org/10.1007/978-981-16-5328-5

This Springer imprint is published by the registered company Springer Nature Singapore Pte Ltd.
The registered company address is: 152 Beach Road, #21-01/04 Gateway East, Singapore 189721, Singapore

About This Book

With the growing need for sustainable energy technologies, advanced characterization methods become more and more critical for optimizing the energy materials and understanding their operation mechanisms. In this book, we focus on the synchrotron-based X-ray imaging technologies and the associated applications in gaining fundamental insights into the physical/chemical properties and reaction mechanisms of energy materials. We will discuss a few major X-ray imaging technologies, including X-ray projection imaging, transmission X-ray microscopy, scanning transmission X-ray microscopy, tender and soft X-ray imaging, and coherent diffraction imaging. Researchers can choose from various X-ray imaging techniques with different working principles based on research goals and sample specifications. With the X-ray imaging techniques, we can obtain the morphology, phase, lattice, and strain information of energy materials in both 2D and 3D in an intuitive way. In addition, due to the high penetration of X-rays, operando/in situ experiments can be designed to track the qualitative and quantitative changes of the samples during operation. We hope this book can broader reader's view on X-ray imaging techniques and inspire new ideas and possibilities in energy materials research.

Contents

About the Editor

Jiajun Wang obtained his doctorate from Harbin Institute of Technology in 2008. He is currently a professor at Harbin Institute of Technology (HIT), one of the Thousand Youth Talents in China. Prior to joining to HIT, he was a synchrotron beamline scientist at National Synchrotron Light Source, Brookhaven National Laboratory, and then Advanced Photon Source, Argonne National Laboratory, to develop in situ/in operando synchrotron X-ray imaging techniques for energy material research. He has published over 80 papers with over 8000 citations. His research activities include synchrotron X-ray technologies, phase transition, electrochemistry, engineering materials, nanoparticles and nanocomposites, energy storage and conversion materials.

Chapter 1
A Category of Synchrotron X-ray Imaging Methods

Shuaifeng Lou, Fang Zhang, Han Wang, and Jiajun Wang

1.1 X-ray Projection Imaging

Roentgen discovered X-rays in 1895 [1]. In 1896, it was used in medical imaging, taking X-ray pictures of women's fingers. Subsequently, the researchers established a projection imaging model based on the particle nature of the X-ray light source. At present, this technology has become a necessary means for clinical medical diagnosis, industrial flaw detection, and quality inspection [2]. To obtain a clear image of the structure of light-element objects, X-ray phase-contrast imaging has been developed. In 1971, X-ray interferometry imaging technology was discovered by using the interference between the reference light and the projected light of the object [3]. In 1997, Chapman proposed the diffraction-enhanced imaging method [4], and in 2003, the grating shear imaging technology was discovered [5]. This method takes advantage of the phase shift of propagating X-rays through the sample. In the past ten years, with the development of X-ray source and detector technology, X-ray imaging technology has become increasingly mature, providing higher spatial and temporal resolution; therefore, X-ray imaging is used in a series of energy conversion and storage materials has been applied.

X-ray projection imaging is mainly composed of X-ray source, sample stage, X-ray area detector, and CCD or scintillator coupled. The composition of the CMOS detector is shown in Fig. 1.1 [6]. In the space very close to the sample, the light travels almost in a straight line, so the geometric projection is displayed on the light screen. X-ray imaging in this area is projection imaging. The spatial resolution mainly depends on the pixel size of the detector (from 1 to 10 μm),

S. Lou (✉) · F. Zhang · H. Wang · J. Wang (✉)
School of Chemistry and Chemical Engineering, Harbin Institute of Technology, Harbin 150001, China
e-mail: shuaifeng.lou@hit.edu.cn

J. Wang
e-mail: jiajunhit@hit.edu.cn

J. Wang (ed.), *Advanced X-ray Imaging of Electrochemical Energy Materials and Devices*, https://doi.org/10.1007/978-981-16-5328-5_1

three-dimensional reconstruction can be performed when rotating the sample. Different from other X-ray imaging systems, the X-ray projection imaging system has no other X-ray optics behind the sample stage. X-rays pass through the sample and then directly project onto the detector. Imaging relies on the different contrast of the sample itself to resolve its structure. Absorption X-ray imaging relies on the attenuation of the incident X-ray beam to form a projected image on the detector. Generally, X-rays pass through light-element objects with little attenuation. With the emerging phase-contrast technique, X-ray microtomography can enhance the contrast between different weakly attenuating materials such as carbon and lithium, though it usually needs more complicated setups and is typically more prone to different types of artifacts.

In terms of application, Shearing et al. [7] obtained the first high-resolution image of the electrode structure of a lithium-ion battery through X-ray projection imaging. A series of parameters such as porosity, tortuosity, and surface parameters were quantified through this image, and the specific characterization of the heterogeneous microstructure of conventional graphite electrodes was obtained for the first time [8]. At present, the X-ray tomography technology derived from X-ray projection imaging has become a conventional method for analyzing battery structure and diagnosing battery failure. At the same time, the combination of X-ray tomography technology and computer three-dimensional modeling can quantify the relationship between the structure and performance of energy materials, and further, expand the application range of X-ray tomography (Fig. 1.2) [9].

Absorption X-ray imaging relies on the attenuation of the incident X-ray beam to form a projection image on the detector. Although X-ray projection imaging has the advantage of non-destructive detection of dynamic structural evolution with high time resolution, it still has certain limitations. First, the high-flux synchrotron X-ray beam may cause radiation damage to the sample. For example, when a Proton Exchange Membrane Fuel Cell (PEMFC) is exposed to an X-ray beam, the degradation of the electrochemical performance of the battery caused by X-ray can be observed in the in-situ experiment [10]. Secondly, the resolution of X-ray projection imaging is insufficient. The best spatial resolution of X-ray projection imaging is about submicron, which is not sufficient to detect the structural changes of the single

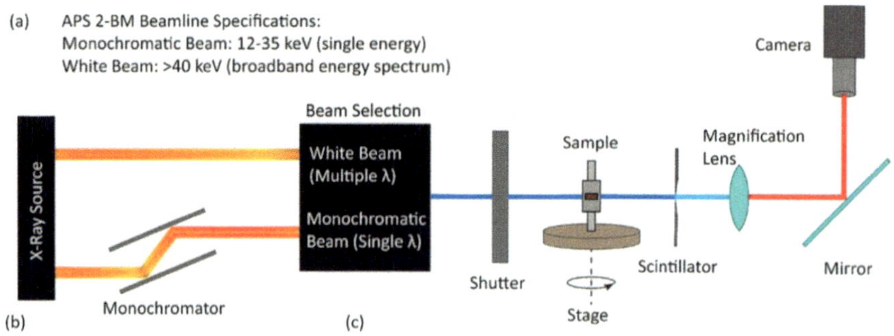

Fig. 1.1 Diagram of the synchrotron X-ray projection imaging setup [6]

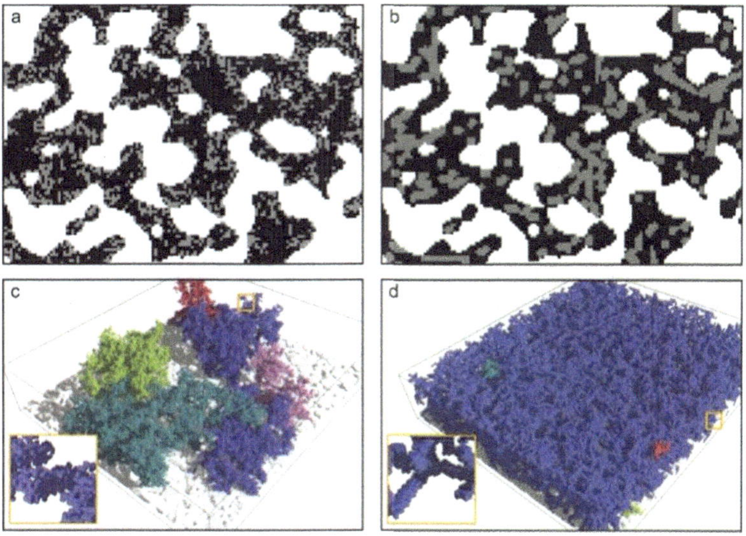

Fig. 1.2 A combination of X-ray tomography and carbon binder modeling: reconstructing the three phases of $LiCoO_2$ Li-ion battery cathodes [9]

electrode particle energy levels of lithium batteries and the crystal structure and chemical state distribution in the sample. Therefore, to compensate for the limitations of X-ray projection imaging, other X-ray imaging systems are also needed.

1.2 Transmission X-ray Microscopy

The first full-field transmission ray microscope (TXM) was built by Gottingen University in 1980 and completed its work in the soft ray band with a resolution of about 150 nm and a focal depth of 3.3 μm [11]. Early X-ray tomography technology was mainly used in geology, so the micron-level spatial resolution can meet the requirements. However, with the development of optical devices and imaging technology, high spatial resolution X-ray tomography technology has been developed and applied to the field of chemical materials [12]. The high chemical phase sensitivity, low-requirement operating environment, and deep non-destructive embedding depth enable this technology to be applied in the field of battery materials. Nowadays, full-field transmission X-ray microscopes have been successfully applied to synchrotron radiation devices all over the world [13]. Thanks to advanced processing technology, the resolution of the TXM imaging system has reached 30 nm in the hard-line band and further improved in the soft-line band to 15 nm [14, 15]. Besides, for light elements, due to its less absorption of hard X-rays, TXM imaging technology has poor image contrast for lighter elements. However, this problem can be solved by adding a phase-contrast mode.

Hard X-rays (>2000 eV) are used as the light source, and the synchrotron X-rays are focused on tens of microns through a rotating ellipsoidal focusing lens. Then, the unfocused light is filtered out through a beam interceptor and a small hole, and then the filtered X-rays are applied to the sample, and the Fresnel zone plate is enlarged to the CCD detector for imaging as shown in Fig. 1.3 [16]. This method can get a larger high-resolution image without taking a single shot. This method of imaging technology will play a very important role in in-situ experiments. For TXM technology, its spatial resolution is determined by the width of the outer ring of the Fresnel zone plate. At present, the best spatial resolution Fresnel zone plate for two-dimensional images is 20–30 nm [17].

The absorption contrast of the image is a very common mode for TXM technology. This contrast is caused by the different degrees of absorption by the sample when X-rays pass through the sample. Because sample thickness and chemical element composition are two important factors that affect X-ray absorption [18]. Therefore, to obtain a higher-quality absorption contrast image, the thickness or particle size of the sample and the chemical composition of the material are the main factors. Normally, the thickness or particle size of the sample ranges from tens of nanometers to tens of microns. Because the specific energy can excite the electrons in the specific shell of the atom, X-rays are absorbed. To obtain a complete absorption spectrum, it is necessary to continuously change the energy of the X-ray before and after the absorption edge of the tested element, to obtain a series of absorption contrast images at different energies. Further, after image post-processing, the chemical Spatial distribution of phases.

Although a variety of in-situ imaging techniques such as in-situ atomic force microscopy (AFM) [19] and transmission electron microscopy (TEM) [20] have been used to study the microstructure changes of electrode materials during charge and discharge. However, TXM imaging technology has its unique advantages, such as thicker electrode material thickness (a true battery, such as a button battery), a large observation range (tens of microns), and detection in the air, etc. [21]. At the Synchrotron Radiation Source Center in Taiwan, my country, the two-dimensional in-situ TXM imaging technology was applied for the first time in battery materials [21, 22].

TXM uses the full-field imaging mode, that is, the sample size is smaller than the spot size at the sample position, so it can complete the two-dimensional transmission image acquisition of the sample in one exposure. Therefore, compared with

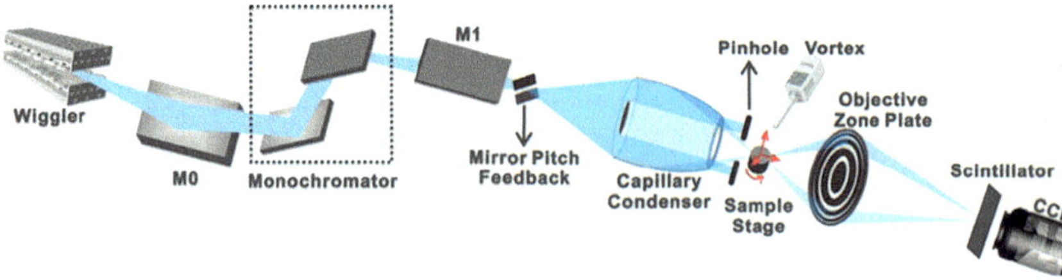

Fig. 1.3 Transmission X-ray microscope (TXM) imaging light path diagram [16]

the scanning imaging method, it has a significant advantage, namely the imaging speed Fast; besides, thanks to the strong penetration of X-rays, the soft X-ray transmission microscope device can use the "water window" to perform three-dimensional non-destructive imaging of cells in the frozen state, and the hard X-ray transmission X-ray microscope is more used Research on micro-scale and nano-resolved three-dimensional non-destructive imaging of energy materials, rare earth magnets, etc. without the need to slice the imaging samples; at the same time, the synchrotron radiation-based TXM device takes advantage of the adjustable X-ray energy and combines the development of X-ray absorption spectrum. The spectroscopic imaging method can perform imaging research on the element distribution and valence distribution of the elements of interest in the sample, to perform the correlation analysis of the function and the morphology of the sample on the mesoscopic scale to realize functional imaging.

The first application of TXM imaging in the field of battery materials is to observe the evolution of topography. As we all know, for high-capacity anode materials (such as Si, Sn, Ge, etc.), large volume changes during charge and discharge are the main factors limiting their wide application. Therefore, it is very necessary to understand and clarify the morphology and volume changes of this type of material during the lithium/sodium ion deintercalation process [23]. Wu et al. [22] observed the evolution of the internal structure of the anode material Sn during charge and discharge through in-situ TXM experiments. In their research, Sn particles are dispersed in a graphite electrode without a current collector, which serves as a negative electrode and is further assembled in an improved coin cell. The study found that during the first discharge, the lithium alloying process presents a core-shell reaction mechanism along the area where cracks are formed during the lithiation process. The formation of cracks is due to the larger volume expansion during the discharge. This volume expansion has a great correlation with the size of Sn particles. The subsequent delithiation process caused the pulverization and fragmentation of the material, resulting in the formation of porous Sn particles, which also provided a new idea for the design of Sn anode materials. Combined with XANES chemical analysis, in in-situ experiments (Fig. 1.4), TXM can analyze chemical elements and phases with higher sensitivity and higher spatial resolution, revealing the chemical valence of the tested sample during the electrochemical reaction process State change [24].

1.3 Scanning X-ray Microscopy

X-ray microscopy lies between electron and light microscopy in terms of specimen size and imaging resolution and is thus suitable for imaging extremely large and complex structures [25]. X ray radiography was discovered in 1896 showed that a position-sensitive detector could detect the locations of photon removal when in contact closely with the objects. Then researchers produced the radiographs of small objects and examined them under the light microscope to gain the early stage

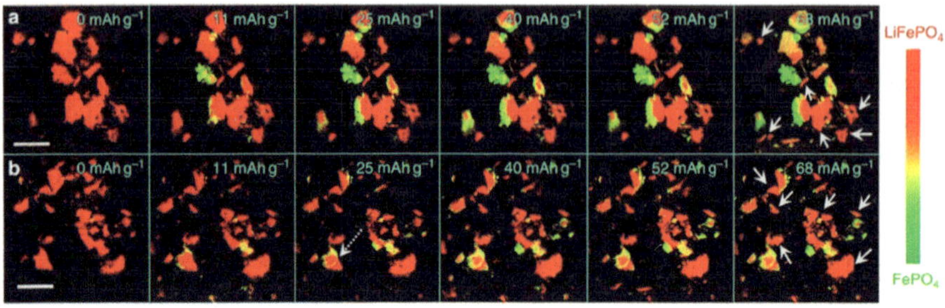

Fig. 1.4 Temporal phase transformation evolution in intraparticle at a fast-charging rate of 1C [24]

of X-ray microscopy. However, because of the insufficient image resolution and many other technological limitations, the early stage of X-ray microscopy was not sufficient to provide precise images. Though there are many difficulties in technology, researchers continued to work hard on technical refinement and eventually obtained more powerful and user-friendly X-ray microscopy which can be applied to many fields [26].

As in other types of microscopy, the construction of scanning X-ray microscopy is allowed because of the existence of focusing devices. Scanning microscopy requires a precision scanning stage with computer control therefore it is more complex than transmission microscopy. But it is by far the simplest technique to analyze the images obtained from the scanning microscope. With the aid of a computer, the image is built up as an absorption map in a serial fashion which makes image processing simple. The elemental map can also be obtained by taking images on either side of an absorption edge and subtracting. Moreover, under comparable signal-to-noise ratios, scanning X-ray microscopy imparts the least dose to the specimen. In scanning X-ray microscopy, X-rays are brought to a small focus in which the specimen is scanned over, and a detector (a gas flow proportional counter) is used to record the transmitted flux [27].

There are currently two types of scanning X-ray microscopy: STXM (scanning transmission X-ray microscopy); and SPEM (scanning photoemission microscopy). In STXM a Fresnel zone plate is used to produce an X-ray probe of 80 nm diameter by focusing an intense low emittance beam of monochromatic soft X-ray. The specimen can be mounted on a 3 mm diameter electron microscope grid and is held outside the vacuum system under normal ambient conditions and then the stationary X-ray probe is used to scan the specimen. X-ray absorption is measured as a function of sample position through the detection of transmitted X-rays. Sample atmospheres include ambient air, He, and low vacuum are all available for STXM. The intensity of the X-rays transmitted by each point on the specimen is detected by the detector which is mentioned above and converted into digital form to build up an image. All the operations are under computer control and the data in digital form greatly convenient for both on-line and subsequent off-line analysis of the images [28]. Figure 1.5 is a photograph of the ambient STXM.

Fig. 1.5 Photograph of the interferometer controlled ambient STXM [28]

Synchrotron-based SPEM is a surface-sensitive and ultra-high vacuum technique. In SPEM a zone plate focusing lens is used to focus X-rays, and then X-rays are impinged on a sample to produce photoelectrons which are detected by a dispersive electron spectrometer. Similar to STXM, synchrotron SPEM offers accurately mechanical scanning and can be combined with angle-resolved photoemission spectroscopy (ARPES). This technique can record the electronic band structure of complex materials at a resolution of nano-lateral (sub-100 nm) [29].

The interaction cross-section of scanning X-ray microscopy is much smaller than for electron scattering, so specimens are much thicker as well without sample preparation or alteration. Samples for STXM can be analyzed under various environmental conditions, including controlled humidity and anoxia. In terms of the system to which it can be applied, STXM is the most universal one. STXM lacks the ultra-high vacuum requirement, provides bulk and surface analyzes rather than merely surface characterization, and entails less radiation damage to the sample for similar spectral resolution.

STXM and SPEM coupled with micro-spectroscopy such as near-edge X-ray absorption fine structure (NEXAFS) can determine the spatial association and speciation of most elements in soils based on maintaining sample integrity. Scanning X-ray microscopy has contributed to the greatest advancements in the understanding of organo-mineral interactions of soil, including mineral control on organic carbon cycling and the mechanisms of biomineral formation. Furthermore, nanoscale C NEXAFS spectra are distinct from the bulk C NEXAFS spectrum which reveals heterogeneity in C speciation between soils and within soils (Fig. 1.6) [30].

There are still challenges in precisely exploring the specification of elements and opening doors to reveal the accurate structure of environmental metrics in the environment due to the highly heterogeneous concerning their composition and conformation of these environmental metrics. The synchrotron-based STXM is becoming a mainstream tool for understanding the environmental processes of elements in complex matrices at the scale of molecular and nanometer [30]. The stabilization of organic carbon and interaction with contaminants in the environment has a lot to do with its structural configurations and functional distributions. Kinyangi et al. [31] resolved the biocomplexity of unaltered soil micro aggregates at the nanoscale with the combination of STXM and C 1s-NEXAFS at a spatial resolution of 50 nm. In that work, researchers proposed that organic matter exists in form of either oxidized C associated with mineral surfaces or particulate aromatic and aliphatic C in soil micro aggregates. Using STXM-NEXAFS to characterize the chemical heterogeneities of carbonaceous materials can greatly help to understand the environmental processes of organic contaminants (Fig. 1.7).

It is encouraging that synchrotron-based scanning X-ray microscopy has been increasingly applied to many emerging areas in recent years. But it is noteworthy that it is still a technical challenge to future apply scanning X-ray microscopy with other chemicals, biological and spectroscopic methods despite the resolution sensitivity of synchrotron has been significantly improved. Fortunately with the reducing cost of both storage and computing, rapidly expanding applications of graphical processors for scientific computing, STXM has already been applied to in-situ studies, such as in-situ spectro-electrochemistry, controlled humidity studies, and catalysis. What's more, SPEM has also been applied to studies of operando in solid oxide fuel cell systems and situ studies of polymer electrolyte membrane fuel cells. With the continual improvements in light source brightness and stability, the future for Scanning X-ray microscopy is bright indeed [32].

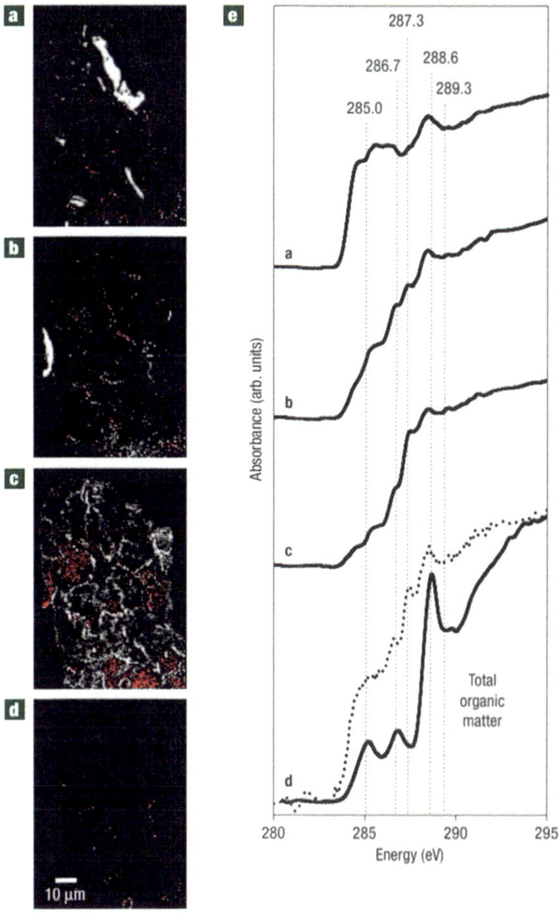

Fig. 1.6 Identification of regions with spectral properties of total organic carbon within a soil micro-assemblage from Nandi Forest (Kenya) determined by STXM- NEXAFS. **a–c** Target maps from clusters obtained by principal component analysis. **d** Target map of total organic carbon. **e** Corresponding spectra. White regions in the target maps indicate areas that are well characterized by the corresponding target spectrum [30]

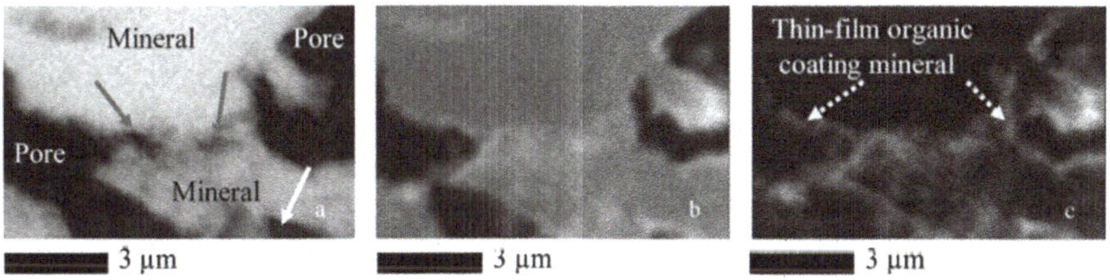

Fig. 1.7 Composition thickness maps of the main thematic regions at the exterior of a micro-aggregate [31]

1.4 X-ray Fluorescence Imaging

X-ray fluorescence analysis is a well-developed technology with a long history. It has been applied in many fields, such as in the archeological, pathobiological, biological, and forensic sciences as well as in the chemical industry. This is because X-ray fluorescence has a unique strength of non-destruction as a kind of quantitative analysis [33]. In electron microanalysis, proposed in 1949, fluorescent radiation is stimulated by specimen irradiation with a focused electron beam. Even though excitation by electron has various advantages, but it still has some serious disadvantages. Firstly, bombarding the specimen with electrons would not only produce the characteristic spectrum but also the intense bremsstrahlung radiation which significantly reduces the sensitivity of the method. Secondly, excitation by electrons is sensitive to the surface which can be applied to study a thin layer of the specimen. Differ from the excitation by electrons, X-ray excitation of fluorescence gives chances to detect much larger specimens [34].

Synchrotron-based X-ray fluorescence imaging is a technique able to quantitatively determine the distribution of trace elements in biological samples with high sensitivity and high resolution. Due to the detection of each element is based on its intrinsic photoelectric effect, the technique does not require labeling steps. A focused X-ray beam is used to scan samples in an x/y pattern and then two-dimensional distributions of elements are produced from the emitted fluorescence spectra at each pixel. Usually, biological samples (>10 μm) are fully penetrated by X-rays and if the thickness of the specimen is known, the concentrations of elements can be calculated in three dimensions. XFI experiments on biological specimens with a micron to submicron resolution are usually operated at synchrotron sources. The reason is that biological specimen has weak absorption contrast to high energy X-rays, high brightness X-ray sources are required to achieve a measurable XFI signal at a nanometer scale [35].

In actual operation, X-ray fluorescence imaging uses a micro-focused X-ray beam incident upon a specimen, an energy detector to surveillance the X-ray fluorescence, and the translation of specimen to ensure the X-ray beam examines different parts of the specimen in sequence to develop an image of the specimen (Fig. 1.8) [36].

Synchrotron-based X-ray fluorescence imaging has become an effective method to investigate the distributions and concentrations of elements in biological specimens. Technological advances now allow imaging with resolution comparable to electron microscopy therefore researchers can visualize the subcellular organelles with appropriate correlative methods [35]. Usually, high exposure to heavy metals and their accumulation is recognized as a reason for many diseases. Using the synchrotron-based X-ray fluorescence microprobe, its high sensitivity and resolution of elements can give a method to study the distribution of heavy metals connected with specific diseases [37]. XFI is an important method in preclinical research that focuses on visualizing the function of cells in a non-invasive way to understand treatment effects [38]. Figure 1.9 shows examples of biological XFI,

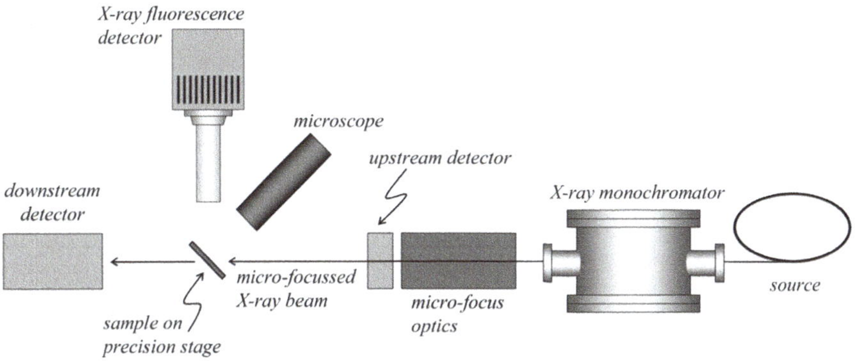

Fig. 1.8 Highly simplified and schematic diagram of the equipment for X-ray fluorescence imaging. The X-ray source is typically synchrotron light with X-ray optics [36]

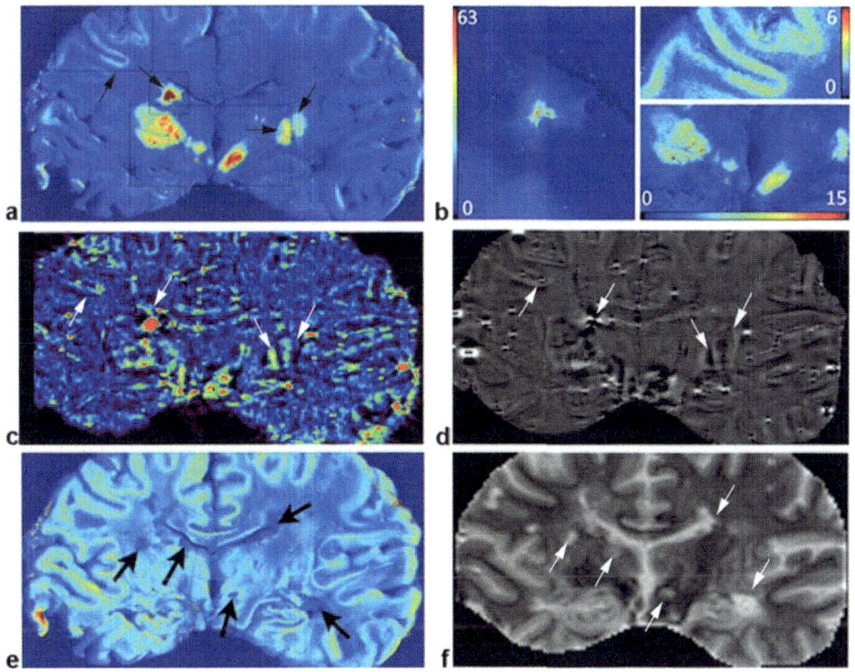

Fig. 1.9 Examples of biological X-ray fluorescence imaging [39]

demonstrating the distribution of Fe and Zn in the human brain taken post-mortem from an individual suffering from multiple sclerosis [39].

XFI of a metal-containing drug in a biological specimen directly detects the metal based on the nature of the element. So that there is no need to sign the specimen, the integrity of the biological structures in the specimen are maintained, various forms of specimen can be detected depending on the required result [40].

Understanding the distribution and concentration of elements in plant culture is important in the plant science field. XFI has some specific abilities which plant scientists are interested in, including analyzes under room pressure and temperature, great analyzes limits, as well as high resolution. Plant scientists anticipate that technological advantages on XFI will also give chances moving into the future, such as screening molecular biology in high-throughput, localizing protein in conjunction with exotic metal tags, and enabling time-resolved analyzes of living plants [41].

Despite the application of XFI systems is now well established, new systems continue to be developed. The researchers are now primarily focusing on the development and expanding application range of portable XFI systems. In the area of cultural heritage, the XFI technique is already a well-rounded imaging technique for the analysis of paintings which is also being applied to the imaging of stained-glass panels. The technique could not only characterize the different shades of blue but also the materials used in the painting [42].

Many X-ray fluorescence related technique was introduced in the research. Scanning micro X-ray fluorescence imaging with a confocal instrument has the advantage of high resolution decided by the size of the micro x-ray beam. The shortcoming of this technique is its long getting time depending on a required specimen size and spatial resolution. By comparison, projection-type X-ray fluorescence imaging has an edge of imaging speed by using a sensitively 2D X-ray detector [43]. Due to its development and applications, the X-ray fluorescence imaging (XFI) technique continues to gain attention. The XFI has evolved to a 3D elemental imaging technique with the emergence of high-speed scanning approaches as well as ED-detectors of high counting rate capability [42].

XFI and its related applications can obtain abundant information about the elements in organs and cells. Actually, in the past two decades, the application of XFI in biology has developed from explorative imaging of elements to the distribution and speciation of platinum in tumor cells as well as iron (Fe) in mouse macrophages. XFI also made a valuable contribution to studies of copper (Cu) that founded highly concentrated Cu pools in biological systems. This found made a great contribution to a better understanding of disease processes with damaged Cu handling and promoted the development of treatment methods for Cu handling disorders [35].

Osaka City University developed a micro X-ray fluorescence instrument using a full lens focused on an X-ray poly-capillary in conjunction with a low-power X-ray tube (Fig. 1.10a) [33]. X-ray fluorescence elemental mapping of a micro SD memory card was demonstrated. The micro X-ray beam was scanned with a step of 100 μm and the X-ray fluorescence analysis was performed for 10 s at the fixed position of the specimen. Detailed structure of the specimen was achieved non-destructively as shown in the maps of Au, Ni, Ti, Cu, and Br. (Fig. 1.10b) [43].

The application of X-ray fluorescence will go on expanding as long as researchers go on to push the characterization, development, and application of X-ray fluorescence imaging. Entering the field of other therapeutic areas and deeper into research will further spread the advantage of X-ray fluorescence [38].

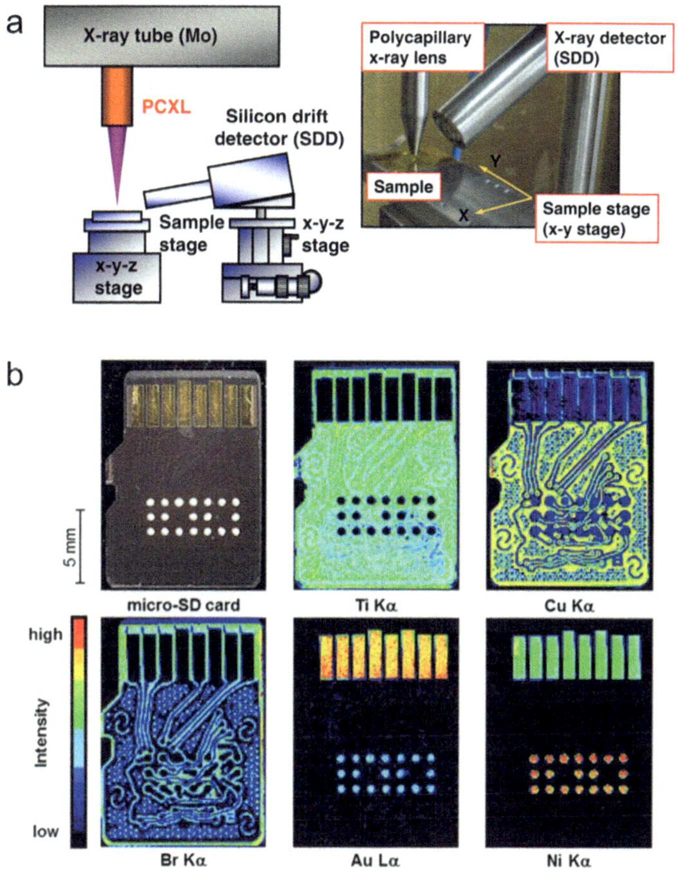

Fig. 1.10 **a** Micro X-ray fluorescence instrument at Osaka City University, [9, 33] **b** Non-destructive elemental imaging in the micro SD memory card [43]

1.5 Coherent Diffraction Imaging

Coherent ray diffraction imaging (CDI) was proposed by Sayer in 1980. It is a "lensless" microscopic imaging method that irradiates the sample with coherent rays and uses a two-dimensional detector directly behind the sample to obtain the distance of the sample. Field diffraction pattern, and finally use the reconstruction algorithm to calculate the absorption and phase information of the sample. The coherent ray diffraction imaging method is the latest in the history of x-ray imaging, and it is likely to realize the dream of three-dimensional spatial resolution imaging at the order of ray wavelength. In other ray microscopy methods, for example, the spatial resolution is limited by factors such as the numerical aperture of the optical element and aberrations, and currently, the best is achieved. The method does not require optical components for imaging, and its spatial resolution is not affected by factors such as numerical aperture, so it is expected to achieve higher spatial resolution than systems such as TXM. Besides, there is no limitation of depth of focus, and a large imaging field of view can be achieved while achieving high spatial resolution. Also, the development of the fourth-generation simultaneous shooting (free-electron laser) provides you with ultra-high-brightness ultra-short ray pulses that can perform time-resolved imaging.

Unlike the lens-based radiographic imaging systems, phase X-ray diffraction imaging (X Coherent X-ray Diffractive Imaging, CXDI) does not require imaging through a lens but uses the coherent characteristics of the incident beam to use the scattered light of the sample to form the image. The core of CDI is to obtain the far-field diffraction pattern of the sample. As shown in Fig. 1.11 [44], the experimental device directly irradiates the more coherent rays on the sample. A detector is placed in the far-field range to collect the far-field diffraction pattern. A light barrier is placed on the front end to remove them through light. The current CXDI is generally carried out under vacuum conditions, the purpose is to eliminate the absorption of air by the air and remove the interference of small particles in the air. Also, a diaphragm is usually placed in front of the sample to obtain a light beam slightly larger than the sample. The function of the front light barrier of the detector is to prevent the high-intensity through light from causing saturation of the detector, otherwise, the diffraction pattern cannot be effectively recorded. The three-dimensional reconstruction image of the sample can be obtained by rotating the sample stage.

CDI imaging technology is the development and extension of X-ray crystallography. Both have similarities in imaging principles and have their characteristics. Because the crystal has a periodic structure, it has a diffraction enhancement effect on rays, and the diffraction pattern is high-intensity and discrete Because there is no periodic structure in the amorphous sample of the Bragg diffraction point, if the sample is irradiated with an incoherent light source, the diffraction pattern obtained is a fuzzy diffraction signal, and there is no definite phase relationship. When an amorphous sample is irradiated with an electron beam, the diffraction pattern obtained is a weak and continuous diffraction signal, which has a definite phase relationship. According to Born approximation conditions, the coherent diffraction pattern at the far-field is the Fourier transform of the sample. Knowing the intensity

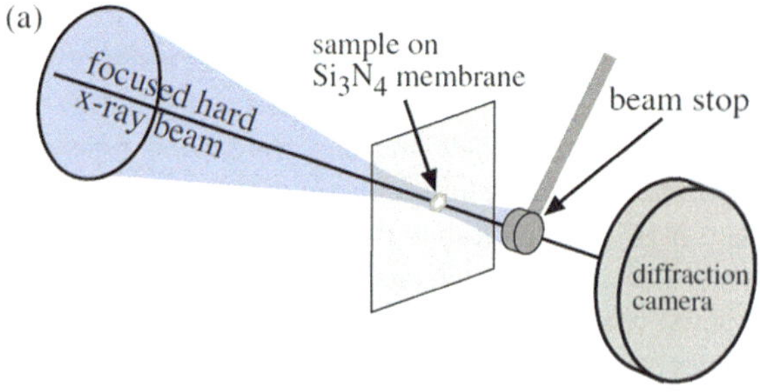

Fig. 1.11 Schematic diagram of the structure of coherent ray diffraction imaging [44]

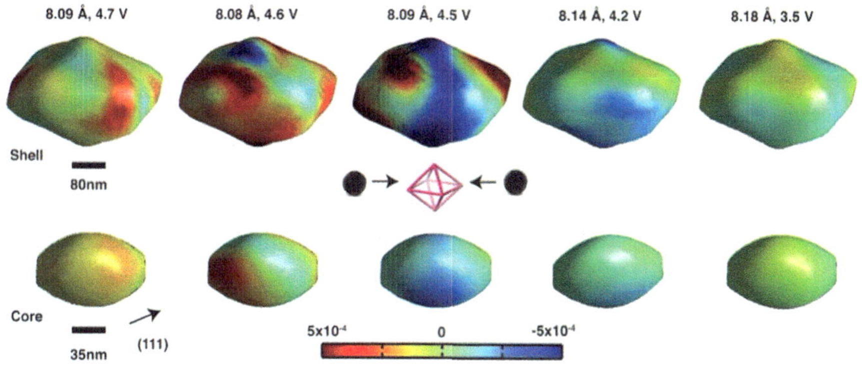

Fig. 1.12 Isosurface projections of strain evolution [49]

and phase information of the diffraction pattern at the same time, you can get the image of the sample in real space.

The emergence of CDI imaging technology overcomes the defects of traditional ray microscopes that are difficult to focus, resolve, and low substrate. It has been widely used in three-dimensional microstructure imaging of materials [45], sample quantitative analysis [46], chemical element distribution imaging [47], Nanocrystal strain field imaging, and other research directions [48]. Coherent light diffraction imaging technology (CDI) has developed into a very effective in-situ test technique for studying the phase transition mechanism of battery materials. This technology can analyze the evolution of nanocrystalline materials (such as stress changes and phase transitions in crystalline materials). Its resolution depends only on the wavelength of incident X-rays, and the current spatial resolution can reach several nanometers. Ulvestad et al. [49] applied the in-situ CXDI operation technology to the study of energy materials for the first time (Fig. 1.12). They tracked the evolution process of 3D defects in $LiNi_{0.5}Mn_{1.5}O_4$ cathode particles, and found uneven stress during the electrochemical cycle distributed.

1.6 Industrial X-ray Microscopy

With the fast development of energy storage devices, extensive requirements have been put forward for the application of X-ray microscopy in industrial equipment. Compared with the requirements of the medical field, the main purpose of X-ray microscopy scans in an industrial environment is very different. Under most conditions, industrial X-ray microscope scanning does not care about the dose of X-rays to the sample as in medicine. Although getting a fast scan is necessarily a clear advantage, the requirement for ultrafast scanning time in an industrial environment is not important. Therefore, industrial X-ray microscopes can use higher-intensity X-ray sources and can extend the scanning time to longer when scanning with high precision. Given mentioned above, industrial X-ray microscopes have many applications in these fields of material characterization, non-destructive testing, and metrology applications, so achieving the maximum possible scan resolution, accuracy, and precision are usually the key goals. Further, the prospective fields of industrial X-ray microscopy are examination of fiber-reinforced composites and multimodal data analysis. A discussion on the specific areas of industrial X-ray microscope focusing will be given later in this book, but this section will determine the specific modifications made to clinical X-ray microscope scanning equipment to meet the needs of the industry in recent years. The fundamental difference in industrial X-ray microscope scanners is the rotating parts of the scanner in material analysis and metrology applications. In the medical field, the scanning device rotates around the patient at a high speed, but in most industrial systems, the scanning device is fixed while the sample is rotating.

Figure 1.13 exhibits the macroscale X-ray CT results before and after the failure of a fully charged battery. The non-destructive nature of the technology allows visual observation of structural changes after battery failure. Figure 1.13a shows the three-dimensional reconstruction of cells before destruction, where orthogonal slices of (*Y*, *Z*) and (*X*, *Z*) planes are also displayed. Meanwhile, orthogonal slices of failed units can also be compared. As shown in Fig. 1.13b, the destruction of the SEI layer is the propagation process of the gas formed due to decomposition. For example, arrow 1 highlights the delamination of electrode layers that may form air pockets. Arrows 2 and 3 indicate the arrangement of the electrode layers and the change of the battery casing. It can be observed from the figure that the ridge line of arrow 2 has expanded, which leads to the increase in the internal pressure of the battery, and the tightly wound electrode layer has also changed accordingly. At 130 °C, the positive electrode reacts with the electrolyte and causes the loss of oxygen at the cathode. When the temperature is higher than 200 °C, the cathode undergoes thermal decomposition, the crystal structure changes, heat is released, and CO_2 and H_2O are also released [41]. Due to the influence of this phenomenon, the positive electrode material has been separated from the aluminum current collector. In addition, the battery casing has also changed, especially at the outer end of the positive electrode. Although the battery swelled, analysis after failure did not reveal a battery rupture. At the same time, Finnegan et al. reported that when Li$(Ni_{0.33}Mn_{0.33}Co_{0.33})O_2$ 2.6 Ah 18,650 cells undergo thermal runaway at 100% SOC, the battery casing and front cover remain intact. Although the gas in the battery will cause the internal pressure to increase, this does not necessarily cause the battery to rupture. Figure 1.13b shows that the internal structure of the battery collapses in this case. According to previous reports, the cylindrical mandrel in the center of the electrode layer plays a vital role in the mechanical failure of the battery. At the same time, it also provides mechanical strength and a path for the gas to reach the exhaust port during failure.

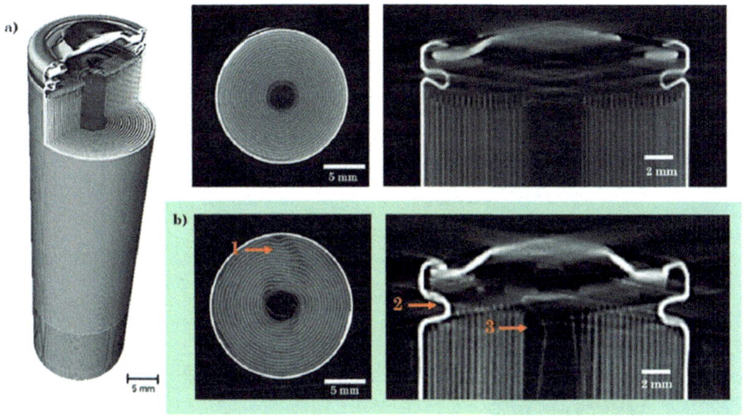

Fig. 1.13 X-ray macro-CT results of a commercial 18,650 cell **a** showing a 3D reconstruction of the whole cell and orthogonal slices in the *XY* and *YZ* planes before thermal runaway and **b** showing orthogonal slices in the *XY* and *YZ* planes after thermal runaway using ARC. Arrows depict areas of deformation within cell architecture

1.7 In-Situ X-ray Imaging

The structural transformation caused by electrochemistry will straightly impact the electrode performance of lithium-ion batteries. Develop advanced characterization methods within the scope of operational ability can intensify the in-depth comprehension of the mechanism, so as to clarify the dynamic stages and structural changes of energy materials. In-situ synchrotron X-ray imaging technology, which can examine energy in real time, has attracted increasing attention in the research of battery energy by virtue of its natural characteristics, non-destructive, sensitive to chemistry and elements, environmentally friendly, and high penetration. Transmission X-ray microscope (TXM), a new full-field hard X-ray imaging technology, was firstly reported from beamline X8C at the National Synchrotron Light Source (NSLS) in Brookhaven National Laboratory (BNL), which was exploited and applied to battery microstructure research. The introduction of this method combined with in-situ XANES can clarify the transitions in the chemical phase and the morphological evolution of active particles during the electrochemical cycle [50]. Unlike others, this technique can provide chemical phase diagram information (oxidation state, electronic, and local environment) with spatial resolution lower than 30 nm, which is a particularly indispensable and distinct function, instead of supplying average information spectra only through conventional X-ray absorption near the edge structure (XANES), as shown in Fig. 1.14a, b. TXM can be used to explain the connection between morphological evolution and chemical reaction under in-situ conditions by gathering morphological information during the cycle and tuning X-ray energy at one cycling state across an absorption edge of an element of interest that emerges a series of X-ray absorption spectra. More

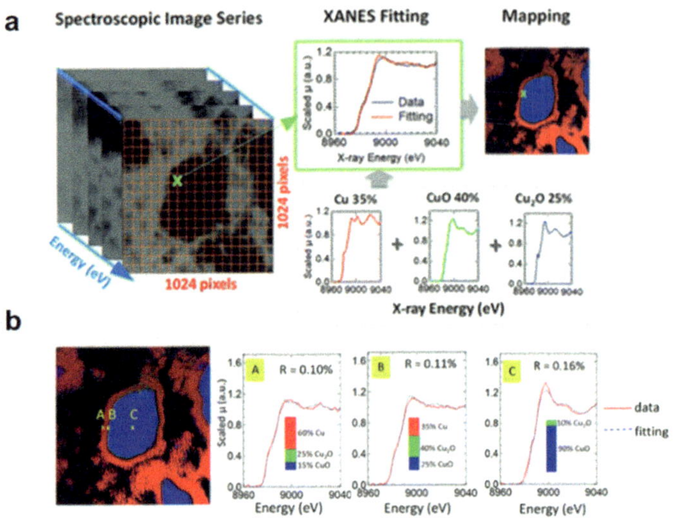

Fig. 1.14 **a** Principles and data processing for TXM-XANES measurements. **b** Phase composition upon the first lithiation at the selected zones [50]

importantly, in-situ X-ray imaging has a wide field of vision for huge and thick samples, breaking the limitation of in-situ transmission electron microscopy (TEM) on nanoscale samples, which means that it can more truly reveal the characteristics of real-time battery response.

In general, the electrochemical cell developed in this study was depended on the universally used button cells, which were revised with two Kapton Windows to accurately monitor the electrochemical reaction in real time, as shown in Fig. 1.15 [51]. In addition, in order to obtain high-resolution and high-quality images, carbon paper can be used as an electrode carrier instead of traditional metal foil because it can better penetrate X-rays, which is especially crucial for subsequent in-situ TXM-XANES acquisition and mapping analysis. A unique ability of the newly developed TXM is to keep constant the magnification of images collected at different energies. This is especially significant for investigating battery materials under in-situ electrochemical situations, ensuring the optimal resolution of all graphics gathered by XANES imaging under multiple X-ray energies.

In conclusion, in-situ X-ray imaging has proved to be a prospective instrument for studying energy materials. The combination of nanoscale spatial resolution and chemical sensitivity has been satisfactory and productive, which is of great significance in the research of some technologically significant battery cathode materials. Systematic quantification of imaging data has generated important cognition of the underlying mechanisms behind battery performance improvement/degradation. The dynamic response of battery particles to distinct reaction conditions and crystal face dynamics can be visualized by the evolution of local charge distribution [52].

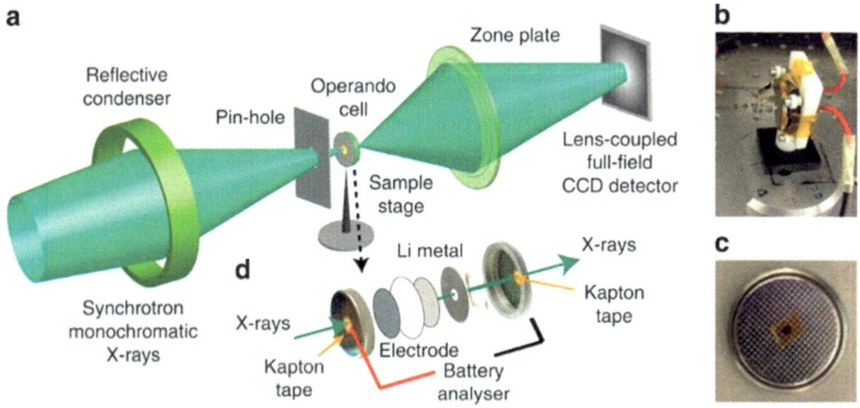

Fig. 1.15 **a** Sketch of the full-field TXM. Photographs of **b** the custom-built cell holder and **c** a perforated coin cell used for the operando studies. **d** Schematic illustration of the operando cell containing the FeF_3 electrode and all the other key components of a realistic battery [51]

While thrilled about these features, we also notice enormous opportunities related to future technology developments. For instance, the spatial and temporal resolution of in-situ X-ray imaging technology may be further advanced along with the growth of X-ray optics and next-generation synchrotron devices with unprecedented coherence characteristics. The further development of X-ray free-electron laser (XFELs) can even research the ultrafast property of materials far from equilibrium. In addition to spectral contrast, it is also appropriate to involve other image contrast mechanisms, such as phase contrast and diffraction contrast. By operating at distinct X-ray energy ranges, we can also probe the surface or overall structure and chemical heterogeneity during the actual battery reaction.

Clearly, the existing in-situ X-ray imaging methods have made certain technological progress, but there is still much room for improvement in energy storage research. In particular, we look forward to further development of technologies that can provide more high-resolution information about the three-dimensional morphology and chemical composition of electrodes, which will be extremely meaningful. In addition, the current challenge for some in-situ tools is the high data collection rate required to detect non-equilibrium states in operational X-ray imaging measurements. More importantly, another key point of in-situ research is to devise electrochemical equipment compatible with application technology [53]. The degree of correlation between these customized batteries and practical batteries should receive more attention and artifacts should be treated with caution. More importantly, the combination of multiple in-situ technologies is formidable enough to provide more overall insight and the great similarity in the design of the respective in-situ cells makes it relatively easy to implement. In view of the recent development status, it can be expected that in-situ research will play an increasingly significant role in the future development of the next generation of lithium-ion batteries.

1.8 Machine Learning X-ray Imaging

Machine learning (ML) is the research of automatic enhancement of computer algorithms via training and experience, which can be regarded as a subset of artificial intelligence study. Generally, based on the sample data (called "training data") to establish a mathematical model, a machine learning algorithm can predict or make decisions without explicit programming. Machine learning algorithms in some areas, such as e-mail filtering and computer vision, to replace the traditional algorithm is not feasible, it is closely related to the calculation and statistics, its focus is to use the computer to predict. The research of mathematical optimization introduces methods, theories, and application fields into the scope of machine learning. Among them, data mining is a related sub content, which mainly studies exploratory data analysis through unsupervised learning. Therefore, machine learning is also called predictive analysis.

X-ray spectro-microscopy has successfully visualized and quantified mesoscale compositional/chemical heterogeneity and revealed the spatial pathways of the reactions. Although we have effectively harvested scientifically valuable structural and chemical information, the limitations of existing workflows are difficult to deal with. In full-field X-ray spectro-microscopy, because we use 1 k $\times$ 1 k or 2 k 2 k pixels area detector to collect spatially resolved spectral data, the data rate is 6 orders of magnitude higher than that of the traditional bulk X-ray spectroscopy, so in fact, it is impossible to interact with every spectrum for detailed analysis. A good degree of automation is required, and several data restoration steps have been implemented. In addition to data reduction, the conversion of "data" into "knowledge" requires a comprehensive interpretation of spectral features, which are usually non-trivial and therefore have been identified as the bottleneck of the procedure.

The mesoscale spectrum-microcosmic study of battery particles relies heavily on a supervised machine learning method, as shown in Fig. 1.16 [54]. This approach is simple but very effective and efficient. The XANES spectra per pixel (Fig. 1.16a) were compared with known quantitative criteria for the local chemical state based on prior knowledge of the list of major chemical species expected in sample Fig. 1.16b, namely NiO and the metal Ni. The local chemical composition is based on the similarity of spectral fingerprints, and the linearity combination is used to fit the citation of Fig. 1.16c. Corresponding pixels are then color-coded to indicate the distribution of local valences. This process can be repeated as a function of perspective, making it possible to reconstruct the three-dimensional chemical heterogeneity (Fig. 1.16d). The results shown in Fig. 1.16 show that, during the cycle, micron-sized NiO electrode pellets break up and are simultaneously reduced to metallic form.

In short, machine learning method has achieved great success in X-ray imaging big data mining. It has been used in X-ray microscopic data collected by different experimental devices, including full-field transmission X-ray microscope, scanning transmission X-ray microscope, and soft X-ray spectroscope. However, two

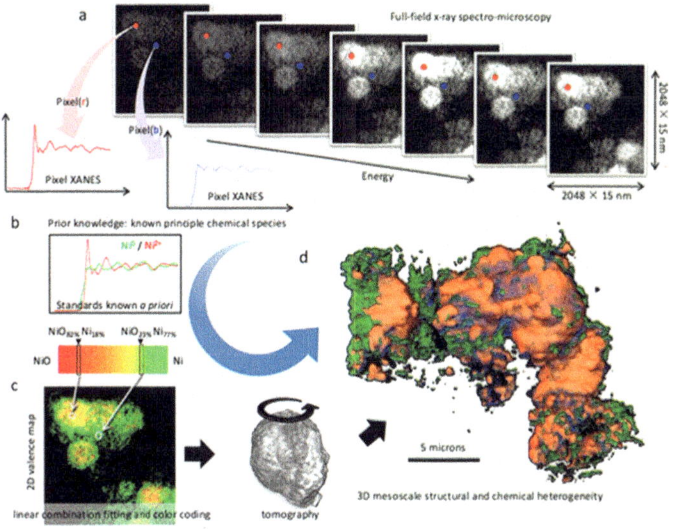

Fig. 1.16 Supervised machine learning approach for extracting the 2D and 3D mesoscale structural and chemical heterogeneity from the full-field X-ray spectro-microscopy data [54]. Panel **a** illustrates the data structure. Panel **b** shows the prior knowledge of a list of anticipated principal chemical species in the sample, which is indispensable for the chemical mapping in this approach. Panel **c** illustrates the quantification procedures including the linear combination fitting, the color-coding, and the tomographic reconstruction. Panel **d** shows the 3D mesoscale structural and chemical heterogeneity of a cluster of partially reduced NiO electrode particles. The scale bar in panel (**d**) is 5 μm

questions were raised during the experiment: (1) Whether the particles studied are representative. (2) Whether prior knowledge can help to supervise the integrity and accuracy of data mining. In fact, if researchers rely too much on previous knowledge, the chances of discovering unknown chemical species may be hindered, which may miss the key information for understanding mesoscale chemistry. Therefore, a large-scale analysis at the electrode level is necessary. Figure 1.17 shows an example illustration of particle crack investigation at the electrode level. Jiang et al. [55] successfully identified more than 650 unique particles of different sizes, shapes, locations, and crack degrees, and separated them from the imaging data in an automatic manner. Figure 1.17a shows the workflow of segmentation based on machine learning. Figure 1.17b shows the comparison between the traditional segmentation results of some representative particles and the machine learning-assisted segmentation results. Obviously, machine learning-assisted segmentation has a higher accuracy rate than traditional segmentation methods.

As a result, we have studied new computational developments in data science. We use and improve the new clustering algorithm, and search through a large number of X-ray energy spectroscopy micro data. In our case, the clusters identified through numerical searches can be further evaluated by researchers with field expertise that demonstrates knowledge of X-ray spectral signatures and battery

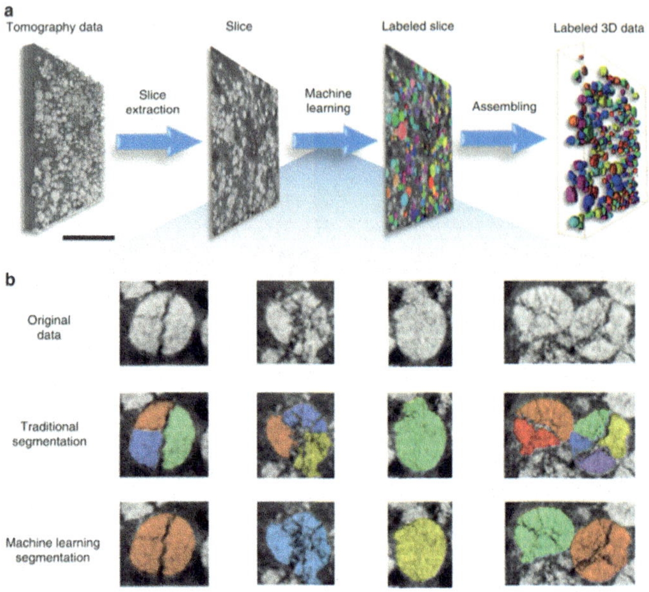

Fig. 1.17 Machine learning-based segmentation and labeling [55]. The scale bar in (**a**) is 50 μm

chemistry to explain the scientific significance of the results. Here, we emphatically introduce the application of the new machine learning method in the research of X-ray spectrophotometry of large-scale imaging data. Monitoring and unsupervised data mining methods can extract scientifically relevant information from large amounts of data efficiently and effectively, thereby greatly complementing the domain expertise of researchers.

References

1. Henderson, L.D.: X rays and the quest for invisible reality in the art of Kupka, Duchamp. Art J **47**(4), 323–340 (1988)
2. Ducros, N., Abascal, J.F.P., Sixou, B., Rit, S., Peyrin, F.: Regularization of nonlinear decomposition of spectral x-ray projection images. Med. Phys. **44**(9), e174–e187 (2017)
3. Pfeiffer, F., Bech, M., Bunk, O., Kraft, P., Eikenberry, E.F., Bronnimann, C., Grunzweig, C., David, C.: Hard-X-ray dark-field imaging using a grating interferometer. Nat. Mater. **7**(2), 134–137 (2008)
4. Dilmanian, F.A., et al.: Computed tomography of x-ray index of refraction using the diffraction enhanced imaging method. Phys. Med. Biol. **45**, 933–946 (2000)
5. Weitkamp, T., Diaz, A., David, C., Pfeiffer, F., Stampanoni, M., Cloetens, P., Ziegler, E.: X-ray phase imaging with a grating interferometer. Opt. Express **13**, 6297 (2005)
6. Shen, F.Y., Dixit, M.B., Xiao, X.H., Hatzell, K.B.: Effect of pore connectivity on Li dendrite propagation within LLZO electrolytes observed with synchrotron X-ray tomography. ACS Energy Lett. **3**(4), 1056–1061 (2018)

7. Shearing, P.R., Howard, L.E., Jørgensen, P.S., Brandon, N.P., Harris, S.J.: Characterization of the 3-dimensional microstructure of a graphite negative electrode from a Li-ion battery. Electrochem. Commun. **12**(3), 374–377 (2010)
8. Ebner, M., Chung, D.-W., García, R.E., Wood, V.: Tortuosity anisotropy in lithium-ion battery electrodes. Adv. Energy Mater. **4**(5) (2014)
9. Zielke, L., Hutzenlaub, T., Wheeler, D.R., Manke, I., Arlt, T., Paust, N., Zengerle, R., Thiele, S.: A combination of X-ray tomography and carbon binder modeling: reconstructing the three phases of $LiCoO_2$ li-ion battery cathodes. Adv. Energy Mater. **4**(8) (2014)
10. Schneider, A., Wieser, C., Roth, J., Helfen, L.: Impact of synchrotron radiation on fuel cell operation in imaging experiments. J. Power Sources **195**(19), 6349–6355 (2010)
11. Schmahl, G., Rudolph, D., Niemann, B., Christ, O.: Zone-plate X-ray microscopy. Q. Rev. Biophys. **13**(3), 297–315 (1980)
12. Wang, R., Xu, C., Sun, J., Gao, L., Yao, H.: Solvothermal-induced 3D macroscopic SnO_2/nitrogen-doped graphene aerogels for high capacity and long-life lithium storage. ACS Appl. Mater. Interfaces. **6**(5), 3427–3436 (2014)
13. Schneider, G., Guttmann, P., Heim, S., Rehbein, S., Mueller, F., Nagashima, K., Heymann, J. B., Müller, W.G., McNally, J.G.: Three-dimensional cellular ultrastructure resolved by X-ray microscopy. Nat. Meth. **7**, 985–988 (2010)
14. Chao, W., Harteneck, B.D., Liddle, J.A., Anderson, E.H., Attwood, D.T.: Soft X-ray microscopy at a spatial resolution better than 15 nm. Nature **435**(7046), 1210–1213 (2005)
15. Chen, J., Wu, C., Tian, J., Li, W., Yu, S., Tian, Y.: Three-dimensional imaging of a complex concaved cuboctahedron copper sulfide crystal by x-ray nanotomography. Appl. Phys. Lett. **92**(23) (2008)
16. Kao, T.L., Shi, C.Y., Wang, J., Mao, W.L., Liu, Y., Yang, W.: Nanoscale elemental sensitivity study of $Nd_2Fe_{14}B$ using absorption correlation tomography. Microsc. Res. Tech. **76**(11), 1112–1117 (2013)
17. Wang, J., Chen-Wiegart, Y.C., Wang, J.: In situ three-dimensional synchrotron X-ray nanotomography of the (de)lithiation processes in tin anodes. Angew. Chem. Int. Ed. Engl. **53** (17), 4460–4464 (2014)
18. Kuppan, S., Xu, Y., Liu, Y., Chen, G.: Phase transformation mechanism in lithium manganese nickel oxide revealed by single-crystal hard X-ray microscopy. Nat. Commun. **8**, 14309 (2017)
19. Beaulieu, L.Y., Hatchard, T.D., Bonakdarpour, A., Fleischauer, M.D., Dahn, J.R.: Reaction of Li with alloy thin films studied by in situ AFM. J. Electrochem. Soc. **150**(11) (2003)
20. Huang, J.Y., Zhong, L., Wang, C.M., Sullivan, J.P., Xu, W., Zhang, L.Q., Mao, S.X., Hudak, N.S., Liu, X.H., Subramanian, A., Fan, H., Qi, L., Kushima, A., Li, J.: In situ observation of the electrochemical lithiation of a single SnO_2 nanowire electrode. Science **330**, 1515–1519 (2010)
21. Chao, S.-C., Yen, Y.-C., Song, Y.-F., Sheu, H.-S., Wu, H.-C., Wu, N.-L.: In situ transmission X-ray microscopy study on working SnO anode particle of Li-ion batteries. J. Electrochem. Soc. **158**(12), A1335 (2011)
22. Chao, S.-C., Song, Y.-F., Wang, C.-C., Sheu, H.-S., Wu, H.-C., Wu, N.-L.: Study on microstructural deformation of working Sn and SnSb anode particles for Li-ion batteries by in situ transmission x-ray microscopy. J. Phys. Chem. C. **115**(44), 22040–22047 (2011)
23. Xie, Y., Wang, H., Xu, G., Wang, J., Sheng, H., Chen, Z., Ren, Y., Sun, C.-J., Wen, J., Wang, J., Miller, D.J., Lu, J., Amine, K., Ma, Z.-F.: In operando XRD and TXM study on the metastable structure change of $NaNi_{1/3}Fe_{1/3}Mn_{1/3}O_2$ under electrochemical sodium-ion intercalation. Adv. Energy Mater. **6**(24) (2016)
24. Li, L., Chen-Wiegart, Y.C., Wang, J., Gao, P., Ding, Q., Yu, Y.S., Wang, F., Cabana, J., Wang, J., Jin, S.: Visualization of electrochemically driven solid-state phase transformations using operando hard X-ray spectro-imaging. Nat. Commun. **6**, 6883 (2015)
25. Sayre, D., Chapman, H.N.: X-ray microscopy. Acta Crystallogr. A. **51**(3), 237–252 (1995)
26. Yamamoto, Y., Shinohara, K.: Application of X-ray microscopy in analysis of living hydrated cells. Anat. Rec. **269** (2010)

27. Luo, L., Zhang, S.Z.: Applications of synchrotron-based X-ray techniques in environmental science. Sci. China Chem. **53**(12), 2529–2538 (2010)
28. Burge, R.E., Beswetherick, J.T., Browne, M.T., Charalambous, P.S., Duke, P.J., Foster, G.F., Hare, A.R., Michette, A.G., Morris, D., Morrison, G.R.: Scanning X-ray microscopy. In: Proceedings X-Ray Instrumentation in Medicine and Biology, Plasma Physics, Astrophysics, and Synchrotron Radiation, International Congress on Optical Science and Engineering, vol. 1140, 528 (1989)
29. Stuckey, J.W., Jianjun, Y., Jian, W., Sparks, D.L.: Advances in scanning transmission X-ray microscopy for elucidating soil biogeochemical processes at the submicron scale. J Environ. Qual. **46**(6), 1166–1174 (2017)
30. Lehmann, J., Solomon, D., Kinyangi, J., Dathe, L., Wirick, S., Jacobsen, C.: Spatial complexity of soil organic matter forms at nanometre scales. Nat. Geosci. **1**(4), 238–242 (2008)
31. Kinyangi, J., Solomon, D., Liang, B., Lerotic, M., Wirick, S., Lehmann, J.: Nanoscale biogeocomplexity of the organomineral assemblage in soil: application of STXM microscopy and C 1s-NEXAFS spectroscopy. Soil Sci. Soc. Am. J. **70**(5), (2006)
32. Hitchcock, A.P.: Soft X-ray spectromicroscopy and ptychography. J. Electron Spectrosc. Relat. Phenom. **200**, 49–63 (2015)
33. Tsuji, K., Matsuno, T., Takimoto, Y., Yamanashi, M., Kometani, N., Sasaki, Y.C., Hasegawa, T., Kato, S., Yamada, T., Shoji, T.: New developments of X-ray fluorescence imaging techniques in laboratory. Spectrochim. Acta Part B: Atomic Spectrosc. **113**, 43–53 (2015)
34. Lider, V.V.: X-ray fluorescence imaging. J. Phys. Usp. **61**(10), 980–999 (2018)
35. Leary, S.C., Ralle, M.: Advances in visualization of copper in mammalian systems using X-ray fluorescence microscopy. Curr. Opin. Chem. Biol. **55**, 19–25 (2020)
36. Pushie, M.J., Pickering, I.J., Korbas, M., Hackett, M.J., George, G.N.: Elemental and chemically specific x-ray fluorescence imaging of biological systems. Chem. Rev. **114**(17), 8499–8541 (2014)
37. Gobe, G.C., Mott, S.A., Jonge, M.D., Hoy, W.E.: Heavy metal imaging in fibrotic human kidney tissue using the synchrotron X-ray fluorescence microprobe. Transl. Androl. Urol. **8** (S2), S184–S191 (2019)
38. Jeffrey, D.: Paradigms in fluorescence molecular imaging: maximizing measurement of biological changes in disease, therapeutic efficacy, and toxicology/safety. Mol. Imag. Biol. **21** (4), 599–611 (2019)
39. Habib, C.A., Zheng, W., Haacke, E.M., Webb, S., Nichol, H.: Visualizing iron deposition in multiple sclerosis cadaver brains. AIP Conf. Proc. **1266**(1), 78–83 (2010)
40. Dillon, C.T.: Synchrotron radiation spectroscopic techniques as tools for the medicinal chemist: microprobe x-ray fluorescence imaging, x-ray absorption spectroscopy, and infrared microspectroscopy. Aust. J. Chem. **65**(3), 204–217 (2012)
41. Kopittke, P.M., Punshon, T., Paterson, D.J., Tappero, R.V., Wang, P., Blamey, F.P.C., van der Ent, A., Lombi, E.: Synchrotron-based x-ray fluorescence microscopy as a technique for imaging of elements in plants. Plant Physiol. **178**(2), 507–523 (2018)
42. Vanhoof, C., Bacon, J.R., Fittschen, U.E.A., Vincze, L.: 2020 atomic spectrometry update—a review of advances in X-ray fluorescence spectrometry and its special applications. J. Anal. At. Spectrom. **35**(9), 1704–1719 (2020)
43. Nakazawa, T., Tsuji, K.J.X.R.S.: Depth-selective elemental imaging of microSD card by confocal micro XRF analysis. **42**(3) (2013)
44. Schroer, C.G., Boye, P., Feldkamp, J.M., Patommel, J., Schropp, A., Schwab, A., Stephan, S., Burghammer, M., Schoder, S., Riekel, C.: Coherent x-ray diffraction imaging with nanofocused illumination. Phys. Rev. Lett. **101**(9), 090801 (2008)
45. Song, C., Bergstrom, R., Ramunno-Johnson, D., Jiang, H., Paterson, D., de Jonge, M.D., McNulty, I., Lee, J., Wang, K.L., Miao, J.: Nanoscale imaging of buried structures with elemental specificity using resonant x-ray diffraction microscopy. Phys Rev Lett. **100**(2), 025504 (2008)

46. Miao, J., Amonette, J.E., Nishino, Y., Ishikawa, T., Hodgson, K.O.: Direct determination of the absolute electron density of nanostructured and disordered materials at sub-10-nm resolution. Phys. Rev. B. **68**(1) (2003)
47. Cherukara, M.J., Cha, W., Harder, R.J.: Anisotropic nano-scale resolution in 3D Bragg coherent diffraction imaging. Appl. Phys. Lett. **113**(20) (2018)
48. Robinson, I.K., Vartanyants, I.A., Williams, G.J., Pfeifer, M.A., Pitney, J.A.: Reconstruction of the shapes of gold nanocrystals using coherent x-ray diffraction. Phys. Rev. Lett. **87**(19), 195505 (2001)
49. Ulvestad, A., Singer, A., Cho, H.M., Clark, J.N., Harder, R., Maser, J., Meng, Y.S., Shpyrko, O.G.: Single particle nanomechanics in operando batteries via lensless strain mapping. Nano Lett. **14**(9), 5123–5127 (2014)
50. Wang, J., Chen-Wiegart, Y.-C.K., Wang, J.: In situ chemical mapping of a lithium-ion battery using full-field hard X-ray spectroscopic imaging. Chem. Commun. **49**(58), 6480–6482 (2013)
51. Li, L., Chen-Wiegart, Y.-C.K., Wang, J., Gao, P., Ding, Q., Yu, Y.-S., Wang, F., Cabana, J., Wang, J., Jin, S.: Visualization of electrochemically driven solid-state phase transformations using operando hard X-ray spectro-imaging. Nat. Commun. **6**, 6883 (2015)
52. Wei, C., Xia, S., Huang, H., Mao, Y., Pianetta, P., Liu, Y.: Mesoscale battery science: the behavior of electrode particles caught on a multispectral x-ray camera. Acc. Chem. Res. **51** (10), 2484–2492 (2018)
53. Harks, P.P.R.M.L., Mulder, F.M., Notten, P.H.L.: In situ methods for Li-ion battery research: a review of recent developments. J. Power Sources. **288**, 92–105 (2015)
54. Meirer, F., Cabana, J., Liu, Y., Mehta, A., Andrews, J.C., Pianetta, P.: Three-dimensional imaging of chemical phase transformations at the nanoscale with full-field transmission X-ray microscopy. J. Synchrotron Radiat. **18**(5), 773–781 (2011)
55. Jiang, Z., Li, J., Yang, Y., Mu, L., Wei, C., Yu, X., Pianetta, P., Zhao, K., Cloetens, P., Lin, F., Liu, Y.: Machine-learning-revealed statistics of the particle-carbon/binder detachment in lithium-ion battery cathodes. Nat. Commun. **11**(1), 2310 (2020)

Chapter 2
Practical Basics and Applications of X-ray Tomography

Xiaogang Yang

2.1 Introduction

Beginning with the advent of X-ray computerized tomography (CT) for routine scanning back in the early 1970s [1], X-ray CT has grown into a powerful imaging modality that can provide the internal three-dimensional (3D) morphology of representative volumes of material science specimens. It is a common tool for the 3D imaging of energy materials and devices. It allows non-destructive and quantitative imaging of complex systems, such as the Lithium-ion battery (LIB) cells. For instance, in the research and development of the LIB, X-ray tomography enables quantification of the structure and the determination of accurate geometric inputs for the cell modeling; observation and analysis of the interplay between electrochemical, mechanical, and thermal effects; and visual confirmation of degradation phenomena [2]. The high-dimensional and high-resolution X-ray tomographic data, which can be used to interpret the complicated thermal-electro-chemo-mechanical interplay that occurs under the operating conditions and collectively determines battery performance [3].

With the third and fourth generations of synchrotron light sources, the X-ray tomography measurement for the energy materials can be done with multiple techniques such as Transmission X-ray Microscopy (TXM) [4], X-ray Phase Contrast Imaging [5], X-ray Diffraction (XRD) [6], X-ray Absorption Near-Edge Structure (XANES) [3, 7, 8], X-ray Fluorescence (XRF) and Ptychography [9]. With these advanced methods, the image contrast is not only limited to the density contrast but also extended to phase contrast and chemical elementary contrast. These applications applied different technologies of X-ray optics to record the individual tomographic projections. However, they use a similar theory of tomography for projection scan and image reconstruction. It is worthwhile to make

X. Yang (✉)
Deutsches Elektronen-Synchrotron DESY, Notkestraße 85, 22607 Hamburg, Germany
e-mail: xiaogang.yang@desy.de

J. Wang (ed.), *Advanced X-ray Imaging of Electrochemical Energy Materials and Devices*, https://doi.org/10.1007/978-981-16-5328-5_2

a general review of tomography for guiding the practice of the experimental measurements.

In this chapter, we introduce the practical basics of X-ray tomography. We start from the origin of the medical CT to explain the fundamental principle of tomography. The mathematical theories are discussed. Following that, three major strategies of tomographic reconstruction are introduced with the example algorithms. The implementations of these algorithms in open-source software are listed for the practice. Then we make an overview of the setup of TXM as an example.

2.2 Tomography Theory

X-ray computerized tomography was developed in medical examinations initially. The basic idea is to scan the specimen in multiple angles to obtain the 2D projections of the X-ray traveling through the specimen. The 3D volume of the internal structure for the specimen can be reconstructed by computational algorithms. The original and most popular way of scanning the projections is to measure the X-ray intensity after the attenuation traveling through the specimen, which is so-called transmission X-ray tomography. However, tomography is not limited to this definition. Any imaging method that measuring the 3D structure of an object by multi-angle scanning can be considered as a branch of tomography.

The first X-ray CT scanner was invented by Godfrey Hounsfield in 1972 and Allan M. Cormack invented a similar system in the same year. They shared the Nobel Prize for their contribution to Medicine research, see Fig. 2.1. Although the modern tomography system became more and more complex, the basic idea behind the measurement is always the same. There are two core steps: scanning of the projection of the coming signal interacted by the object in different angles; tomographic reconstruction from the scanned projections. Besides the medical CT (see Fig. 2.2), tomographic measurements are widely used in scientific research. It is an essential tool for the study of energy materials.

Fig. 2.1 Godfrey Hounsfield and Allan M. Cormack with their original sketch of the first tomography system

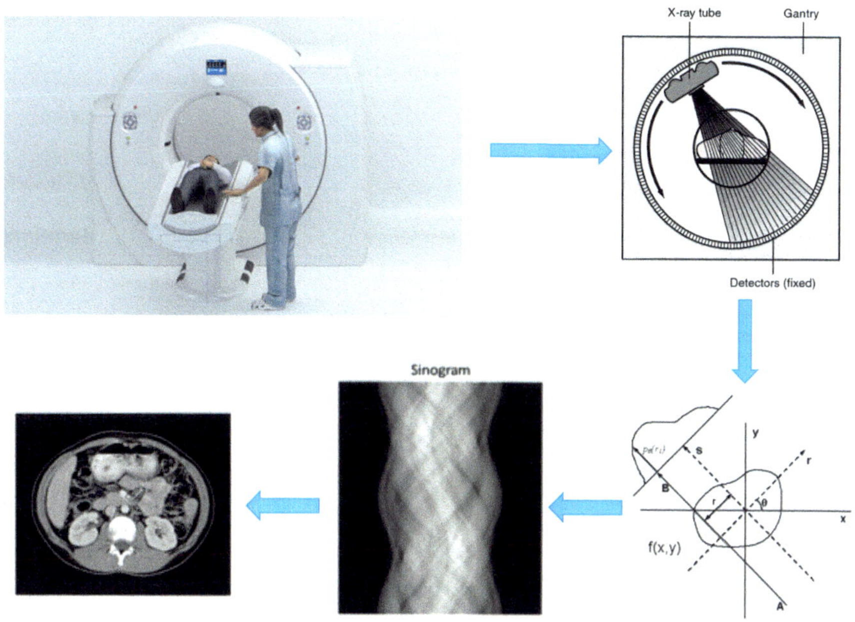

Fig. 2.2 General sketch of the medical X-ray tomography

For an easy start, we introduce the tomography theory with the full field transmission X-ray tomography. There are different geometries of the tomography setup, see Fig. 2.3. Here we introduce the case of parallel beam scan, which is popular in the TXM measurement of synchrotron light sources. The following model is shown base on a 2D slice from the 3D volume. In the parallel beam case, the 3D volume is simply from the heaping up of the 2D slices. Thus, most tomographic models and algorithms start from the 2D case.

The monochromatic traveling through matters following the Lambert–Beer law. The intensity of the X-ray attenuates as:

$$\frac{I_d}{I_o} = \exp(-\mu x) \tag{2.1}$$

where I_o is the incident X-ray intensity, which is the number of photons registered per second when the system is in a vacuum; $I(x)$ is the X-ray intensity after passing a distance through the material; μ is the linear absorption coefficient of the material; x is the distance of the X-ray passing through the material.

In the actual scanning scheme, as shown in Fig. 2.4, we measure

$$g(l, \theta) = -\ln\left(\frac{I_d}{I_o}\right) \tag{2.2}$$

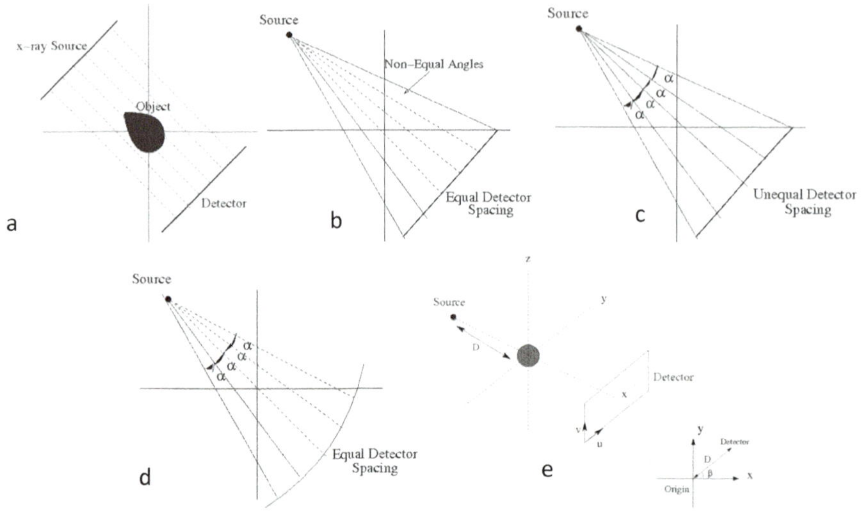

Fig. 2.3 Measurement geometries of the tomography. **a** is the parallel beam tomography; **b**–**d** are different arrangements of the detector for fan-beam tomography; **e** is the corn beam tomography

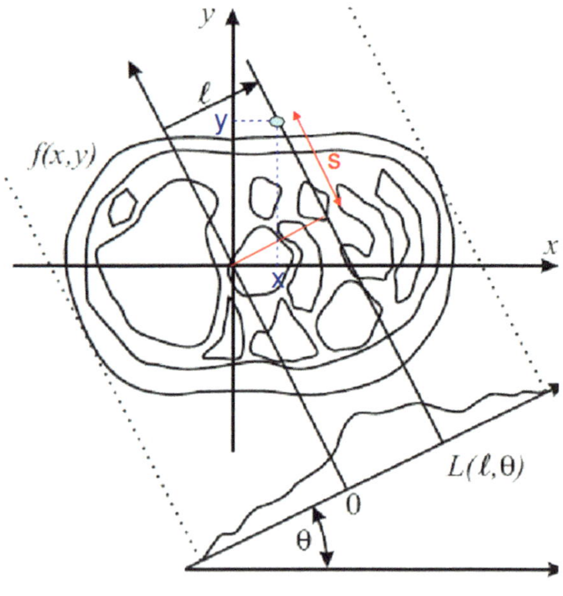

Fig. 2.4 Geometrical sketch of the projection and object for the tomographic scanning

and need to find $f(x, y) = \mu(x, y)$ by using

$$\begin{aligned} g(l, \theta) &= \int_{-\infty}^{\infty} f(x(s), y(s))\mathrm{d}s \\ x(s) &= l \cdot \cos(\theta) - s \cdot \sin(\theta) \\ y(s) &= l \cdot \sin(\theta) + s \cdot \cos(\theta) \end{aligned} \tag{2.3}$$

Therefore, we get

$$g(l, \theta) = \int_{-\infty}^{\infty} \int_{-\infty}^{\infty} f(x, y) \cdot \delta(x \cos\theta + y \sin\theta - l)\mathrm{d}x\mathrm{d}y \tag{2.4}$$

Equation 2.4 is the Radon transform. It calculates the forward projection of the parallel beam traveling through the objects. The tomographic models can be simulated by the forward calculation of the Radon transform. By computing the projections in different angles, the objects are transformed as a sinogram, see Fig. 2.2. The tomographic reconstruction is the inverse of the Radon transform. The form of the equation is simple. However, it is computationally difficult due to the complex conditions of experimental data. Therefore, tomographic reconstruction is a key procedure to obtain the final structural information of the measurements.

2.3 Tomography Reconstruction

There are three major categories of tomographic reconstruction algorithms: the analytical methods, the algebraic methods, and the optimization algorithms, see Fig. 2.5. There is not a perfect reconstruction algorithm overall. Each has its advantage and disadvantage. It would be helpful for the researchers to understand the basics of these algorithms.

2.3.1 Fourier Slice Theorem

Fourier slice theorem is the direct mathematical attempt to find the inversion of Eq. 2.4. It is the foundation of the analytical methods. We first take the 1D Fourier transform for a projection $g(l, \theta)$:

$$G(\rho, \theta) = \Im_{1D}\{g(l, \theta\} = \int_{-\infty}^{\infty} g(l, \theta)\mathrm{e}^{-j2\pi\rho l}\mathrm{d}l \tag{2.5}$$

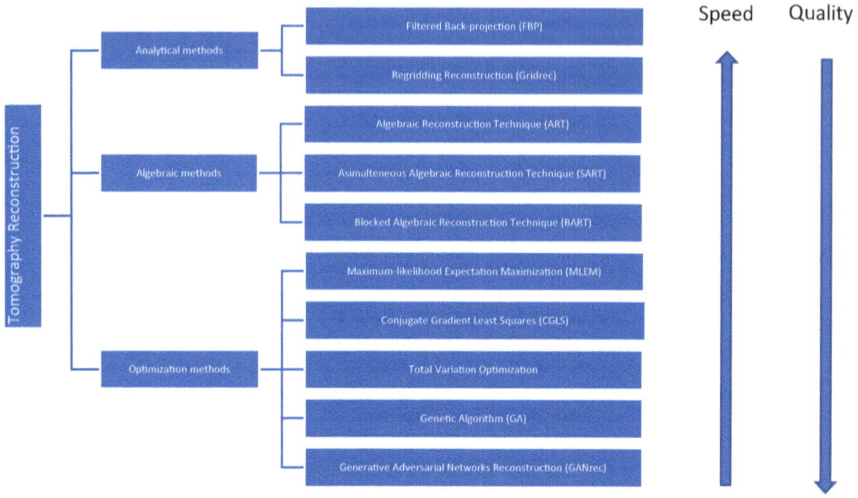

Fig. 2.5 The general list of different strategies of the tomographic reconstruction with some examples of popular algorithms. In practice, computational speed decreases from top to bottom, and the reconstruction quality increases

Next, we substitute the Radon transform for $g(l, \theta)$:

$$g(l, \theta) = \int_{-\infty}^{\infty} \int_{-\infty}^{\infty} f(x, y) \cdot \delta(x \cos \theta + y \sin \theta - l) \mathrm{d}x\mathrm{d}y \tag{2.6}$$

$$G(\rho, \theta) = \int_{-\infty}^{\infty} \int_{-\infty}^{\infty} \int_{-\infty}^{\infty} f(x, y) \cdot \delta(x \cos \theta + y \sin \theta - l) e^{-j2\pi\rho l} \mathrm{d}x\mathrm{d}y\mathrm{d}l \tag{2.7}$$

Then, we do a little rearranging:

$$G(\rho, \theta) = \int_{-\infty}^{\infty} \int_{-\infty}^{\infty} f(x, y) \left\{ \int_{-\infty}^{\infty} \delta(x \cos \theta + y \sin \theta - l) e^{-j2\pi\rho l} \mathrm{d}l \right\} \mathrm{d}x\mathrm{d}y \tag{2.8}$$

Now we apply the properties of the delta function:

$$\begin{aligned} G(\rho, \theta) &= \int_{-\infty}^{\infty} \int_{-\infty}^{\infty} f(x, y) \left\{ e^{-j2\pi\rho[x \cos \theta + y \sin \theta]} \right\} \mathrm{d}x\mathrm{d}y \\ G(\rho, \theta) &= \int_{-\infty}^{\infty} \int_{-\infty}^{\infty} f(x, y) \left\{ e^{-j2\pi[x\rho \cos \theta + y\rho \sin \theta]} \right\} \mathrm{d}x\mathrm{d}y \end{aligned} \tag{2.9}$$

It looks like

$$F(u,v) = \int_{-\infty}^{\infty}\int_{-\infty}^{\infty} f(x,y)e^{-j2\pi[xu+yv]}\mathrm{d}x\mathrm{d}y \tag{2.10}$$

with $u = \rho\cos(\theta), \quad v = \rho\sin(\theta)$.

Equation 2.8 becomes the form like:

$$G(\rho,\theta) = \int_{-\infty}^{\infty}\int_{-\infty}^{\infty} f(x,y)e^{-j2\pi[x\rho\cos\theta + y\rho\sin\theta]}\mathrm{d}x\mathrm{d}y \tag{2.11}$$

$$G(\rho,\theta) = F(\rho\cos(\theta),\rho\sin(\theta)) \tag{2.12}$$

It is exactly the form for the 2D Fourier transform of the objective equation $f(x, y)$. This means that if we take the 1D Fourier transform of the projections and rearranged them as the order of their angles, we can get the Fourier transform of the objective equation $f(x, y)$. Then we can do the tomographic reconstruction only by making the inverse of the Fourier transform, see Fig. 2.6.

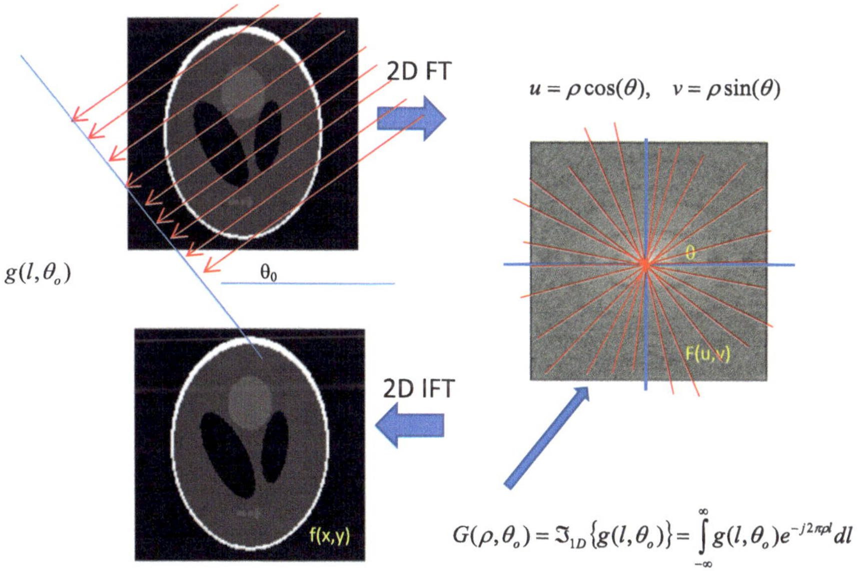

Fig. 2.6 The diagram of the Fourier slice theorem

2.3.2 Analytical Methods

Although there are all kinds of novel algorithms developing for tomography reconstruction, the analytical methods are still the most popular algorithms both in commercial and in research [10]. The most reason is computational efficiency. As we discussed in Sect. 2.3.1, the analytical methods are the direct application of the Fourier slice theorem.

The Filtered back-projection (FBP) algorithm was the earliest tomography algorithm. It is one step in addition to the back-projection process of the Fourier slice theorem. It works as the following steps:

1. 1D Fourier transform for the projections of the in each angle;
2. Rearrange the transformed 1D data in the polar coordinate as the angular orders;
3. Apply the high pass filter to the rearranged data;
4. Performing the inverse Fourier transform to the filtered data to get the final reconstruction.

The reason for the additional step 3 is that the sampling density of the data is not uniform in Fourier space, see the right figure of Fig. 2.6. The low-frequency part is more densely sampled than the high-frequency part, which results in a blurred reconstruction. The high pass filter compensates for the sparse sampling at high frequency, this is a problem of directly using the Fourier slice theorem for tomographic reconstruction. Another way to compensate for this problem is to interpolate the pixels of the high-frequency part in the Fourier domain, which is called Regridding reconstruction (Gridrec).

The analytical reconstruction algorithms are based on the continuous space of the Fourier transform. However, the data of real measurements are always discrete. Considering the geometry of the data rearrangement in polar coordinate, the full angle scan (0–180°) is always required. The sampling density of the scanning angle should fit the number of pixels of the final reconstruction ($N_\theta \approx N_P \frac{\pi}{2}$, where N_θ is the number of projections, N_P is the image width/height of the reconstruction in pixels). This requirement can be satisfied in the commercial X-ray tomography system. However, under-sampled scanning always happens in laboratory measurements. The analytical methods can easily generate reconstruction artifacts for these cases. Advanced algorithms are then needed.

2.3.3 Algebraic Methods

The algebraic algorithms consider the relation between the projections and the attenuation coefficient as the linear equations, see Fig. 2.7.

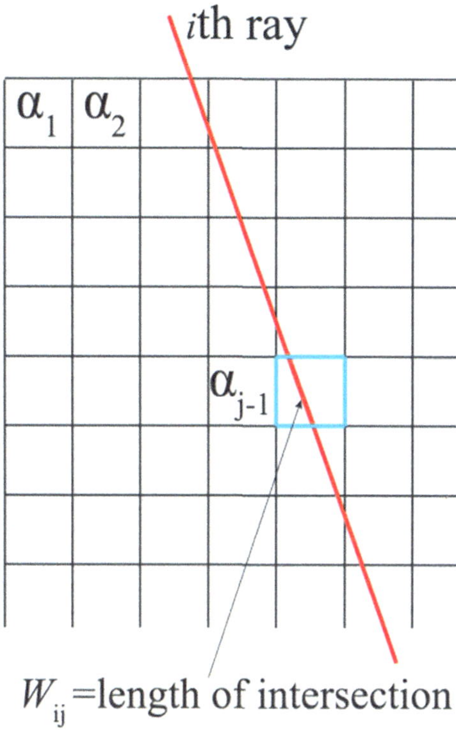

Fig. 2.7 The diagram of the linear equations system for the X-ray attenuation

The process of the reconstruction is to solve the ill-posed linear equations:

$$\tilde{g}_i = \sum_{j=1}^{N} W_{ij} f_j \tag{2.13}$$

The Algebraic Reconstruction Technique (ART) is the first development of the algebraic algorithm family [11]. It follows the formula as:

$$f_j^{(k+1)} = f_j^{(k)} + \frac{g_i - \sum_{j=1}^{N} f_{ji}^{(k)}}{N} \tag{2.14}$$

It is the iterative process to correct the error of f_j gradually from an initial guess of the solution, see Fig. 2.8:

ART proved its improvements on the reconstruction quality on the under-sampled data. However, the correction of the projection is done ray-by-ray locally. The criteria of iteration are also evaluated locally for each angle of projection. The error of the abnormal values such as noise and measurement artifacts, can be accumulated step-by-step, leading to the convergence to the local minimum. In the end, the reconstruction artifacts are easily generated.

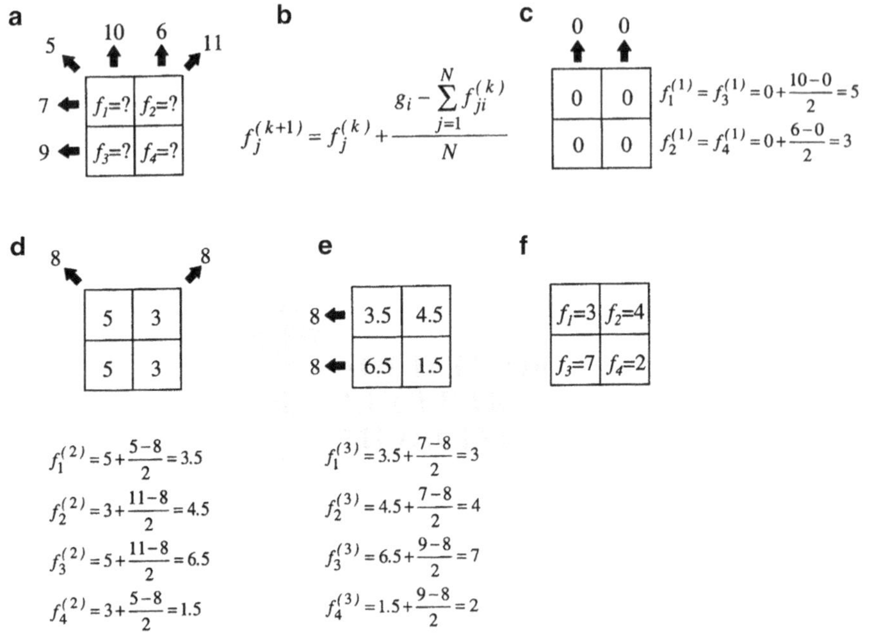

Fig. 2.8 Simple example of the iterations for the ART algorithm

To dispel the problem of the local error accumulation, one further improvement of ART was developed, the Simultaneous Algebraic Reconstruction Technique (SART) [12]. It simultaneously applies to a pixel the average of the corrections generated by all rays, instead of updating the pixels on a ray-by-ray basis sequentially. This is an improvement from the standard ART to reduce the accumulated error in the ray-by-ray iteration. The SART algorithm is written as:

$$f_j^{n+1} = f_j^n + \frac{1}{W_{+,j}} \sum_i^N \left(\frac{g_i - g_i^*}{W_{i,+}} W_{ij} \right) \tag{2.15}$$

where $W_{+,j} = \sum_{i=1}^{N} W_{ij}$, $W_{i,+} = \sum_{j=1}^{M} W_{ij}$, M is the total number of rays and N is the total number of pixels.

There are a lot of different developments of the algebraic family for tomographic reconstruction. The core idea behind them is the same as ART. We would not list all of them. In the same computational condition and coding condition, the algebraic algorithms are slower than the analytical methods. They can get better reconstruction accuracy than the analytical methods for the under-sampled measurements. However, the measurement noise and the artifacts from the data acquisition can significantly influence the convergence of the iterations. They are good alternatives, but not better solutions.

2.3.4 Optimization Methods

Optimization methods using the global optimization algorithms to solve the forward projection of Radon transform directly. The common idea of these algorithms is to minimize the difference between measurement projection and the guessing projection, which is computed by the forward projection of the guessing solution. The final solution is obtained when the difference becomes small enough, ideally 0, but not possible for the real data.

The basic formula of the optimization methods is:

$$\min\|g - g^*\| \tag{2.16}$$

where g is the measurement projection, g^* is the projections computed from the candidate solution.

In the actual computing, the formula normally written as:

$$f = \arg\min_f \left\{ \|g - Wf\|^2 + \lambda\|f\|^2 \right\} \tag{2.17}$$

In principle, any global optimization algorithm can solve this formula. The popular applications are Maximum-likelihood Expectation Maximization (MLEM) [13], Conjugate Gradient Least Squares (CGLS) [14], and Total Variation Optimization (TV) [15], but not limited to them. Some of the novel algorithms of global optimization have also been developed for tomography reconstruction, for example, the Genetic Algorithm (GA) [16].

The deep learning algorithms are a good option for the optimization process to compute the global minimization of the mathematical model. One recent development is to integrate the Generative Adversarial Networks [17] for the iterative reconstruction of tomography (GANrec) [18], see Fig. 2.9. The input of the GANrec is the sinogram to be reconstructed. The sinogram is transformed into a candidate reconstruction by the generator of the GAN algorithm. The candidate reconstruction is projected to a model sinogram by the Radon transform. The model sinogram is compared with the input sinogram by the discriminator of the GAN. A GAN loss is obtained from this comparison. The weights of the generator and discriminator of the GAN are updated by optimizing the GAN loss.

The optimization methods evaluate the reconstruction error globally for each iteration, which is the overall error between the measurement projections and the model projections. They are not easily disturbed by the local errors of the measurement data. In principle, these algorithms can reconstruct the result with better accuracy than the analytical and algebraic algorithms. The gain of the accuracy is traded from the sacrifice of computational efficiency. The optimization algorithms are naturally slower than the analytical and algebraic algorithms. They are normally applied to special measurement setups with sparse scanning density and relatively low-quality data. Fortunately, the computational power is improving rapidly.

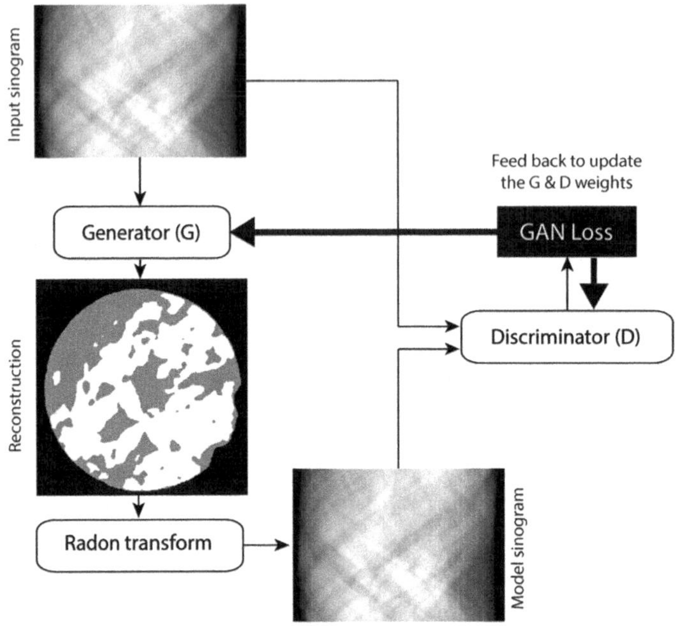

Fig. 2.9 The flowchart of the GANrec algorithm

High-Performance Computing (HPC) and Graphics Processing Unit (GPU) enable parallel computing for tomographic algorithms. The optimization algorithms can also be implemented for large datasets if the experimental condition is limited for the tomographic scan.

2.3.5 *Tomography Software*

The commercial CT systems always include the reconstruction software and their data are encrypted. The image reconstruction and process are not the issues. In scientific research, the data process software is key-important for the self-developed tomographic system. In the synchrotron light source facilities, each beamline has its in-house codes for the tomographic reconstruction. Because of the common needs from different labs, the open-source software's for tomography becomes popular. Tomopy toolbox [19] and Astra toolbox [20] are two softwares that are widely used in scientific research.

Tomopy was developed specifically for synchrotron facilities. It is a python interface and core functions are coded with C, see Fig. 2.10. It works following the order of pre-processing, reconstruction, and post-processing. It includes frequently required tools for these processes. It collected the reconstruction algorithms of Gridrec, ART, and MLEM. Astra toolbox was developed for electron tomography.

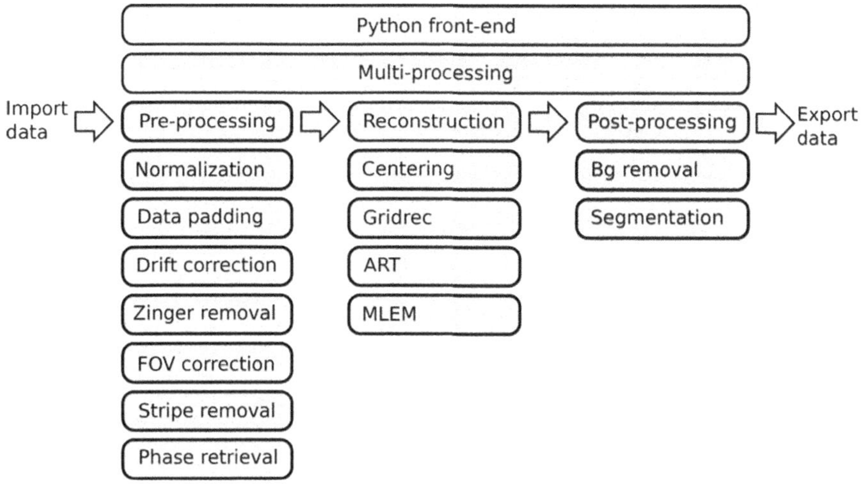

Fig. 2.10 The framework of Tomopy. The figure is extracted from Ref. [19]

However, it is also widely applied for X-ray tomography in both desktop X-ray facilities and Synchrotron X-ray facilities. It has both Matlab and python interface and core functions are coded with C++, see Fig. 2.11. It is optimized for GPU computing, which is a cheap but efficient alternative when the HPC is not available. The reconstruction algorithms of the Astra toolbox are also integrated into the Tomopy toolbox [21]. Both toolboxes made significant upgrades after their initial release. They currently include most of the major and classical reconstruction algorithms. They are valuable tools for the data of the tomographic measurements in energy materials research.

2.4 Experimental Setup of X-ray Tomography

We introduce the experimental setup of Transmission X-ray microscopy (TXM) in the synchrotron light source as an example of the measurement practice on energy materials applications. The hardware of the TXM in the synchrotron mainly contains three parts: the X-ray source, the rotational sample stage, the detector system, and the X-ray optics system if necessary, see Fig. 2.12.

2.4.1 The X-ray Source

The availability of synchrotron X-ray sources, where high fluxes of controllable monochromatic rays are readily generated through the use of dual crystal

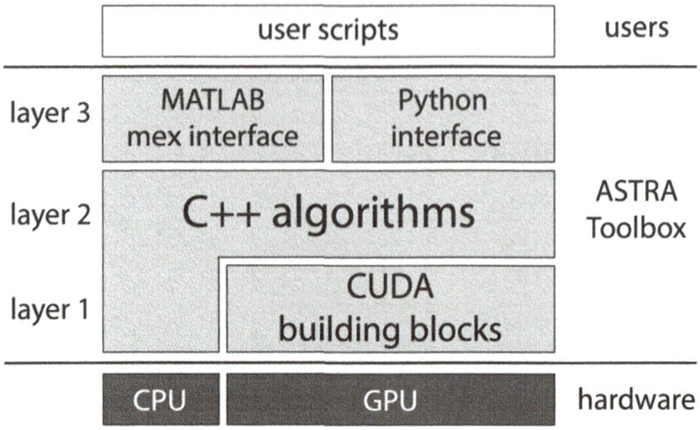

Fig. 2.11 The framework of Astra toolbox. The figure is extracted from Ref. [20]

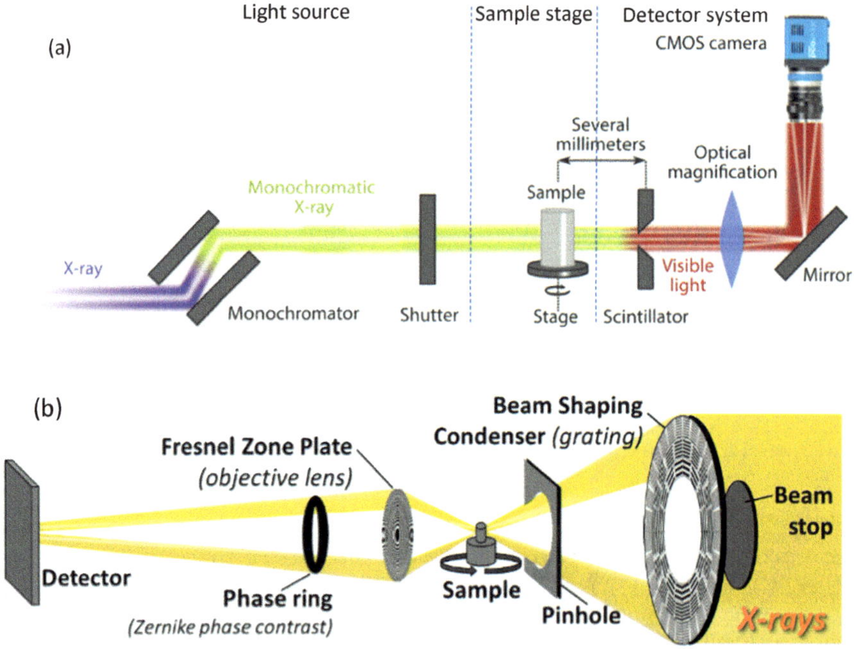

Fig. 2.12 **a** Sketch of a TXM experimental setup [25]. Coherent X-ray penetrates the sample, which is mounted on a rotating stage. The transmitted beam is converted to visible light by the scintillator. A CCD or COMS camera is used to record the converted image. **b** Optics scheme for a nano-tomography of a transmission x-ray microscope (TXM) [26]

monochromators, has greatly transformed the prospects of utilizing TXM. There are TXM beamlines in different synchrotron facilities providing soft X-ray (100 eV ~ 1 keV) and hard X-ray (>5 keV). Normally the hard X-ray is used for TXM imaging of the battery materials.

The tunable energy of X-ray in the synchrotron enables the high flexibility of TXM measurements. Choosing optimum energy for the measurement is important to visualize the object with the best contrast. Following the Lambert-Beer law ($I/I_0 = e^{-\mu d}$), the X-ray intensity of attenuation is the function of the attenuation coefficient and object size. The I/I_0 is suggested approximately 0.135 for good contrast [22]. X-ray mass attenuation coefficients are the function of X-ray energy and the elementary property of the materials. There is not a direct theoretical formula to calculate it. An open-source database such as NIST Standard Reference Database 126 [23] is a useful tool for calculation. Besides, each synchrotron facility has its software for the design of the measurement parameters [24].

2.4.2 The Sample Rotation

The sample stage with rotational capability is the only difference of the tomographic imaging compared with the simple 2D X-ray imaging. The rotation stage is normally the standard commercial hardware. There are two major challenges of the sample rotation: the alignment of the sample and the stability of the stage during the rotation. These problems are especially critical for the nano-tomography, in which the scale of stage motion is beyond the accuracy of the mechanical limits.

In nano-tomography, the rotation axis of the stage can easily be off-centered when mounting the sample. In the mesh grid of reconstruction algorithms, the rotating center should be aligned to the center of the detector, see Fig. 2.13. If it is shifted, the reconstruction will show the deformation of the pattern and fake shadows. Given an accurate value of rotation axis to the reconstruction algorithm is important to obtain a correct reconstruction. There are automatic algorithms to calibrate the rotation axis [27, 28]. These traditional algorithms are not robust for noisy and poor-condition data. Manual calibration is the most reliable way for these cases, but is not feasible for large datasets. Deep learning is thus developed to mimic the manual calibration of the rotation axis [29].

The unstable sample stage during thetomographic scan causes another problem for the data process. It results in similar fake reconstruction as the off-centered rotation axis, but is much more difficult to calibrate. This problem is not so common for TXM measurement. However, it is still a challenge for some beamlines with strong environmental vibrations. In nano-probe based tomography, such as XRF tomography and tomographic ptychography, the alignment of projection is crucial. Although the early development of the alignment algorithms was developed for electron microscopy tomography, they are also valid for the TXM data. The most common approach is to use cross-correlation between projections acquired at

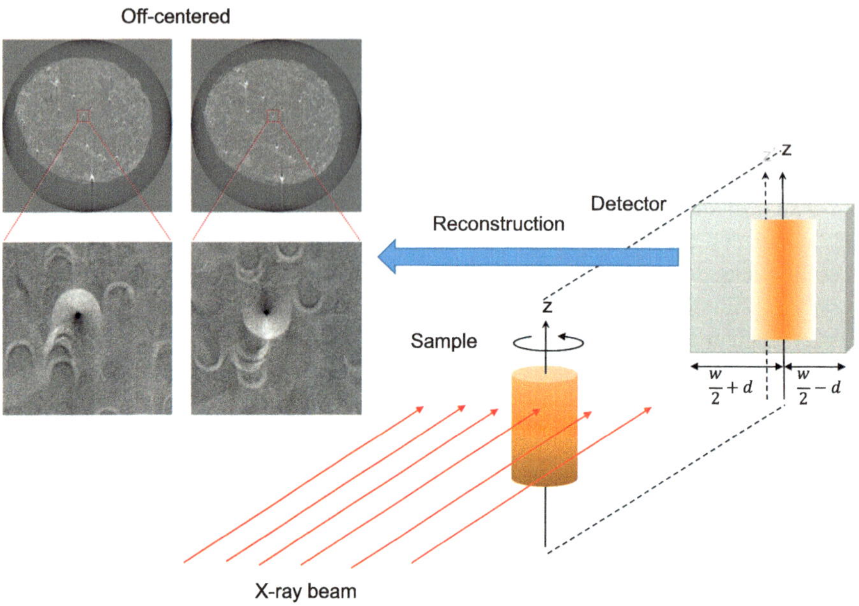

Fig. 2.13 The reconstruction when the rotation axis is off-centered

adjacent rotation angles [30, 31], or correlation of vertical variations in the mass of the sample [32]. The iterative forward and backward projection algorithm is another option for this case [33]. All the current computational alignments are limited to the requirement of good data quality. Perform the measurement properly to avoid misalignment is the best choice to avoid such a data problem.

References

1. Hounsfield, G.N.: Computerized transverse axial scanning (tomography): Part 1 Description of system. Br. J. Radiol. **46**, 1016–1022 (1973)
2. Wood, V.: X-ray tomography for battery research and development. Nat. Rev. Mater. **3**, 293–295 (2018)
3. Yu, Z., Wang, J., Liu, Y.: High-dimensional and high-resolution x-ray tomography for energy materials science. MRS Bull. **45**, 283–289 (2020)
4. Wang, J., Chen-Wiegarta, Y.C.K., Wang, J.: In situ chemical mapping of a lithium-ion battery using full-field hard X-ray spectroscopic imaging. Chem. Commun. **49**, 6480–6482 (2013)
5. Eastwood, D.S., et al.: The application of phase contrast X-ray techniques for imaging Li-ion battery electrodes. Nucl. Instruments Methods Phys. Res. Sect. B **324**, 118–123 (2014)
6. Jensen, K.M.Ø., et al.: X-ray diffraction computed tomography for structural analysis of electrode materials in batteries. J. Electrochem. Soc. **162**, A1310–A1314 (2015)
7. Meirer, F., et al.: Three-dimensional imaging of chemical phase transformations at the nanoscale with full-field transmission X-ray microscopy. J. Synchrotron Radiat. **18**, 773–781 (2011)

8. Yang, F., et al.: Nanoscale morphological and chemical changes of high voltage lithium-manganese rich NMC composite cathodes with cycling. Nano Lett. **14**, 4334–4341 (2014)
9. Yu, Y.S., et al.: Three-dimensional localization of nanoscale battery reactions using soft X-ray tomography. Nat. Commun. **9**, 1–7 (2018)
10. Pan, X., Sidky, E.Y., Vannier, M.: Why do commercial CT scanners still employ traditional, filtered back-projection for image reconstruction? Inverse Prob. **25**, 1230009 (2009)
11. Herman, G.T., Lent, A., Rowland, S.W.: ART: mathematics and applications. A report on the mathematical foundations and on the applicability to real data of the algebraic reconstruction techniques. J. Theor. Biol. **42**, 1–32 (1973)
12. Andersen, A.: Simultaneous algebraic reconstruction technique (SART): a superior implementation of the ART algorithm. Ultrason. Imaging **6**, 81–94 (1984)
13. Nuyts, J., Michel, C., Dupont, P.: Maximum-likelihood expectation-maximization reconstruction of sinograms with arbitrary noise distribution using NEC transformations. IEEE Trans. Med. Imaging **20**, 365–375 (2001)
14. Zhu, W., et al.: Iterative total least-squares image reconstruction algorithm for optical tomography by the conjugate gradient method. J. Opt. Soc. Am. A **14**, 799 (1997)
15. Ritschl, L., Bergner, F., Fleischmann, C., Kachelrieß, M.: Improved total variation-based CT image reconstruction applied to clinical data. Phys. Med. Biol. **56**, 1545–1561 (2011)
16. Yang, X., van Ommen, J.R., Mudde, R.F.: Comparison of genetic algorithm and algebraic reconstruction for X-ray tomography in bubbling fluidized beds. Powder Technol. **253**, 626–637 (2014)
17. Goodfellow, I. et al.: Generative adversarial networks. Commun. ACM 63, 139–144 (2020)
18. Yang, X., et al.: Tomographic reconstruction with a generative adversarial network. J. Synchrotron Radiat. **27**, 486–493 (2020)
19. Gürsoy, D., De Carlo, F., Xiao, X., Jacobsen, C.: TomoPy: a framework for the analysis of synchrotron tomographic data. J. Synchrotron Radiat. **21**, 1188–1193 (2014)
20. van Aarle, W., et al.: The ASTRA toolbox: a platform for advanced algorithm development in electron tomography. Ultramicroscopy **157**, 35–47 (2015)
21. Pelt, D.M., et al.: Integration of TomoPy and the ASTRA toolbox for advanced processing and reconstruction of tomographic synchrotron data. J. Synchrotron Radiat. **23**, 842–849 (2016)
22. Grodzins, L.: Optimum energies for x-ray transmission tomography of small samples: Applications of synchrotron radiation to computerized tomography I. Nucl. Instruments Methods Phys. Res. **206**, 541–545 (1983)
23. Hubbell, J., Seltzer, S.: Tables of X-Ray Mass Attenuation Coefficients and Mass Energy-Absorption Coefficients 1 keV to 20 MeV for Elements Z = 1 to 92 and 48 Additional Substances of Dosimetric Interest, http://physics.nist.gov/PhysRefData/XrayMassCoef/cover.html. (1995)
24. Henke, B., Gullikson, E., Davis, J.: X-Ray Interactions: Photoabsorption, Scattering, Transmission and Reflection E = 50-30,000 eV, Z = 1-92. Atomic Data and Nuclear Data Tables, 54(2). LBNL Report #: LBL-33908. Retrieved from https://escholarship.org/uc/item/9wh2w9rg (1993)
25. Pietsch, P., Wood, V.: X-ray tomography for lithium ion battery research: a practical guide. Annu. Rev. Mater. Res. **47**, 451–479 (2017)
26. De Andrade, V., et al.: A new transmission x-ray microscope for in-situ nano-tomography at the APS. SPIE 9967, 11 (2016)
27. Azevedo, S.G., Schneberk, D.J., Fitch, J.P., Martz, H.E.: Calculation of the rotational centers in computed tomography sinograms. IEEE Trans. Nucl. Sci. **37**, 1525–1540 (1990)
28. Yang, Y., et al.: Registration of the rotation axis in X-ray tomography. J. Synchrotron Radiat. **22**, 452–457 (2015)
29. Yang, X., De Carlo, F., Phatak, C., Gürsoy, D.: A convolutional neural network approach to calibrating the rotation axis for X-ray computed tomography. J. Synchrotron Radiat. **24**, 469–475 (2017)

30. Guckenberger, R.: Determination of a common origin in the micrographs of tilt series in three-dimensional electron microscopy. Ultramicroscopy **9**, 167–173 (1982)
31. Hayashida, M., Terauchi, S., Fujimoto, T.: Automatic coarse-alignment for TEM tilt series of rod-shaped specimens collected with a full angular range. Micron **41**, 540–545 (2010)
32. Guizar-Sicairos, M., et al.: Phase tomography from x-ray coherent diffractive imaging projections. Opt. Express **19**, 21345 (2011)
33. Gürsoy, D., et al.: Rapid alignment of nanotomography data using joint iterative reconstruction and reprojection. Sci. Rep. **7**, 1–12 (2017)

Chapter 3
Advanced Transmission X-ray Microscopy for Energy Materials and Devices

Qingxi Yuan, Xiqian Yu, Hongyi Pan, and Kai Zhang

3.1 Transmission X-ray Microscopy

Transmission X-ray microscopy (TXM) can acquire a full-field projection image with spatial resolution of tens of nanometers using one-time exposure. Thus, it is easy to combine this imaging method with computed tomography and to get three-dimensional (3D) morphology information of samples. In this chapter, we will introduce the methods of TXM and corresponding operation principles, the application examples of energy materials and devices, and the possible development of these imaging technologies.

3.1.1 Methods of Transmission X-ray Microscopy

There are two kinds of methods currently used for full-field TXM, one is zone plate-based TXM, which uses zone plate as imaging optics to magnify the sample information to detector and get the magnified sample image at detector. The other is KB (Kirkpatrick-Baez) mirror-based TXM, which uses KB mirror as focusing optics to form a focal spot with tens of nanometers and uses the beam magnifying effect after focal spot to get magnified image information at the detector. The details of these two methods are given as the following.

Q. Yuan (✉) · K. Zhang
Institute of High Energy Physics, Chinese Academy of Sciences, Beijing, China
e-mail: yuanqx@ihep.ac.cn

X. Yu · H. Pan
Institute of Physics, Chinese Academy of Sciences, Beijing, China

J. Wang (ed.), *Advanced X-ray Imaging of Electrochemical Energy Materials and Devices*, https://doi.org/10.1007/978-981-16-5328-5_3

3.1.1.1 KB-mirror-Based Transmission Microscopy

KB-mirror-based TXM using synchrotron source was first developed at beamline ID16A at European Synchrotron Radiation Facility (ESRF), with the imaging scheme shown in Fig. 3.1 [1]. Using this setup, the synchrotron radiation X-ray beam from the beamline is focused by the KB mirror to form a very small focal spot and at the same time a divergent beam after a focal point. The sample is set at a small distance Z_s (tens of millimeters) downstream of the focus and Fresnel diffraction patterns with variable magnification are recorded on the detector set at a large distance Z_d (several meters). The magnification factor M is given by $(Z_s + Z_d)/Z_s$. Changing Z_s allow us to vary the magnification of the system. The imaging field of view (FOV) is defined by the beam divergence and the focus-sample distance. At a fixed setup, if we want to use a larger FOV, we have to reduce the magnification or set a longer focus-detector distance. The imaging spatial resolution of this imaging method depends on the size of the focal spot. With the development of the KB-mirror manufacturing process and mirror bending technology, the available focal spot size formed by the KB mirror becomes smaller and smaller. A focal spot of less than 13 nm size (FWHM) was realized at beamline ID16A of ESRF using 33.6 keV X-rays and graded multilayer-coated KB mirrors [2], meaning that about 10 nm spatial resolution can be got with this method if the stability of the whole system is good enough.

For KB-mirror-based TXM system, one of the important advantages is its X-ray energy can go higher than 30 keV without needing to consider the efficiency loss, meaning that dynamical studies can be carried out at higher energies with considerable photon flux. Since a Fresnel diffraction radiograph can be easily acquired due to the propagation between sample and detector, another advantage is that phase-contrast imaging mode can be easily performed using this approach, especially for those samples with very weak X-ray attenuation. However, this at the same time becomes a limitation for dynamic studies. Since the boundary

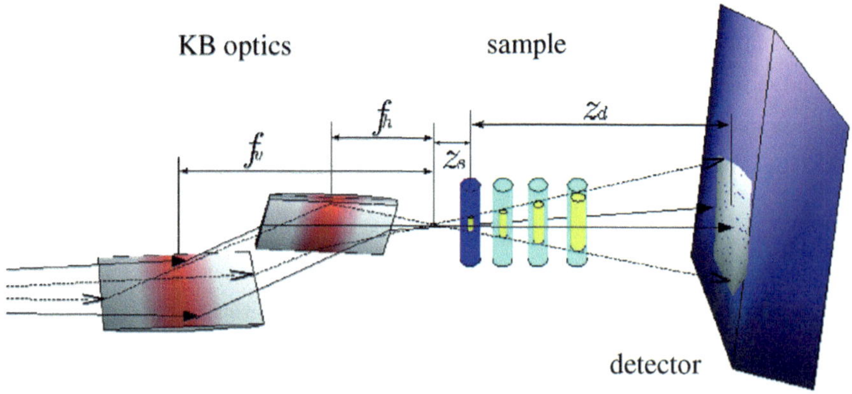

Fig. 3.1 The schematic diagram of KB-mirror-based TXM [1]

enhancement effect from the phase effect will disturb the projected image, at least three projection images with different sample-detector distance are needed to reconstruct phase-contrast image and absorption contrast image of the specimen, which limits possible in-situ/operando application. Another limitation of the KB-mirror-based TXM system is the stability needed is at a very high level. The smaller the focal spot size is, the higher imaging spatial resolution we can get, the higher stability for the whole system is needed. Typical beamline at synchrotron facility for this kind of TXM system is ID16A [3] at ESRF and P10 [4] at PETRA III. This imaging approach is still under development and is not the focus of this chapter. Interested readers can refer to the relevant literature.

3.1.1.2 Zone Plate-Based Transmission Microscopy

Zone plate-based full-field X-ray microscopes were developed over the past half-century, first by a group at the University of Gottingen [5] and later by the center for X-ray optics in Berkeley [6]. Due to the development of manufacturing technology for zone plates, microscopes of this type are rapidly gaining popularity with instruments operating soft X-ray (180 eV–2 keV, mainly 200 eV–600 eV), tender X-ray (2 keV–5 keV), and hard X-ray (5–12 keV). Figure 3.2 shows the schematic diagram of the zone plate-based TXM system developed at Advanced Light Source (ALS). Zone plate-based TXM system operated at tender X-ray or hard z-ray region uses a similar setup.

As shown in Fig. 3.2, the condenser optics can be zone plate for soft X-ray, while ellipsoidal mono-capillary condenser [7] is often used for hard X-ray due to its high reflection efficiency (usually higher than 80%) and to avoid the lower efficiency of zone plate condenser. Absorption contrast images can be acquired using these optics. If an appropriately designed phase ring is inserted into the optical path at the back focal plane of the zone plate, the Zernike phase-contrast image can be acquired with the same resolution level at designed X-ray energies.

For zone plate-based TXM system, X-rays from beamline or laboratory source is focused by the condenser to sample position, and the front focus of the imaging

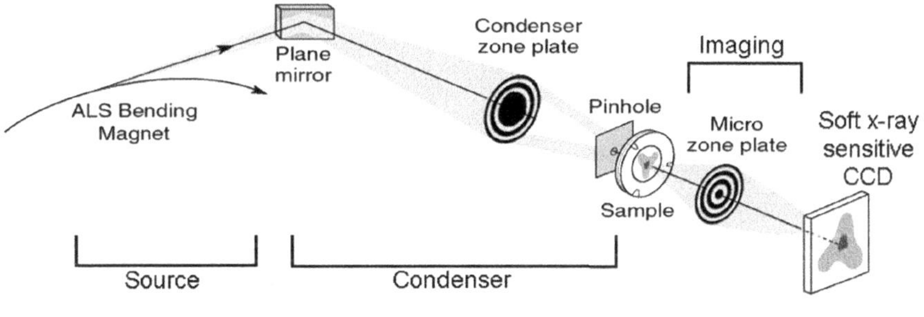

Fig. 3.2 Schematic diagram of zone plate-based TXM system developed at ALS

zone plate is set a little downstream of sample, resulting in that imaging zone plate can form a magnified image of the sample to the detector. The magnification is defined by the value of image distance divided by object distance, which is similar to that in optical lens imaging. When a lens coupled CCD or CMOS detector is used, the whole magnification is the product of zone plate magnification and lens magnification. In this system, the imaging FOV of zone plate-based TXM depends on focal spot size defined by a condenser, and imaging spatial resolution depends on the outermost zone width of the zone plate. The diffraction efficiency of the zone plate depends on X-ray energy and the aspect ratio of its zone. The higher the X-ray energy used, the higher the zone aspect ratio needed. Since the high aspect ratio zone plate for hard X-ray region is not easy to manufacture, zone plate-based TXM system cannot be operated with X-rays higher than 15 keV now due to very low zone plate diffraction efficiency. Typical spatial resolution of zone plate-based TXM system is about 10 nm [8, 9] at soft X-ray region and 20–30 nm [10–13] for hard X-ray.

There are two important parameters for zone plate imaging. One is the beam size at the sample position, which determines the imaging FOV. The other is the beam divergency downstream of the sample, which is usually two times of zone plate numerical aperture to fully illuminate the whole zone plate area. Beam size times beam divergence at the sample position, which is usually limited by the emittance of the source and the acceptance of the system, is the phase space required by zone plate imaging. For third-generation synchrotron sources, the source emittance is usually in tens of nm·rad, meaning that the phase space at the source point is less than what is needed by zone plate imaging. One approach that has been implemented to overcome this problem is shaking the capillary condenser in horizontal and vertical directions to expand the beam size at the sample position and to fully illuminate the zone plate. But this reduces the possibility for imaging in Zernike phase-contrast mode. Another approach is using a newly developed optics beam shaper condenser (BSC) [14]. The idea of BSC design is to divide a conventional zone plate into sectors, keeping the local spatial frequency within each sector constant, and the first diffraction order of every sub grating will form exactly coinciding illumination in its focal plane with the illuminated area equal to the size of the gratings. As a result, each sector will produce a flat-top illumination in the focal plane. Since BSC is a kind of Köhler illumination optics, the illumination size and angle on the sample can be designed according to the requirement of zone plate imaging. The illumination size on a sample is determined by the size of the sector, and the illumination angle on the sample is determined by the size of the beam shaper condenser and its focal length. The manufacturing process of the beam shaper condenser is similar to that of the zone plate, therefore the diffraction efficiency mainly depends on the grating aspect ratio and the X-ray energy to be used. Although there will be some limitations to use this kind of optics in higher-energy X-rays, BSC does provide a solution for zone plate-based TXM at low emittance synchrotron source, especially for those nearly diffraction-limited 4th generation storage ring photon source. Several BSC based TXM system [15–18] with absorption contrast and Zernike phase-contrast imaging modes have been built at

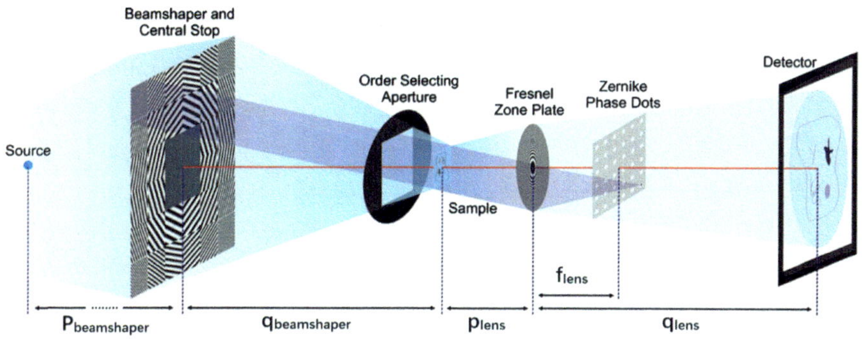

Fig. 3.3 Schematic optical setup for BSC based full-field transmission microscope at TOMCAT beamline of Swiss Light Source [14]

the currently operating synchrotron radiation facility. Figure 3.3 shows the typical optical setup of BSC based TXM system, which is similar to that of a zone plate-based TXM system.

Besides the low diffraction efficiency from zone plate and/or beam shaper condenser, the zone plate-based TXM is not easy to achieve at X-ray energy higher than 15 keV now due to manufacturing difficulties of the corresponding zone plate. This is the limitation of this imaging method. While the advantages of this method can be summarized that it is easy to realize absorption imaging mode or Zernike phase-contrast imaging mode, meaning that zone plate-based TXM can give absorption image of specimen without phase retrieval process which is necessary for KB-mirror-based TXM. With absorption images, the attenuation difference from different parts of a specimen can be acquired at 2D or 3D space. With a synchrotron radiation source, when two images are taken just above and below the K-edge energy of a specific element and logarithmical subtraction is done with these two images, the element contrast will emerge. This technique is K-edge subtraction (KES) imaging, whose contrast comes from a very big attenuation difference of a specific element while there is only very little attenuation difference comes from other elements at these two X-ray energies. Figure 3.4 is an example showing the KES imaging contrast from a porous Ni–YSZ (yttria-stabilized zirconia) specimen. Since Ni K-edge is 8333 eV as shown in Fig. 3.4a, the Ni element contrast is obviously shown with two sets of images taken below Ni K-edge (8300 eV) and above Ni K-edge (8400 eV) shown in Fig. 3.4b. The absorption contrast-based KES imaging is an interesting approach to acquire 2D or 3D distribution of those phases containing specific elements.

Another important approach based on absorption contrast image with zone plate-based TXM is TXM XANES imaging [19, 20], which can give morphological and chemical state distribution information at nanoscale resolution by combing zone plate-based TXM with X-ray absorption near-edge structure (XANES) spectroscopy. Figure 3.5 shows the concept of TXM XANES imaging. With tuning the monochromator of beamline with a fairly small step near K-edge region of a

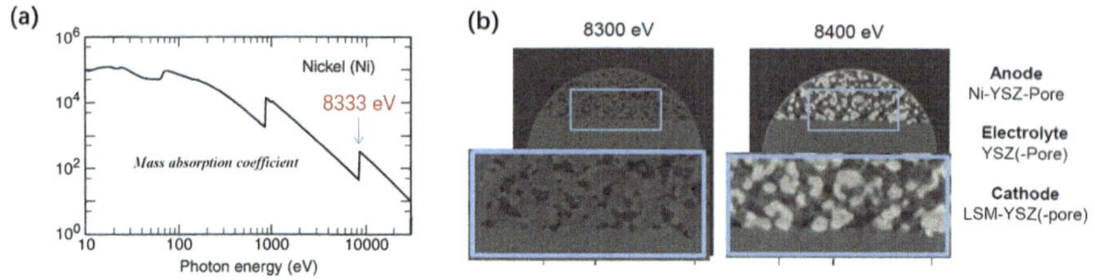

Fig. 3.4 Nickle contrast in Ni–YSZ using two images taken just below and above Ni K-edge energy

specific element (in Fig. 3.5 Ni), a series of projection image or tomography data is collected with the energy scanning, resulting in 2D or 3D XANES imaging data. Since zone plate focusing is energy-dependent, zone plate z position needs to be linked to each energy during the data collection process. After reference correction and image correlation, the intensity change of each pixel as a function of energy gives XANES spectra. The chemical phase information from each pixel XANES spectra can be expressed in a 2D colorful map, and then a 3D chemical state map can be got using tomographic reconstruction of these 2D maps at all projection angles. As a non-destructive 2D or 3D high-resolution chemical state imaging technique, TXM XANES shows great potential in those studies of energy materials and devices, especially combined with the study of charge/discharge or other processes.

3.2 Applications of TXM in Energy Materials and Devices

With the continuous development of high-brightness, high-energy X-ray sources and the improvement of the efficiency of optical components, nanometer resolution X-ray microscopic imaging device has been successively built at many synchrotron radiation facilities (such as SSRL, NSLS, APS, ESRF, etc.) and imaging laboratories in the world. The construction of these imaging devices has added new research methods and is opening vast scientific opportunities for mankind to study the microscopic world. Compared with visible light and electron beams, X-rays have short wavelengths and penetration capability, it also has diverse imaging mechanisms (such as absorption, fluorescence, valence state, spin, phase, etc.), and the characteristics of rich contrast sources, so nanometer resolution X-ray microscopy imaging technology can greatly improve the resolution of optical microscopes and can also be applied to map the distribution of several chemical elements. Besides, nanometer resolution X-ray microscopic imaging technology is also different from electron microscopy. When combined with computer tomography technology, nanometer resolution X-ray microscopy can get the non-destructive 3D structure information of the sample with tens of nanometers resolution. All these unique characteristics allow X-ray microscopy to effectively make up for the

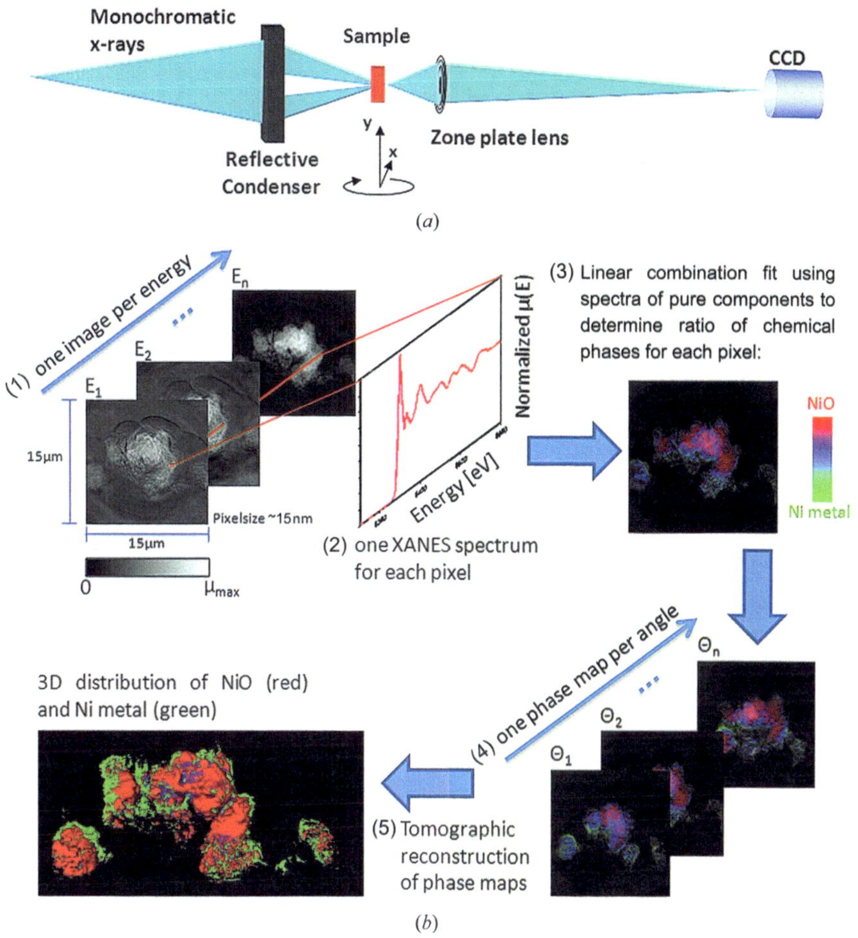

Fig. 3.5 Microscope setup and principles of data processing for 3D XANES microscopy

deficiencies of optical microscopes and electron microscopes. And it also made nanometer resolution X-ray microscopic imaging technology a great success in the explorative research of medicine, biology, material science, chemistry, and other disciplines.

In this section, some representative application examples will be discussed in detail. These application examples can reflect the extremely rich amount of sample information in nanometer resolution X-ray imaging data. Whether this information can be effectively explored determines the role that nanometer resolution X-ray microscopic imaging technology can play in specific scientific research topics. We hope that the discussion of these scientific research cases in this section can inspire readers and promote the development and application of nanometer resolution X-ray microscopic imaging technology.

3.2.1 Application of TXM on Structural Imaging

The most direct information that could get from TXM is the bulk structure and the composition of the sample. In 2018, Xia et al. [21] use nanometer resolution X-ray microscopy (nano-XCT) to observe $LiNi_{0.6}Mn_{0.2}Co_{0.2}O_2$ secondary particles. Zone plate-based TXM is performed after cycling NMC cathode at different cycle rates (1 C, 2 C, 5 C, and 10 C) between 2.5 V and 4.5 V for 50 cycles. From the 2D and 3D images, it can be found that particles display different degrees of damage, which could infer high cycle rate could cause severer cathode particle degradation (Fig. 3.6a). In 2020, Jiang et al. [22] also collect nanoscale TXM tomography of NMC cathode. As shown in Fig. 3.6b, NMC particles and carbon/binder can be simply segmented by threshold based on intensity. Apart from structure detection and phase segmentation, in 2020, Lu et al. [23] introducing external compression into TXM characterization that enables in-situ quantification of cathode internal change. This external compressing process called calendaring is believed a key step in the industrial production of lithium-ion batteries because it reduces the electrode thickness and enables higher-energy density. Calendering step achieves these benefits by enhancing the contact between the components inside the electrode, thus improving electrical and thermal conductivity. And it might also modify electrolyte wettability by changing the pore structure contributing to better long-cycle stability and electrochemical performance. However, over the incremental calendering steps, unexpected deformation and rotation of NMC particles are observed (Fig. 3.6c), which could be the inducement that leads to a negative influence on batteries. Panel I–K in Fig. 3.6c shows the distance between particles decreases while calendering ratio increases. Except for cathode materials, nanoscale TXM is also used to characterize composite polymer electrolytes. Zaman et al. [24] observes PEO-based composite polymer electrolyte with different amount of Al–LLZO ($Li_{7.5}La_3Zr_2Al_{0.25}O_{12}$) as inorganic filler. The curve of normalized surface area to the volume fraction of Al–LLZO, which can easily be measured by TXM, fits well with the corresponding ionic conductivity of the composited polymer electrolyte (Fig. 3.6d).

The information mentioned above is all intuitive. Nevertheless, unlike other qualitative and semi-quantitative techniques, TXM is powerful for its quantification ability. In Xia's work [21], 4–6 particles are selected from each sample electrode, afterward, the porosity and specific surface area are analyzed, the volume ratio of unaffected regions and cracks induced diffusion deterrent (α, self-defined parameter) of them are compared (Fig. 3.7a). Accordingly, porosity increases with the rising of cycle rate, so as the specific surface area, which indicates cycle rate positively correlates with the degree of particle damage (see Fig. 3.6a). In the meantime, the inversely correlated α and the unaffected regions prove the cognition that cracks can deteriorate the Li^+ pathway inside NMC particles. Particles of different sizes are also studied. Results show that the degree of particle damage positively correlated with particle size, which is another factor inducing heterogeneous failing of cathode. Moreover, surroundings could affect the target material's

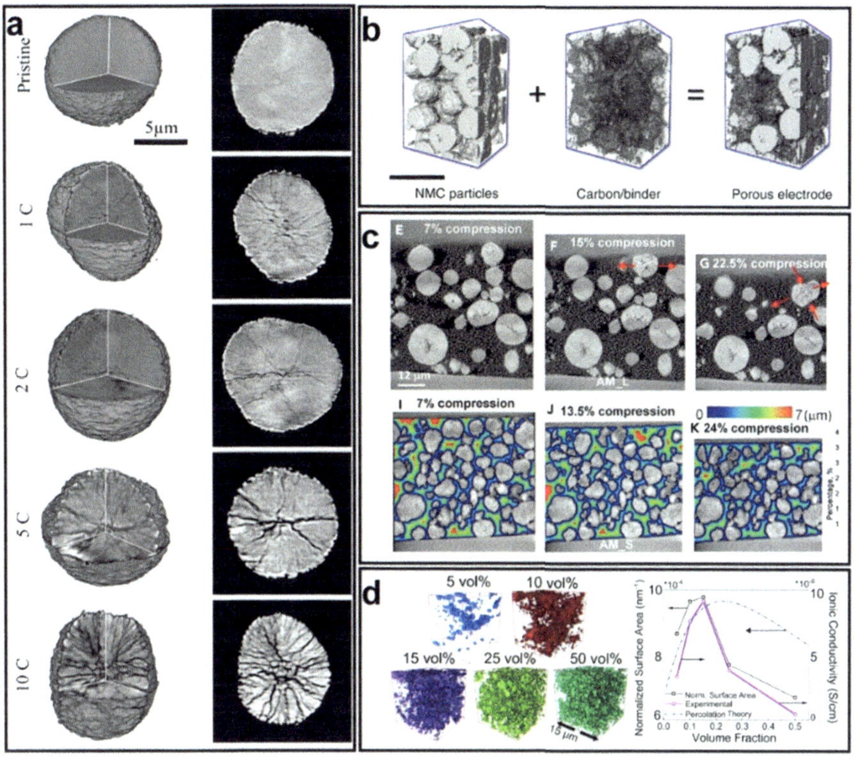

Fig. 3.6 **a** The 3D rendering of the particles and the corresponding 2D slices through the centers of the particles at the pristine state and after 50 cycles at 1 C, 2 C, 5 C, and 10 C [21]; **b** the active NMC particles and inactive carbon/binder domains rendered separately and jointly. The scale bar is 20 μm [22]; **c** in-situ electrode calendering experiment using X-ray nano-CT and the microstructural evolution [23]; **d** reconstructed images for 5, 10, 15, 25, and 50 vol% composite polymer electrolytes, and the measured normalized surface area and experimentally measured ionic conductivity [24]

properties. According to Fig. 3.7b, for NMC particles partially covered by carbon and binder, the coverage is strongly related to the surface local electrical resistance. The more area NMC particles detached from carbon/binder is, the higher the relative electrical resistance would be. Although it cannot be an accurate reflection of real surface electrical resistance, it can still be a good reference for the heterogeneity of surface electronic conductivity. On the other hand, the tomography could also include the distribution and posture information. Take Fig. 3.7c for example, Lu performed a detailed particle morphology analysis [23]. The size and posture of each NMC particle are represented by cylinders with different volumes and orientations, and the color refers to the ovality of that particle. Comparing the size and orientation before and after calendering, it can be told that the ovality distribution of the larger particles (AM_L) is quite wide (Fig. 3.7c, Panel A), indicating a greater

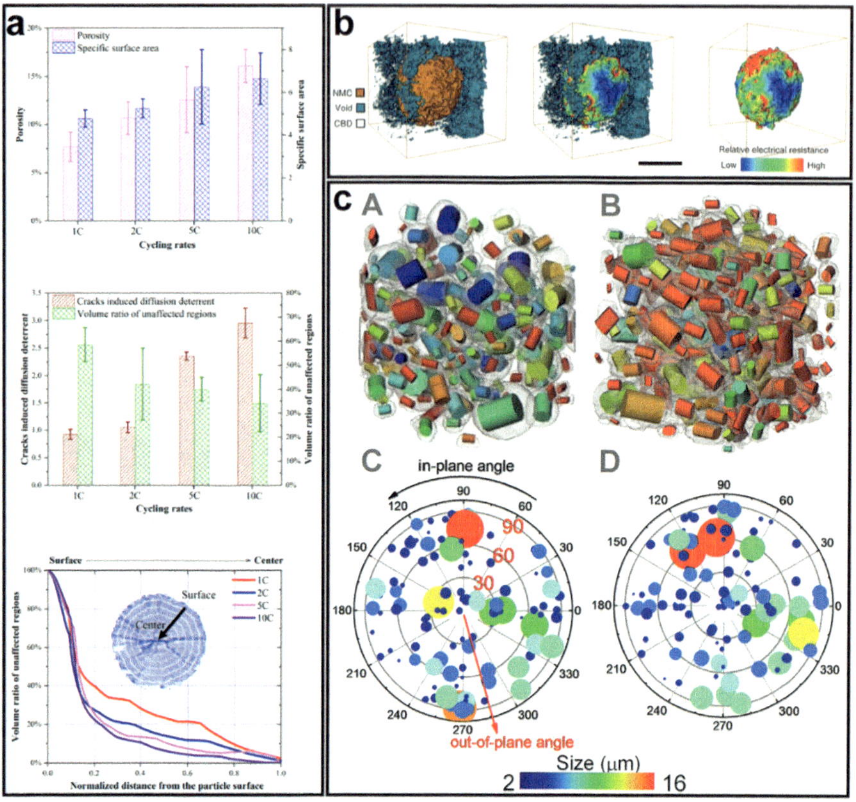

Fig. 3.7 **a** Quantification of the porosity, the specific crack surface area, the crack induced diffusion deterrent, and the unaffected regions, with the depth profile of the spatial distribution of the unaffected regions [21]; **b** visualization and modeling of the NMC particle detachment on the relative local electrical resistance [22]; **c** shape and orientation of NMC particles under calendering [23]

heterogeneity of its microstructure compared to the smaller particles (AM_S). The particles are randomly distributed in AM_L with no meritocratic orientation, even after calendering. AM_S, on the other hand, shows a strong preferential arrangement of particles even before calendering. After calendering, the particles initially located in the inner ring tend to move outward and the preferential orientation becomes more pronounced.

Data collected by TXM seems only able to provide 3D geometrical information and correlated statistical analysis, however, the 3D image is the best model for Multiphysics simulation. Xia [21] simulates electrolyte infiltration into the newly formed cracks to study electrolyte influence on the electronic and ionic conductivity inside the cracked particle (Fig. 3.8a). The result illustrates electrolyte infiltration of the damaged particle can only reduce the ionic diffusion length but not electronic diffusion length, which finally leads to the mismatch between the local electronic and ionic conductivity resulting in heterogeneous distribution of ionic and electronic

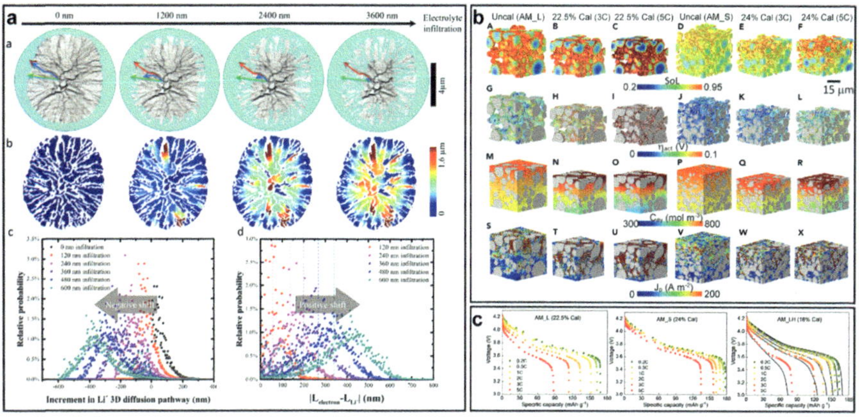

Fig. 3.8 **a** Simulation of the electrolyte infiltration effect on NMC particle [21]; **b** electrochemical parameters obtained at 60% DOD, at discharge rates of 3 C and 5 C with incremental calendering; **c** simulated and experimental discharge performance [23]

traffic load over the particle surface. Lu et al. [23] also simulate electrolyte wetting as well as the spatial distribution of state of lithiation, activation overpotential at the active particle/electrolyte reacting interface, electrolyte concentration, and Li^+ ion flux. The simulation result infers that the wide range of pore sizes in AM_L, attributed to the wide distribution of particle size and shape, affects the homogeneity of the carbon binder area and pore size distribution, while the microstructure in AM_S is more uniform. For the electrode without calendering, AM_L showed overall slower transfer kinetics with only 70% of the permeability of AM_S. However, the permeability of AM_S with calendering is lower than that of AM_L, suggesting that electrolyte wetting is hindered by high surface area and viscous resistance, and thus requires longer wetting times in calendered electrodes composed of fine particles (Fig. 3.8b). Using the rendered 3D model as a cathode, they simulate the discharge performance of each type of cathode. The simulated result displays quite similar regularity with experimental electrochemical performance (Fig. 3.8c).

3.2.2 Application of TXM with XANES Imaging

The combination of XANES with TXM, makes in-situ element valence tracing possible, which can help to explore the real chemical and electrochemical process of the material during battery operation. Wei et al. [25] develop a method that uses the more reliable peak energy of XANES spectra to determine the valence of transition metal. Here, the charged state and the pristine LCO particles are detected and compared. From Fig. 3.9a, it appears that the pristine particles are more homogeneous, while the charged state particles in exhibit oxidative heterogeneity. In addition to the variation in the average oxidation state, the Co K-edge XANES spectra of the entire particle manifest that the charged particles exhibit a broader distribution of Co valence states, as shown by the increasing width of the probability distribution. XANES is performed on the more severely damaged charged state particle, which is filled with cracks and domain boundaries (Fig. 3.9b). This severely damaged particle coexists with normal intact particles in the same electrode, highlighting heterogeneity of the electrode-level reaction. Thus, TXM XANES provides local chemical information of $LiCoO_2$ particles that can illustrate the inhomogeneity of the charge. While Wei uses peak energy as standard, Xu et al. [26] use the energy value at 0.5 of the normalized absorption step height to determine the Mn K-edge positions for each individual pixel of the Li-rich layered cathode materials ($Li_2Ru_{0.5}Mn_{0.5}O_3$). The change in the 2D TXM XANES slices means Mn oxidation is gradually reduced during cycling Fig. 3.9c). Xu divide the cathode particle into several layers and discovered that depth-related Mn reduction heterogeneity happens from the 10th cycle, which could be attributed to the oxygen release. However, this depth dependency is diminished at the 20th cycle, which might be because of the electrolyte infiltration into the newly formed interconnected pore network. Hong et al. [27] detect a bare $LiCoO_2$ particle and found an obvious grain boundary dividing it into two domains, which leads to higher Co valence in both domains than the one without this kind of grain boundary (Fig. 3.10a).

Similar to the capability of TXM to divide different materials through their internal X-ray absorption ability, TXM XANES can map the distribution of multiple elements inside the sample. Zhang et al. [28] collect Al, Co, and Ti signals of the Ti, Mg, Al trace doped $LiCoO_2$ (TMA–LCO). In contrast to the uniformly distributed Al, Ti shows a large degree of separation, which forms a complex interconnection network that divides the $LiCoO_2$ particles into regions (Fig. 3.10b). Active subdomains are defined as the regions with an equal or lower Ti-to-Co ratio than the nominal value. The increase of the specific surface area of those active subdomains is believed to ensure fast Li^+ diffusion. Furthermore, the subdomains separated by the Ti-rich phase can effectively reduce the lattice shrinkage and expansion during cycling and are more resistant to lattice strain and particle fracture, which could improve the long-term cycling stability of TMA-LCO.

3.2.3 Application of TXM with Phase-Contrast Imaging

In 2019, Vanpeene et al. [29] uses phase-contrast TXM to observe large single Si particles in standard and mature electrodes relatively. For standard electrodes, it can be seen that small Si particles around the largest one disappear at ~70% DOD with the decline of their density, and pores start to fill gas at 45% DOD (Fig. 3.11). When DOD reaches 100%, the largest Si particle is also hidden by the surroundings. Cracks and pores enlarge with SOC increase, which is not observed in the mature electrode. On the other hand, Si particles in both the electrodes change their brightness and position after discharge and charge process indicating the unideal reversibility of Si particles and the electrode.

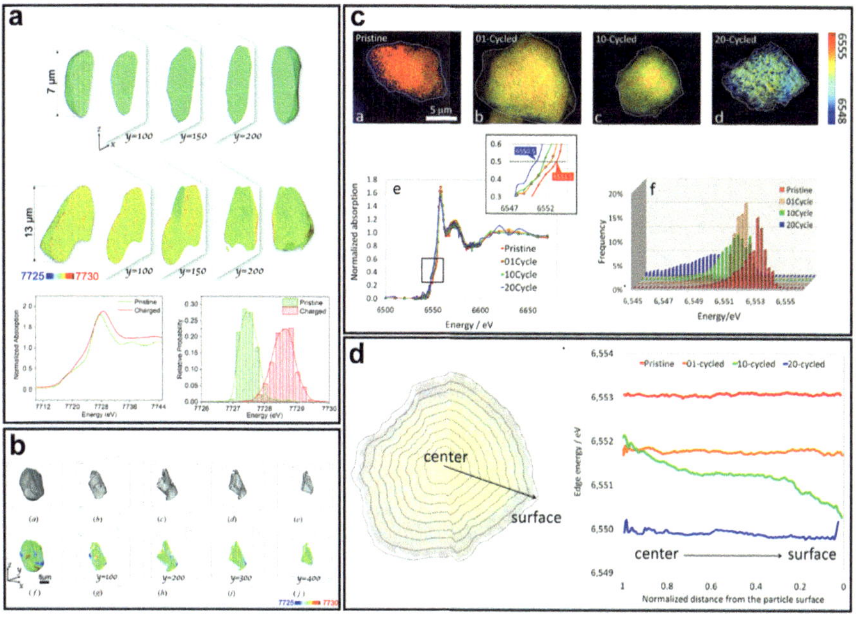

Fig. 3.9 **a** 3D chemical mapping of the LCO particles in pristine and charged states using peak energy as the descriptor; **b** 3D rendering of a charged LCO particle with complicated morphology [25]; **c** 2D nanoscale mapping of the distribution of the Mn oxidation states in particles of different cycles; **d** evaluation of the Mn oxidation distribution's depth dependency [26]

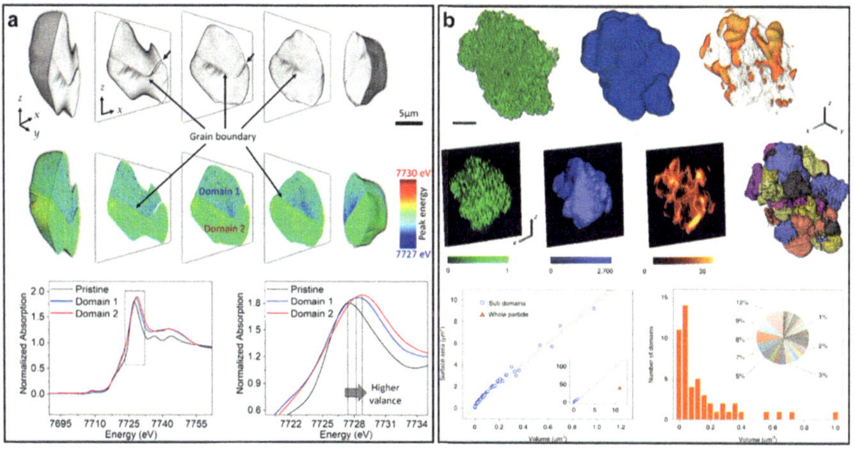

Fig. 3.10 **a** The crystal grain separation and charge heterogeneity in a charged bare LCO particle [27]; **b** 3D X-ray tomography reconstruction and element distribution in TMA–LCO [28]

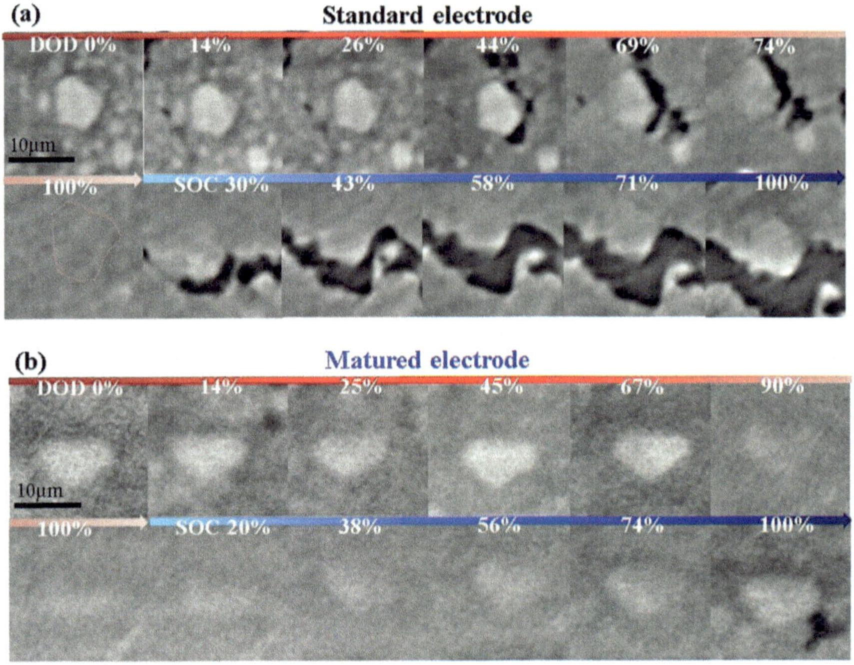

Fig. 3.11 Evolution during the first cycle of XCT 2D images focused on a large Si particle for the standard and matured electrodes [29]

Jiang et al. [22] use phase-contrast nano-tomography and spectro-microscopy simultaneously. By calculating the obtained data, electron density and absorption edge energy map can be acquired from phase-contrast nano-tomography and spectro-microscopy respectively (Fig. 3.12). They further compare a great number of particles to explore the correlation between the two maps finding a reasonably good degree of similarity. The Pearson correlation coefficient of 0.54 confirms the positive correlation between the electron density and the Ni valence state, as well as the local SOC. Thus, with phase-contrast TXM detection, coarse valence state distribution might be able to speculate.

3.2.4 *Application of TXM with KES Imaging*

Solid-oxide fuel cells (SOFCs) are the most widely used fuel cells because they exhibit flexibility, power generation efficiency, and low pollution formation. And many scientists, researchers, and engineers around the globe carried out a variety of research studies to improve the performance of SOFC. Researchers found that SOFC's performance will drop sharply in the long-term operation of SOFC, which is an urgent problem and limit SOFC's widespread application. Therefore, it is very

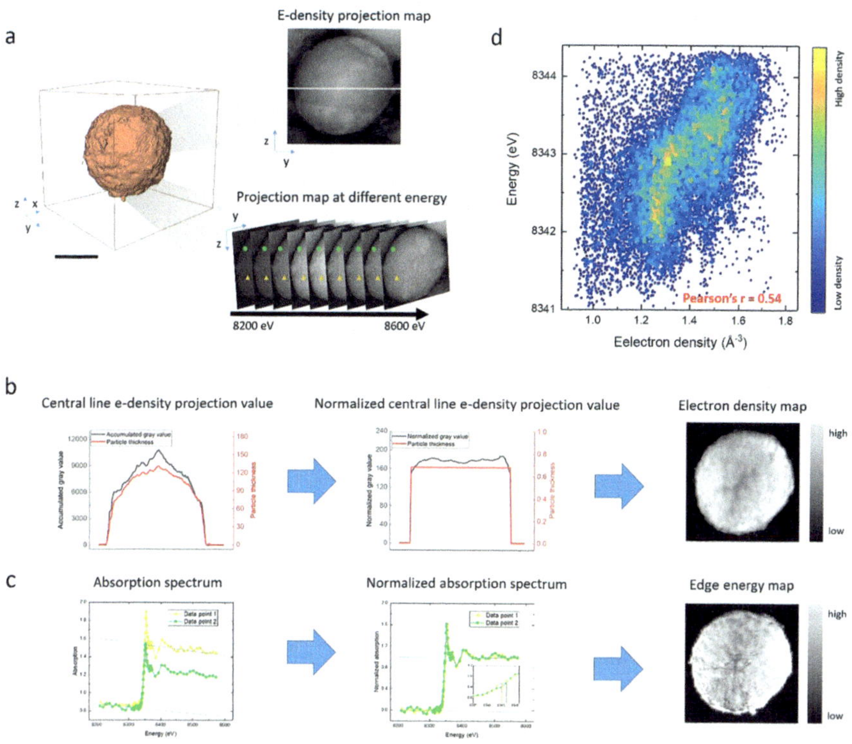

Fig. 3.12 Electron density and Ni valence of an NMC particle [22]

important to understand the mechanism of fuel cell performance degradation by studying the changes in the electrode microstructure after fuel cell operation. For this research, nanometer resolution X-ray microscopy is an ideal research technique with many advantages. The best resolution ranges (20–50 nm) are available in many facilities, providing similar 3D structural feature detectability to SEM. In addition to high spatial resolution, the penetrating power of X-rays enables imaging of large volumes. Moreover, when nanometer resolution X-ray microscopy was combined with absorption edge imaging technique, the 3D element information can be identified by differences in absorption intensities between the images taken above and below the absorption edge of the element under investigation. The acquisition of these 3D element distribution information and structural information will provide important guidance for improving the performance and efficiency of SOFC.

Researchers from the Hefei National Synchrotron Radiation Laboratory used the nanometer resolution X-ray microscopy to collected two 3D scans in absorption contrast mode below (8300 eV) and above (8380 eV) the Ni K-shell absorption edge to get the nanometer resolution 3D element distribution information and structural information of the solid-oxide fuel cell Ni-YSZ anode. Two radiographs of the sample taken at 8300 and 8380 eV are shown in Fig. 3.13a, b, in which

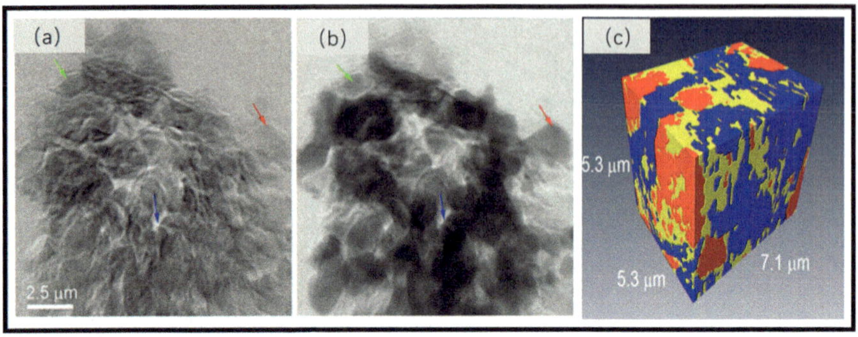

Fig. 3.13 2D radiographs of the anode imaged **a** at 8300 eV (below the K-absorption edge of Ni) and **b** at 8380 eV (above the K-absorption edge of Ni). The darker pixels represent regions of higher X-ray absorption. The region with a strong contrast variation (red arrow) is the Ni phase; the region with a moderate grey level (green arrow) is the YSZ phase, and the brightest region (blue arrow) is the pore space. **c** A 3D rendering of Ni–YSZ sample 1, where Ni, YSZ, and the pore phase are indicated in red, yellow, and blue, respectively

darker features represent regions of higher X-ray absorption. Because the contrast of Ni at 8380 eV was higher than at 8300 eV. It was possible to determine the absolute location of the Ni within the 3D matrix of the anode by comparing the difference between the reconstructed volumes from these radiographs. The other two phases could also be identified by comparison of the grey levels that did not change significantly over the 80 eV photon energy change [30].

Because the shape of the Ni–YSZ sample was irregular, two cuboids were cropped from the two reconstructed 3D volume data separately. The volume data then were segmented into Ni, YSZ, and pore phases based on their grey levels. The processed 3D volume of Ni–YSZ sample 1 was rendered and is shown in Fig. 3.13c (Ni in red, YSZ in yellow, and pore space in blue). Then 3D volume of Ni–YSZ sample 1 was used to calculate several key parameters for quantitative analysis, such as volume fraction, surface area, three-phase boundary length, and electrical conductivity. Table 3.1 shows the obtained results for the two samples. These calculated parameters are critical for understanding the electrochemical conversion efficiency, studying the electrode reliability, and improving manufacturing processes.

Based on the above study, researchers also studied the variation in the 3D structure of the Ni–YSZ anode support layer under thermal cycling conditions. Through the quantitative analysis and calculation based on the 3D reconstructed image shown in Fig. 3.13c, it is found that the 3D structure of Ni has changed, which leads to the decrease of its surface area and connectivity, and the decrease of Ni connectivity further leads to the decrease of anode conductivity. Calculations also show that the total length of the three-phase interface and three-phase boundary length is gradually decreasing during the thermal cycling. But the YSZ phase in the anode structure remains almost stable in a high-temperature environment. This is because the formed network structure plays a role in supporting the skeleton in the

Table 3.1 Summary of microstructural parameters measured from 3D reconstructions of Ni–YSZ samples 1 and 2 [30]

	Sample 1	Sample 2
Volume fractions of Ni (%)	23.2	24.0
Volume fractions of YSZ (%)	35.8	35.6
Volume fractions of pore (%)	41.0	40.4
Surface area/ volume of Ni (μm^{-1})	1.7	1.6
Surface area/ volume of YSZ (μm^{-1})	4.1	3.9
Surface area/ volume of pore (μm^{-1})	3.6	3.7
Interface area/ volume of Ni/YSZ (μm^{-1})	1.2	0.9
Connected porosity (%)	40.7	40.0
Average Ni diameter (nm)	820	900
Average YSZ diameter (nm)	520	550
Average pore diameter (nm)	680	650
TPB length (m/cm^3)	4.44×10^6	3.10×10^6
Conductivity (S/cm)	188.99–502.74	233.10–597.07

anode and prevents the aggregation and growth of Ni particles. Besides, based on these quantitative analysis data, the researchers obtained the spatial distribution of effective three-phase boundary, and then further developed a gas diffusion and electrochemical reaction coupling model to simulate the gas diffusion and electrochemical reaction process in the electrode [31], and provided theoretical models and necessary data for optimizing battery performance. Using nanometer resolution X-ray microscopy and electrical performance testing techniques, researchers correlated the changes in the anode structure with its performance, revealing the mechanism of the fuel cell performance degradation.

3.3 Possible Development of TXM

For KB-mirror-based TXM system, since it can operate at a very large X-ray energy range with relatively high available photon flux, this imaging method is the best choice for 3D X-ray imaging study on dynamic process at more than 10 Hz temporal resolution using equally sloped tomography (EST) [32] data acquisition or other approaches, especially at X-ray energy range higher than 20 keV. Besides EST, it is also a possible development for fast CT scan to use convolutional neural networks or other machine learning approaches for improvement of slice image quality [33, 34].

In the case of zone plate-based TXM, EST and machine learning are definitely helpful to reduce data acquisition time, meaning fast CT imaging. Flenner et al. [18] realized 6 s CT imaging with acceptable results using a convolutional neural network to improve image quality. Other possible developments might be using redshift radiation of undulator to increase the available photon flux from synchrotron beamline at nearly diffraction-limited storage ring [35], designing Kohler illumination reflection condenser [36] to get high efficiency and to acquire a suitable imaging field of view, combining XAFS (X-Ray Absorption Fine Structure) technique to do XAFS imaging to get spatial distribution of atomic coordination information.

References

1. Mokso, R., Cloetens, P., Maire, E., Ludwig, W., Buffiere, J.Y.: Nanoscale zoom tomography with hard X-rays using Kirkpatrick-Baez optics. Appl. Phys. Lett. **90**, 144104 (2007)
2. Da Silva, J.C., Pacureanu, A., Yang, Y., Bohic, S., Morawe, C., Barrett, R., Cloetens, P.: Efficient concentration of high-energy X-rays for diffraction-limited imaging resolution. Optica **4**(5), 492–495 (2017)
3. Hubert, M., Pacureanu, A., Guilloud, C., Yang, Y., da Silva, J.C., Laurencin, J., Lefebvre-Joud, F., Cloetens, P.: Efficient correction of wavefront inhomogeneities in X-ray holographic nanotomography by random sample displacement. Appl. Phys. Lett. **112**, 203704 (2018)
4. Bernhardt, M., Nicolas, J.D., Osterhoff, M., Mittelstadt, H., Reuss, M., Harke, B., Wittmeier, A., Sprung, M., Koster, S., Salditt, T.: Correlative microscopy approach for biology using X-ray holography, X-ray scanning diffraction and STED microscopy. Nat. Commun. **9**, 3641 (2018)
5. Niemann, B., Rudolph, D., Schmahl, G.: Soft X-ray imaging zone plates with large zone numbers for microscopic and spectroscopic applications. Opt. Commun. **12**(2), 160–163 (1974)
6. Chao, W., Anderson, E., Denbeaux, G.P., Harteneck, B., Liddle, J.A., Olynick, D.L., Pearson, A.L., Salmassi, F., Song, C., Attwood, D.T.: 20-nm-resolution soft x-ray microscopy demonstrated by use of multilayer test structures. Opt. Lett. **28**(21), 2019–2021 (2003)
7. Zeng, X., Duewer, F., Feser, M., Huang, C., Lyon, A., Tkachuk, A., Yun, W.: Ellipsoidal and parabolic glass capillaries as condensers for X-ray microscopes. Appl. Opt. **47**(13), 2376–2381 (2008)
8. Chao, W., Harteneck, B.D., Liddle, J.A., Anderson, E.H., Attwood, D.T.: Soft X-ray microscopy at a spatial resolution better than 15 nm. Nature **435**, 1210–1213 (2005)
9. Chao, W., Fischer, P., Tyliszczak, T., Rekawa, S., Anderson, E., Naulleau, P.: Real space soft X-ray imaging at 10 nm spatial resolution. Opt. Express **20**(9), 9777–9783 (2012)
10. Chu, Y., Yi, J., De Carlo, F., Shen, Q., Lee, W.K., Wu, H.J., Wang, C.L., Wang, J., Liu, C., Wang, C., Wu, S., Chien, C., Hwu, Y., Tkachuk, A., Yun, W., Feser, M., Liang, K., Yang, C., Je, J., Margaritondo, G.: Hard-x-ray microscopy with Fresnel zone plates reaches 40 nm Rayleigh resolution. Appl. Phys. Lett. **92**, 103119 (2008)
11. Liu, Y., Andrews, J.C., Meirer, F., Mehta, A., Gil, S.C., Sciau, P., Mester, Z., Pianetta, P.: Applications of hard X-ray full-field transmission X-ray microscopy at SSRL. AIP Conf. Proc. **1365**, 357–360 (2011)
12. Yuan, Q., Zhang, K., Hong, Y., Huang, W., Gao, K., Wang, Z., Zhu, P., Gelb, J., Tkachuk, A., Hornberger, B., Feser, M., Yun, W., Wu, Z.: A 30 nm-resolution hard X-ray microscope with X-ray fluorescence mapping capability at BSRF. J. Synchrotron Rad. **19**, 1021–1028 (2012)
13. Wang, J., Chen, Y. K., Yuan, Q., Tkachuk, A., Erdonmez, C., Hornberger, B., Feser, M.: Automated markerless full field hard X-ray microscopic tomography at sub-50 nm 3-dimension spatial resolution. Appl. Phys. Lett. **100**, 143107 (2012)

14. Jefimovs, K., Vila-Comamala, J., Stampanoni, M., Kaulich, B., David, C.: Beam-shaping condenser lenses for full-field transmission X-ray microscopy. J. Synchrotron Rad. **15**, 106–108 (2008)
15. Stampanoni, M., Mokso, R., Marone, F., Vila-Comamala, J., Gorelick, S., Trtik, P., Jefimovs, K., David, C.: Phase-contrast tomography at the nanoscale using hard X rays. Phys. Rev. B. **81**, 140105(R) (2010)
16. Andrade, V.D., Deriy, A., Wojcik, M., Gürsoy, G., Shu, D., Mooney, T., Peterson, K.M., Glowacki, A., Yue, K., Yang, X., Vescovi, R., Carlo, F.D.: A new transmission x-ray microscope for in-situ nano-tomography at the APS. SPIE Proc. **9967**, 99670H (2016). https://doi.org/10.1117/12.2239449
17. Longo, E., Sancey, L., Flenner, S., Kubec, A., Bonnin, A., David, C., Muller, M., Greving, I.: X-ray Zernike phase contrast tomography: 3D ROI visualization of mm-sized mice organ tissues down to sub-cellular components. Opt. Express **11**(10), 5506–5517 (2020)
18. Flenner, S., Storm, M., Kubec, A., Longo, E., Doring, F., Pelt, D.M., David, C., Muller, M., Greving, I.: Pushing the temporal resolution in absorption and Zernike phase contrast nanotomography: enabling fast in situ experiments. J. Synchrotron Rad. **27**, 1339–1346 (2020)
19. Meirer, F., Cabana, J., Liu, Y., Mehta, A., Andrews, J.C., Pianetta, P.: Three-dimensional imaging of chemical phase transformations at the nanoscale with full-field transmission X-ray microscopy. J. Synchrotron Rad. **18**, 773–781 (2011)
20. Wang, J., Chen-Wiegart, Y.C., Wang, J.: In situ chemical mapping of a lithium-ion battery using full-field hard X-ray spectroscopic imaging. Chem. Commun. **49**, 6480–6482 (2013)
21. Xia, S., Mu, L., Xu, Z., Wang, J., Wei, C., Liu, L., Pianetta, P., Zhao, K., Yu, X., Lin, F., Liu, Y.: Chemomechanical interplay of layered cathode materials undergoing fast charging in lithium batteries. Nano Energy **53**, 753–762 (2018)
22. Jiang, Z., Li, J., Yang, Y., Mu, L., Wei, C., Yu, X., Pianetta, P., Zhao, K., Cloetens, P., Lin, F., Liu, Y.: Machine-learning-revealed statistics of the particle-carbon/binder detachment in lithium-ion battery cathodes. Nat. Commun. **11**, 2310 (2020)
23. Lu, X., Daemi, S.R., Bertei, B., Kok, M.D.R., O'Regan, K.B., Rasha, L., Park, J., Hinds, G., Kendrick, E., Brett, D.J.L., Shearing, P.R.: Microstructural evolution of battery electrodes during calendering. Joule **4**, 1–23 (2020)
24. Zaman, W., Hortance, N., Dixit, M.B., Dc Andrade, V., Hatzell, K.B.: Visualizing percolation and ion transport in hybrid solid electrolytes for Li–metal batteries. J. Mater. Chem. A **7**, 23914–23921 (2019)
25. Wei, C., Hong, Y., Tian, Y., Yu, X., Liu, Y., Pianetta, P.: Quantifying redox heterogeneity in single-crystalline $LiCoO_2$ cathode particles. J. Synchrotron Rad. **27**, 713–719 (2020)
26. Xu, Y., Hu, E., Yang, F., Corbett, J., Sun, Z., Lyu, Y., Yu, X., Liu, Y., Yang, X., Li, H.: Structural integrity—Searching the key factor to suppress the voltage fade of Li-rich layered cathode materials through 3D X-ray imaging and spectroscopy techniques. Nano Energy **28**, 164–171 (2016)
27. Hong, Y., Huang, X., Wei, C., Wang, J., Zhang, J., Yan, H., Chu, Y., Pianetta, P., Xiao, R., Yu, X., Liu, Y., Li, H.: Hierarchical defect engineering for $LiCoO_2$ through low-solubility trace element doping. Chem **6**, 2759–2769 (2020)
28. Zhang, J., Li, Q., Ouyang, C., Yu, X., Ge, M., Huang, X., Hu, E., Ma, C., Li, S., Xiao, R., Yang, W., Chu, Y., Liu, Y., Yu, H., Yang, X., Huang, X., Chen, L., Li, H.: Trace doping of multiple elements enables stable battery cycling of $LiCoO_2$ at 4.6 V. Nat. Energy **4**, 594–603 (2019)
29. Vanpeene, V., Villanova, J., King, A., Lestriez, B., Maire, E., Roue, L.: Dynamics of the morphological degradation of Si-based anodes for Li-ion batteries characterized by in situ synchrotron X-ray tomography. Adv. Energy Mater. **9**, 1803947 (2019)
30. Guan, Y., Li, W., Gong, Y., Liu, G., Zhang, X., Chen, J., Gelb, J., Yun, W., Xiong, Y., Tian, Y., Wang, H.: Analysis of the three-dimensional microstructure of a solid-oxide fuel cell anode using nano X-ray tomography. J. Power Sources **196**, 1915–1919 (2011)

31. Guo, P., Guan, Y., Liu, G., Liang, Z., Liu, J., Zhang, X., Xiong, Y., Tian, Y.: Modeling of gas transport with electrochemical reaction in Nickel–Yttria-stabilized zirconia anode during thermal cycling by Lattice Boltzmann method. J. Power Sources **327**, 127–134 (2016)
32. Miao, J., Forster, F., Levi, O.: Equally sloped tomography with oversampling reconstruction. Phys. Rev. B. **72**, 052103 (2005)
33. Pelt, D.M., Sethian, J.A.: A mixed-scale dense convolutional neural network for image analysis. Proc. Natl. Acad. Sci. USA **115**(2), 254–259 (2018)
34. Pelt, D.M., Batenburg, K.J., Sethian, J.A.: Improving tomographic reconstruction from limited data using mixed-scale dense convolutional neural networks. J. Imaging **4**, 128 (2018)
35. Yuan, Q., Zhang, K., Huang, W., Li, M., Hu, L., Yang, F., Zhu, P.: Conceptual design of TXM beamline at high energy photon source. AIP Conf. Proc. **2054**, 050002 (2019)
36. Suzuki, Y., Takeuchi, A., Uesugi, K., Terada, Y., Nakazawa, H., Ohzawa, S., Aoyama, T., Nii, H., Handa, K.: Hard X-ray imaging microscopy using X-ray guide tube as beam condenser for field illumination. J. Phys.: Conf. Ser. **463**, 012028 (2013)

Chapter 4
Principles of Transmission X-ray Microscopy and Its Applications in Battery Study

Zhao Wu, Xu Ding, Chao Zhang, Gang Liu, Yangchao Tian, and Yong Guan

4.1 Introduction

With the decrement of non-renewable resources, renewable energy materials turn out to be imperious demands for social sustainable development [1]. The research of energy materials, especially battery materials, becomes a hot topic in materials science. Investigations of the morphologic change, the elemental distributions and phase change of an electrode during charge and discharge process may reveal the operation mechanism and improve the performance of battery materials [2]. Transmission X-ray microscopy (TXM), which is characterized by its high resolution compared with optical microscopy and its high penetration in contrast to electron microscopy, has been proved to be a preferred imaging tool [3, 4].

TXM performed at single energy has been utilized to observe the morphologic change of anode and cathode [5]. Several significant geometrical parameters, such as porosity, tortuosity, surface area, pore and particle size distributions, could be calculated [6]. The different structural failure mechanisms of Li-ion battery and Na-ion battery were also revealed. Furthermore, Zernike phase contrast technique was also combined with TXM to observe the growth of low contrast substance.

TXM obtains morphology of the imaged object, while X-ray absorption near-edge structure (XANES) spectroscopy can reveal the chemical composition; therefore, the combination of TXM and XANES (called as "TXM-XANES imaging" below) is usually adopted to investigate battery charge and discharge performance and high-temperature performance. With TXM-XANES imaging, Meirer et al. proposed the core–shell lithiation–delithiation mechanism, according to a typical anode material for Li-ion batteries and the resulting chemical phase transition [7]; Wang et al. disclosed only two phases in the partially charged $LiFePO_4$

Z. Wu · X. Ding · C. Zhang · G. Liu · Y. Tian · Y. Guan (✉)
National Synchrotron Radiation Laboratory, University of Science and Technology of China, Hefei, Anhui 230026, People's Republic of China
e-mail: yongg@ustc.edu.cn

J. Wang (ed.), *Advanced X-ray Imaging of Electrochemical Energy Materials and Devices*, https://doi.org/10.1007/978-981-16-5328-5_4

particles [8]; Yang et al. studied the chemical changes of high capacity lithium manganese-rich nickel cobalt oxide (LMR-NMC) composite cathodes with cycling [9]; Xu et al. considered the change of the oxidation state of Mn in the cycle is probably driven by the release of oxygen [10]; Wei et al. attributed the local oxidation of Ni to the thermally driven redistribution of Li-ions [11].

In this chapter, we introduce the principle of TXM and its variant method, such as its combination with X-ray absorption near-edge structure spectroscopy at first. Then we mainly review how researchers employ TXM technology to investigate the battery materials. Last, based on these reports, we try to discuss the future development of transmission X-ray microscopy.

4.2 Principle of Transmission X-ray Microscopy

Figure 4.1 illustrates the principle of transmission X-ray microscopy. Entrance light from source is focused by condenser to enhance the illumination on the sample, which is imaged by an objective zone plate. Ellipsoidal mirror is usually employed as the condenser, due to its high efficiency. Occasionally, zone plate replaces the ellipsoidal mirror for smaller focal spot. The critical imaging optical element in transmission X-ray microscopy is objective zone plate, which plays the same role of convex lens in visible microscopy. Zone plate is composed of concentric alternating transparent and opaque zones. For the first diffractive order, the optical path lengths of light through centers of adjacent zones differ half wavelength, thus the light interferes constructively at the focus from transparent zones. When the object distance and image distance are, respectively, a and b, we can get the following equation with definition of r_n as nth zone.

$$\sqrt{a^2+r_n^2}+\sqrt{b^2+r_n^2}=a+b+mn\lambda/2 \tag{4.1}$$

where m represents the diffractive order and only equals odd with the same width of adjacent transparent and opaque zone.

Neglecting the second-order terms of small quantity, we can obtain

$$r_n=\sqrt{mn\lambda f} \tag{4.2}$$

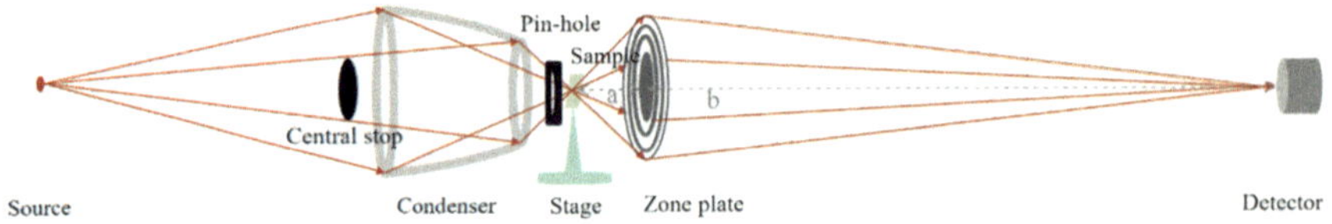

Fig. 4.1 Schematic diagram of transmission X-ray microscopy

with the focus distance $f = ab/(a+b)$.

We can calculate the width of nth zone by differential operation of Eq. (4.2)

$$\Delta r_n = \frac{1}{2}\sqrt{m\lambda f/n} \quad (4.3)$$

Therefore,

$$f = 2r_n\Delta r_n/(m\lambda) \quad (4.4)$$

$$NA = r_N/f = m\lambda/(2\Delta r_N) \quad (4.5)$$

$$R = 0.61\lambda/NA = 1.22\Delta r_N/m \quad (4.6)$$

$$DOF = \lambda/NA^2 = 4\Delta r_N^2/(m^2\lambda) \quad (4.7)$$

where NA is the numerical aperture, N the total number of zones, Δr_N the width of the outmost zone and R the resolution of the imaging system according to Rayleigh criterion and DOF the depth of field.

From Eqs. (4.4–4.7), we can conclude these items below:

a. The focal length is inversely proportional to the wavelength; therefore, the work distance is larger with increasing energy.
b. The numerical aperture of the zone plate is determined by its outmost zone, wavelength and the diffractive order, but not related to its diameter.
c. The resolution of the imaging system depends on the width of the outmost zone and the diffractive order.
d. The depth of field decreases fleetly with the increment of the resolution and is inversely proportional to the imaging wavelength.
e. The resolution becomes high, and the depth of field reduces at high-order imaging.

Zone plate is a monochromator; therefore, the monochromaticity of illumination light is required. When the change of focus distance caused by the wavelength is larger than the depth of field, the obtained image will be dim. According to Eq. (4.4), we can calculate the change of focus distance due to the monochromaticity.

$$\Delta f = 2r_N\Delta r_N\Delta\lambda/(m\lambda^2) \quad (4.8)$$

Comparing Eqs. (4.7) and (4.8), we can get monochromaticity of illumination light should be larger than the total number of half-period zones multiplied by the diffractive order, which is formulated by

$$\lambda/\Delta\lambda > mN \tag{4.9}$$

Three-dimensional (3D) tomography based on TXM, obtaining 3D structure of the imaged object from plenty of two-dimensional (2D) projection images, can be performed with reconstruction algorithms. TXM only probe morphology of the imaged object, while X-ray absorption near-edge structure (XANES) spectroscopy can reveal the chemical composition; therefore, the combination of TXM and XANES is adopted to investigate two-dimensional and three-dimensional change of chemical valence. A relatively simple double-energy TXM, which collects the images above and below the absorption edge of the element of interest, can be employed to obtain specific elemental distributions among different elements. In TXM-XANES imaging, zone plate needs to be moved to refocus with the scanned energy; hence, the magnification changes and misalignment of field of view due to the motor precision should be taken into consideration in the image registration algorithms.

4.3 Applications of TXM in Energy Materials

TXM is a powerful and outstanding tool to investigate energy materials, and we especially focus on the battery materials in this section. The performance of the battery mainly depends on the efficiency and reversibility of the electrochemical phase change on the solid electrode, so it is extremely important to monitor and understand the changes in the morphology and chemical level of battery materials. TXM can provide morphological information of the investigated battery materials by absorption contrast or phase contrast at single energy, as well as the elemental distributions and even the valence change by scanning the energy and combining with the XANES spectroscopy technique. Here, we will review some of the relevant typical research works.

A. **Morphological Imaging of Battery Microstructure by Absorption TXM**

As early as 2010, Shearing et al. adopted TXM tomography to acquire three-dimensional (3D) morphological structure of a graphite negative porous electrode in a Li-ion battery, as shown in Fig. 4.2. Several significant geometrical parameters, such as porosity, tortuosity, surface area and pore and particle size distributions, were extracted from the 3D information to provide opportunities to learn the correlation between the electrode structure and the performance of the battery [12].

Several years later, Wang et al. employed in situ X-ray nano-tomography to investigate and compare the Li-ion battery (LIB) [13] and Na-ion battery (NIB) [14], and an interesting phenomenon was revealed and displayed visually in Fig. 4.3. Structural failure of LIB mainly occurs in the process of Li-ion extraction, on the contrary, the expansion and cracking of Sn particles in NIB occur during Na-ion insertion but no significant pulverization takes place during Na-ion

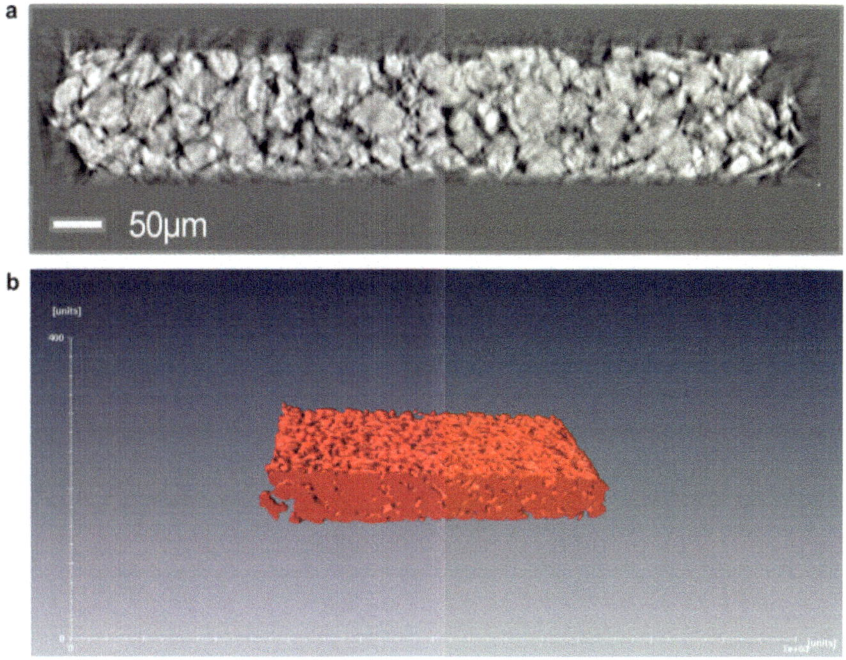

Fig. 4.2 **a** Slice of reconstructed tomography sequence; **b** rendered solid graphite of 300 slices

extraction. The 3D microstructural evolution leads to material degradation and can be quantified with the linear attenuation coefficient, surface curvature and morphological complexity. The decrease of the linear attenuation coefficient implies sodiation process from Sn to Na_xSn in NIB. At the beginning of the sodiation process, Na_xSn ($x \approx 1.3$) is generated, and the volume change is negligible. However, as the Na-ion is further inserted, solidation-induced stress increases and leads to volume expansion. Furthermore, the distribution of the convex and concave surface curvature of all particles is varied during the sodiation process. With the insertion of Na-ion, a lattice mismatch between Sn and Na_xSn leads to an evident increase of concave surface, which plays a major role in affecting the stability of the microstructure. In addition, the morphological complexity, defined by R_s/R_v, can effectively characterize the degradation in pores, cracks, fracture and pulverization of particles, where R_s refers to the radius calculated from the actual surface area S and R_v refers to radius from the volume V.

Similar to the research topics introduced above, recently, Zhao et al. conducted synchrotron X-ray nano-CT experiments to reveal the failure mechanism of the Li-ion battery by characterizing the 3D morphological evolution in different capacity cycling modes in the nanoporous silicon (Np-Si) anode [15]. The results, displayed in Fig. 4.4, point out that the failure of the battery is caused by the

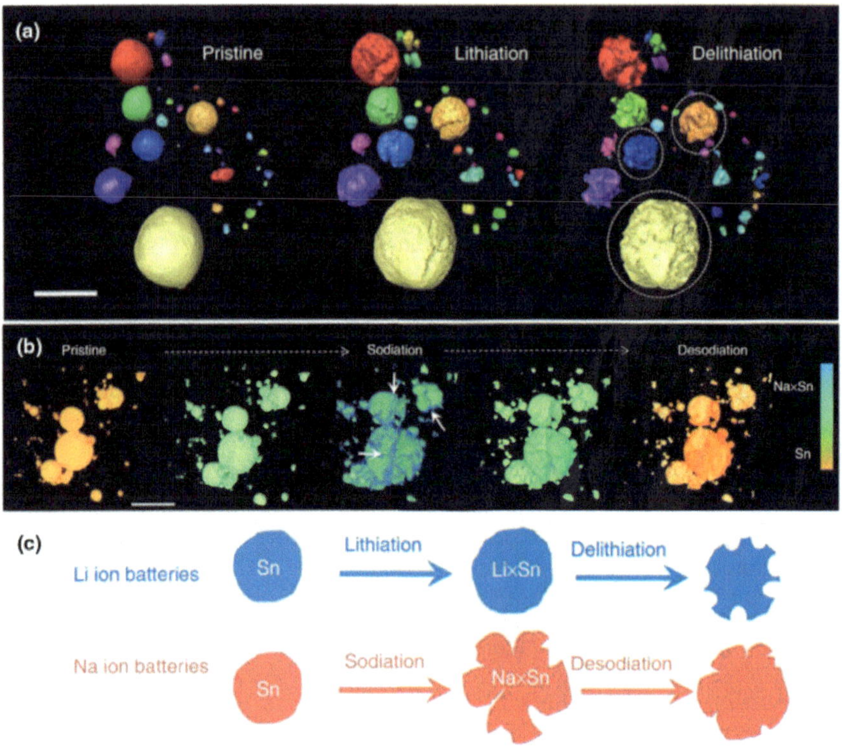

Fig. 4.3 **a** 3D visualization of Sn particles during the first lithiation–delithiation process; **b** 3D visualization of Sn particles during the first sodiation–desodiation process; **c** comparison between LIB and NIB in morphological evolution

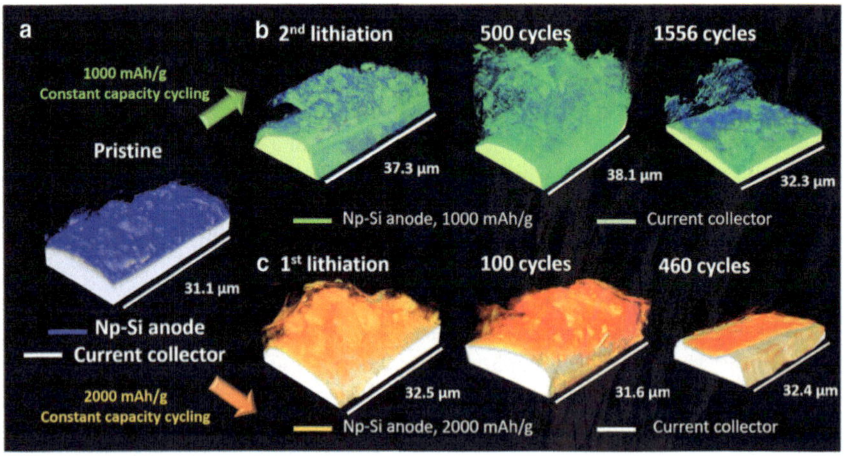

Fig. 4.4 3D morphological evolution in different capacity cycling modes in the nanoporous silicon (Np-Si) anode

volume expansion and gradual delamination of np-Si anode. Particularly, the particle agglomeration leads to a shorter cycling life in the higher capacity cycling mode.

Li et al. used high-resolution synchrotron TXM to examine the morphological evolution of Ge and $Ge_{0.9}Se_{0.1}$ anodes [16]. According to the experimental results of *in operando* 2D TXM and 3D in situ tomography, as shown in Fig. 4.5, $Ge_{0.9}Se_{0.1}$ particle has sudden changes in morphology and optical density during the first lithiation, which is different from the Ge. In addition, $Ge_{0.9}Se_{0.1}$ particle has

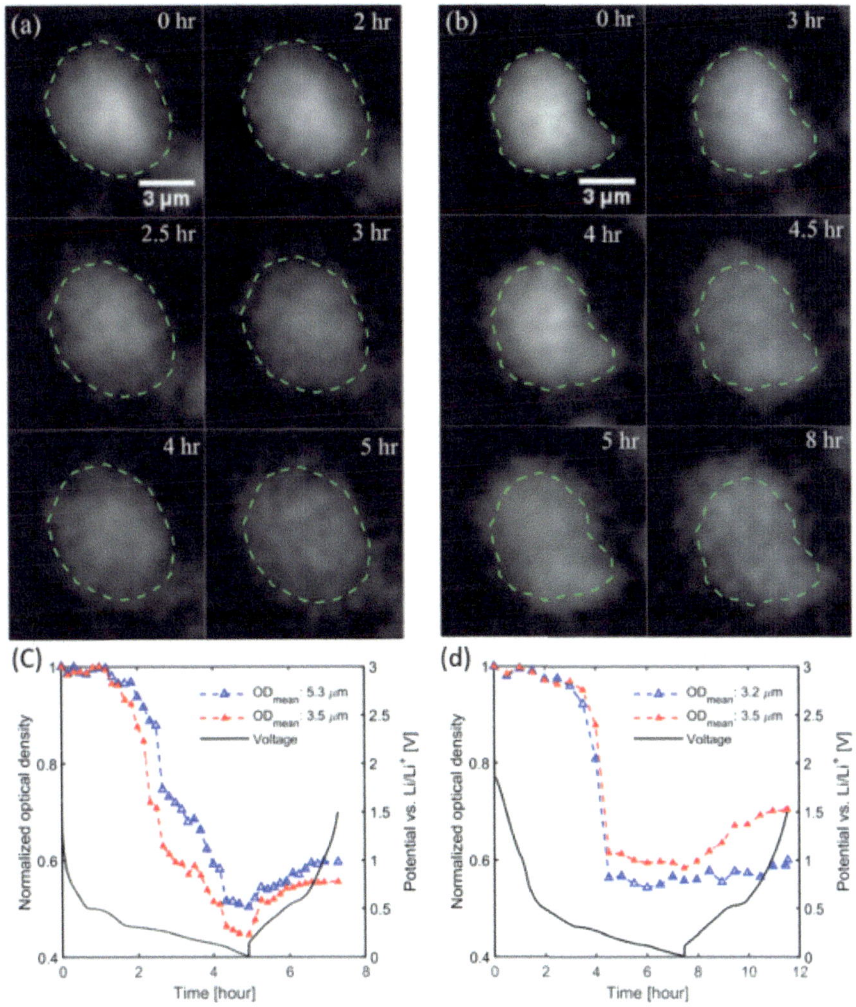

Fig. 4.5 *Operando* 2D TXM of Ge **a** and $Ge_{0.9}Se_{0.1}$ **b** particles as well as normalized optical density dynamics of Ge **c** and $Ge_{0.9}Se_{0.1}$ **d** during the first lithiation

a more homogeneous volume change and smoother surface than Ge, which makes $Ge_{0.9}Se_{0.1}$ particle show a better mechanical stability during the cycling.

TXM can also afford an insight into morphology change of cathode and its mechanism. Nelson et al. conducted *operando* TXM experiments at 6 keV photon energy to study the morphology change of a cathode of sulfur/super P carbon composite particles during the electrochemical cycle of Li–S battery [17]. *In Operando,* TXM can track the individual particles and monitor the loss of polysulfides from the electrolyte in real time by calculating the contrast between composite particles and background. Figure 4.6 shows that there is no significant dissolution of the sulfur/super P carbon composite particles.

In addition, many battery cathodes, composed of transition metal (TM) oxides, have also been explored by TXM. Xu et al. investigated the morphological and chemical changes of Li-rich $Li_2Ru_{0.5}Mn_{0.5}O_3$ cathode caused by electrochemical cycling [10]. The rendering results in Fig. 4.7 show that internal pore grows as the number of electrochemical cycles increases.

Recently, Li et al. realized a direct visualization of structural and chemical complexity of a single $LiNi_{0.8}Mn_{0.1}Co_{0.1}O_2$ (NMC811) secondary particle by using a combination of nano-resolution X-ray probes in both soft and hard X-ray regimes [18]. Based on their experimental results and finite element modeling, a mechanism of mutual modulation between the surface chemistry and the bulk microstructure is formulated.

B. Morphological Imaging of Battery Microstructure by Phase Contrast TXM

Absorption contrast TXM has been widely used in the investigation of battery materials; nevertheless, it does not work well in some cases, due to the low absorption contrast. Lithium is a light element with an atomic number of three, which has a low contrast under X-ray absorption imaging. Thanks to the different physical mechanism, the phase contrast may provide high contrast. The first direct visualization of lithium growth from nucleation to development of dendrite was achieved by Cheng et al. in 2017 [19] by using phase contrast TXM. As Fig. 4.8 shows, the lithium nucleates on the Cu surface at first and makes lithium plates easier on lithium than on Cu. As the formation of nuclei, lithium grows gradually and then overlaps until a layer structure of 5–7 μm thick is formed (Fig. 4.8b–h). Moreover, from 3350 to 4570 s, the stripping process of Li dendrite begins, resulting in a slow increase in voltage. Finally, there are nothing more lithium can be extracted (Fig. 4.8i–l), resulting in irreversible capacity shrinkage. There are more researches of mechanistic understanding of Li dendrites growth by TXM imaging technique, which can be found in a comprehensive review written by Foroozan et al. [20].

In addition to the microstructure properties of anode and cathode have a great influence on the performance Li-ion or Na-ion battery, separator may have an important influence on the performance too. Due to the low X-ray attenuation coefficients of polymer separators, Finegan et al. used Zernike phase contrast technique applied on TXM to capture the microstructures of Li-ion polymer

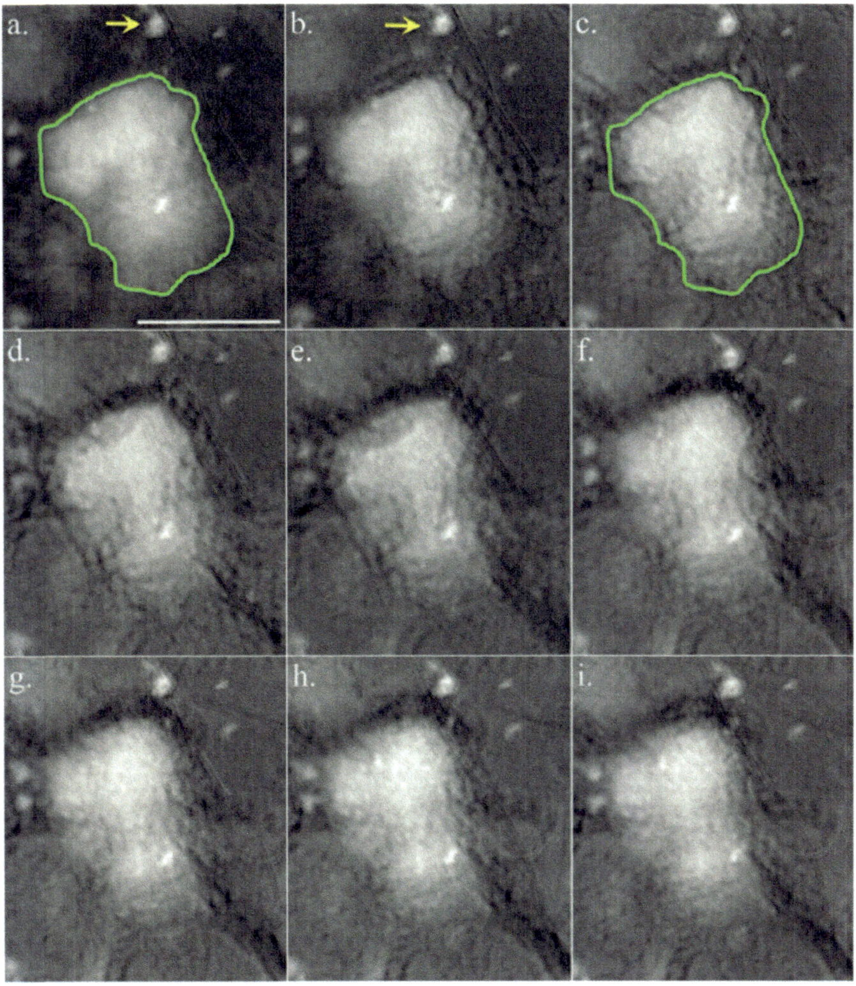

Fig. 4.6 Morphology changes of a cathode of sulfur/super P carbon composite particles during the electrochemical cycle of Li–S battery, the scale bar is 10 μm

separators [21], as shown in Fig. 4.9. The structural properties such as anisotropy of the porosity and average pore diameter are determined by 3D quantification techniques and stereology, which are related to the influence on safety and rate capability of batteries.

III. **Elemental Distributions by Dual-Energy TXM Imaging**

Dual-energy imaging is a simplified multiple X-ray energy imaging method. By recording TXM images at two energies below and above, the absorption edge of the element of interest can obtain elemental distributions without oxidation state

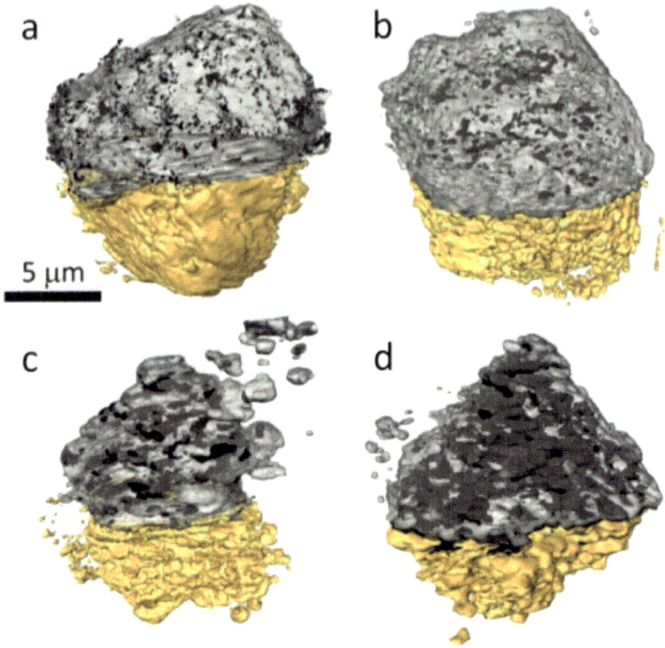

Fig. 4.7 Morphological change of Li-rich $Li_2Ru_{0.5}Mn_{0.5}O_3$ cathode particles at different stages **a** pristine, **b** 01-cycled, **c** 10-cycled and **d** 20-cycled. Internal pore grows as the number of electrochemical cycles increases

information. Grew et al. developed a nondestructive dual-energy X-ray nano-tomography technology to examine the phase and pore network in the solid oxide fuel cells (SOFCs) anode and to reveal the contributions of heterogeneous microstructure to the origins of transport-related losses [22]. After that, Guan et al. used 3D X-ray nano-tomography to investigate the effects of thermal cycling on the microstructure of a porous nickel-yttria-stabilized zirconia (Ni-YSZ) anode [23]. As Fig. 4.10 shows, there is an agglomeration of the nickel particles due to Ni migration after six thermal cycles. By conducting quantitative analysis of microstructure and conductivity evolution, they infer that the conductivity is correlated with the connectivity of the Ni phase.

Furthermore, Rahman et al. adopted dual-energy TXM imaging to study the highly heterogeneous 3D transition metal elemental distributions in sodium-layered cathode [24]. $Na_{0.9}Cu_{0.2}Fe_{0.28}Mn_{0.52}O_2$ (CFM-Cu) has excellent electrochemical properties and a depth-dependent charge compensation mechanism, but it still cannot avoid capacity fading after repeated cycles. So they get the 3D elemental association maps of CFM-Cu by transmission X-ray tomography after 400 cycles at 1C rate in a Na half-cell. As Fig. 4.11 shows, the surface of the particle is basically Mn only association. And the decrease of Mn/Fe/Cu association implies the failure

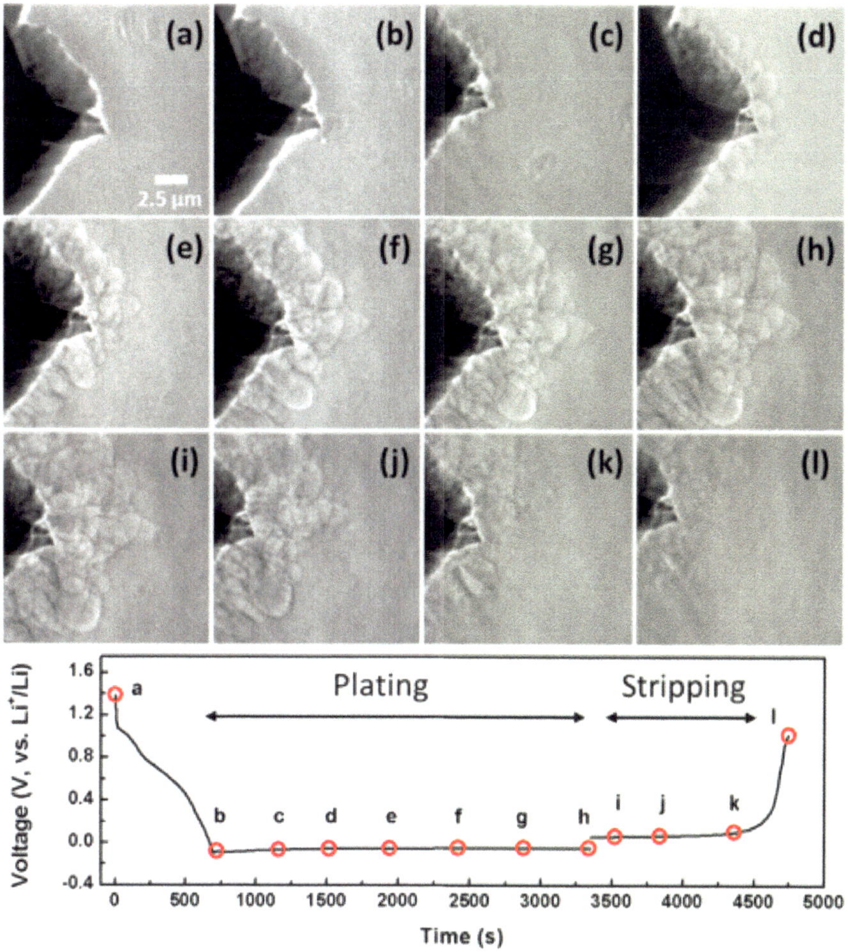

Fig. 4.8 *In operando* 2D TXM images of lithium plating **a–h** and stripping **i–l** and the bottom is corresponding voltage changes during the lithium plating/stripping

mechanism in CFM-Cu is likely related to Mn segregation from Cu–Fe–Mn association and its consequential deposition and reduction at the surface of cathode particles.

Dual-energy nano-tomography can also be used to study Li-ion battery. Wei et al. performed nano-tomography at six different X-ray energies which is below and above the absorption K-edges of Mn, Co, and Ni, respectively [11]. It is helpful to explore the thermally driven mesoscale chemomechanical interplay of $Li_{0.5}Ni_{0.6}Mn_{0.2}Co_{0.2}O_2$ cathode materials.

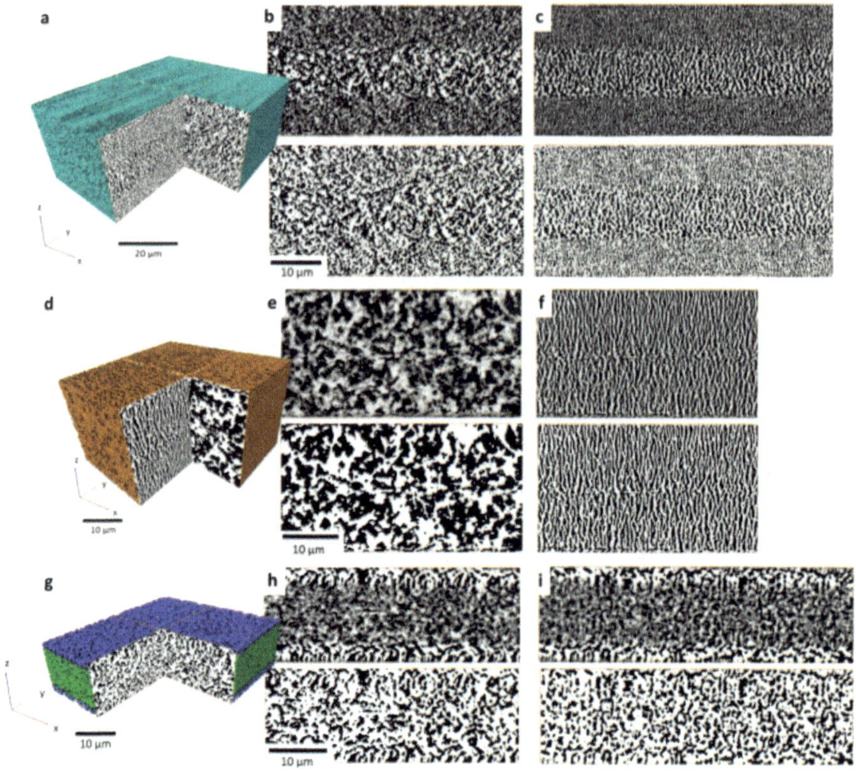

Fig. 4.9 Microstructures of Li-ion polymer separators obtained by phase contrast TXM: **a** volume rendering of Celgard 235 and its binary slices in the **b** *YZ* and **c** *XZ* planes, **d–f** and **g–i** are the counterparts of Celgard 2500 and MTI ceramic-coated membrane

IV. Investigation of Valence Change by TXM-XANES Imaging

In this part, we will introduce the characterization of the changes of chemical information in the materials. X-ray absorption near-edge structure (XANES) spectroscopy is sensitive to the chemical and local electronic changes of the detected elements and is widely used to characterize fine structural changes in a variety of materials. In order to understand the chemical evolution of electrode particles in the reaction process, we use the combination technique of TXM and XANES to draw and track chemical mappings [25, 26]. This technology can visualize the valence state distribution of elements as a function of the state of charge at the nanometer level. The full-field imaging function in a large (tens of microns) field of view (FOV) also allows multiple particles to be inspected in the same measurement [27].

For energy materials, the current electrode materials can already provide a large capacity, but their performance in terms of cycle performance and safety is not satisfactory [11, 28]. Real-world battery operating conditions may introduce

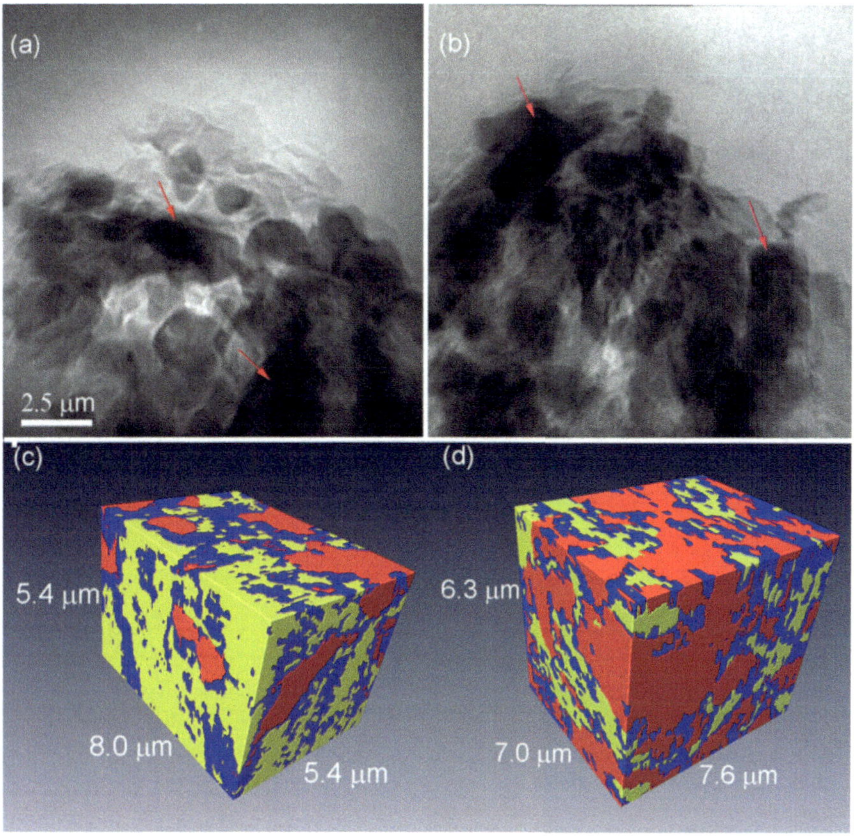

Fig. 4.10 2D radiographs of the anode imaged at 8.38 keV **a** non-thermal cycling test and **b** after six thermal tests. The regions indicated are the Ni phase. The 3D rendering of the **c** non-tested anode and the anode **d** after six thermal cycles, where the red is Ni, the blue is YSZ, and yellow is pore space

complex local chemical events. First, the occurrence of certain chemical events, such as the heterogeneity distribution of element valence states in the cycle, may seriously damage the performance of electrochemical cycles. The decrease in the capacity retention rate of the cathode is an important evaluation index that determines whether the material can be used in commercial applications. Second, the negative effects of thermal abuse caused by local chemical effects, such as local oxygen release, are undesirable, and extreme situations may even bring safety problems [11]. In the following, we will introduce applications of TXM-XANES imaging to investigate battery charge–discharge performance and high-temperature performance.

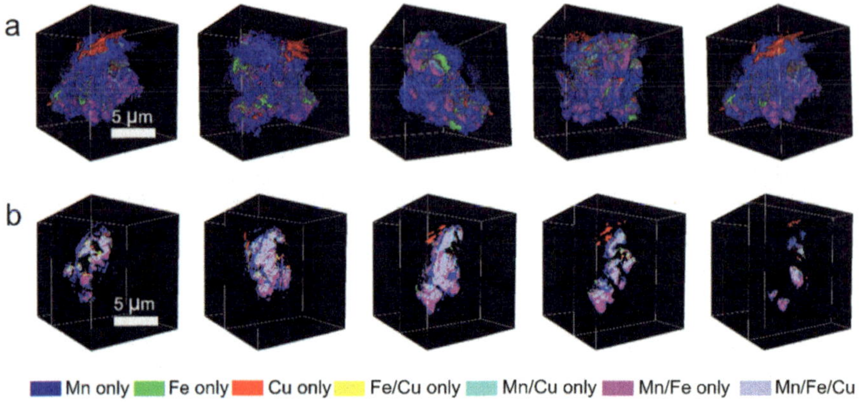

Fig. 4.11 **a** 3D resolved elemental associations at the surface of CFM-Cu in five nanodomains by using TXM tomography after 400 cycles at 1C rate; **b** 2D perception of the elemental association in planes cut through the particle from the surface to the bulk at different depths

D.1 Charge–Discharge Performance

In this subsection, we will review that researchers used TXM-XANES to characterize the valence changes of elements in the electrode particles and infer the occurrence of phase changes. Based on these observed phase maps, researchers analyzed some reasons for the failure of the cathode particles during the charge–discharge cycle, including voltage drops caused by the irreversible phase change and the release of oxygen, and the large chemical gradients that prevent lithium-ion transmission.

In 2011, Meirer et al. demonstrated the phase change of the NiO anode during the first discharge (reduced) and charge (re-oxidized) [7]. The sample cannot return to its original state (sample P and sample O in Fig. 4.12a) after a charge–discharge cycle. From three-dimensional imaging of the partially reduced sample R1 (Fig. 4.12b), we can find Ni is widely reduced on the surface of larger aggregates, but it may also occur through cracks inside the grains. Next, Wang et al. elucidated the conversion reaction mechanism of CuO [25], a typical negative electrode material for lithium-ion batteries, and the chemical phase transition produced strongly indicates the core–shell lithiation–delithiation mechanism.

Wang et al. also used the in situ 3D XANES method to perform imaging experiments on the anisotropic crystal $LiFePO_4$ [8]. It can be observed in Fig. 4.13a that the direction of propagation of the phase boundary (the junction of $LiFePO_4$ and $FePO_4$) is anisotropic, toward an optimal direction at the beginning. It is obvious that the phase boundary is inclined and uneven, which is caused by the anisotropic migration of Li-ions. With the degree of delithiation increases during charging, the larger lithium chemical potential gradient may provide sufficient driving force to make the two-phase boundary move along multiple directions, so that the propagation direction becomes isotropic. In addition, after the phase

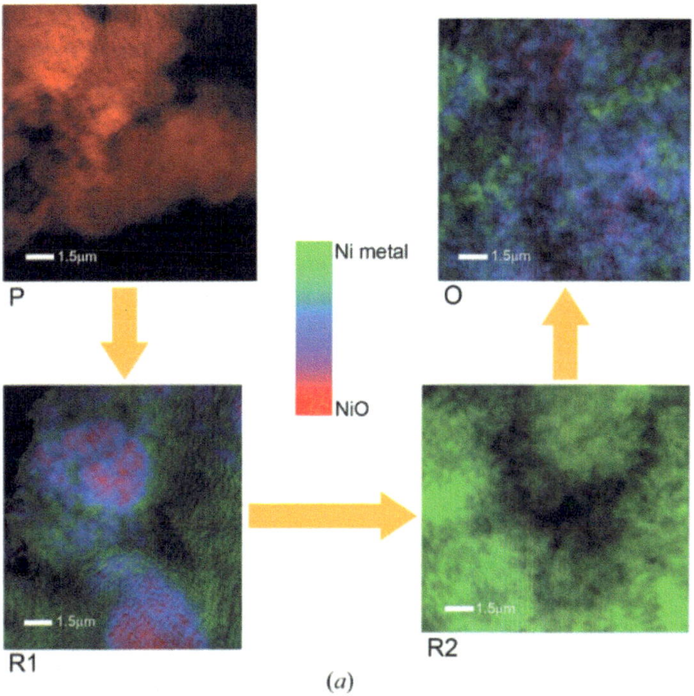

(*a*)

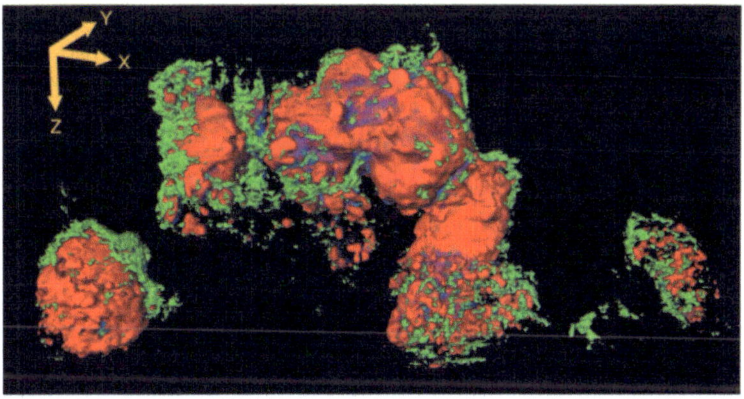

(*b*)

Fig. 4.12 **a** 2D phase maps of NiO particles in the first charge–discharge cycle (red green blue, respectively stands for NiO, Ni and mixed states) for samples P (pristine electrode), R1 (partially reduced), R2 (fully reduced to Ni) and O (re-oxidized) **b** 3D visual reconstruction of sample R1 propagation direction changes from anisotropy to isotropy, the core–shell structure is clearly seen. The XANES spectra of the cross section of the particle drawn in Fig. 4.13b, c, indicate that there are only two phases in the partially charged $LiFePO_4$ particles at the microscopic scale, because only two valence states of Fe cation with energy difference of 2 eV can be observed.

The above researches illustrate that TXM-XANES provides great helps for researchers to analyze the valence changes and phase transitions of the elements inside the electrode during the reaction. Actually, TXM-XANES are also usually employed to understand the cause of the irreversible decrease in specific capacity during electrochemical cycling, which is a very hot research topic in recent years. Yang et al. studied the chemical changes of high capacity lithium manganese-rich nickel cobalt oxide (LMR-NMC) composite cathodes with cycling [9]. Figure 4.14a–d shows the chemical phase and distribution of Mn at different stages of battery lifetime, where the cathode particles composed of $Li_{1.2}Mn_{0.525}Ni_{0.175}Co_{0.1}O_2$. 2D XANES chemical maps of Mn show that the

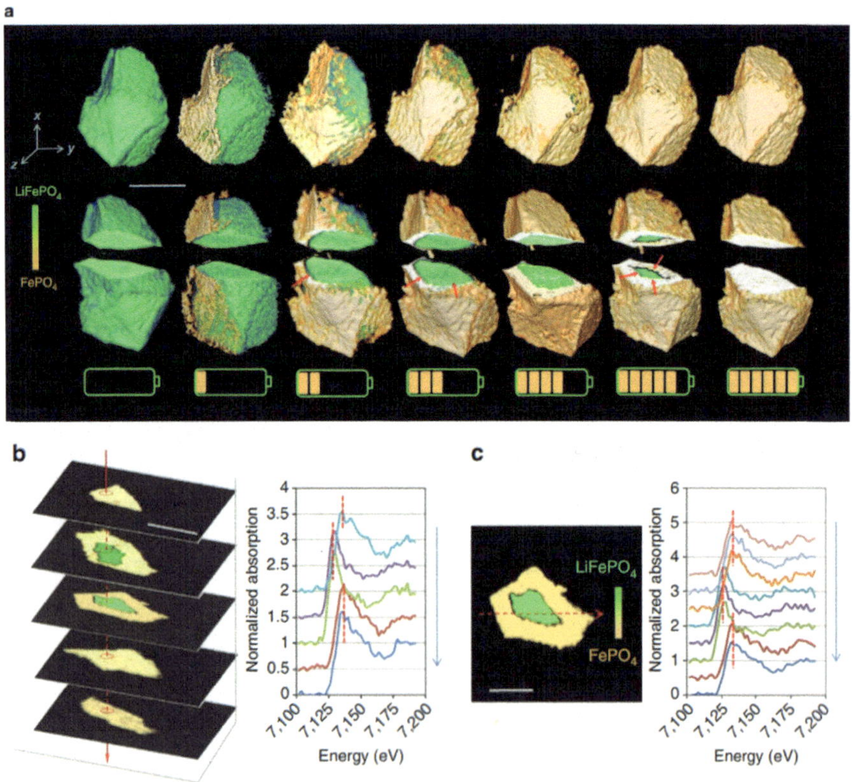

Fig. 4.13 **a** 3D visualization of the evolution of $LiFePO_4$ to $FePO_4$ during charging. **b** Cross-sectional view of partially charged particles along the vertical axis and corresponding XANES spectra. **c** XANES spectra along the horizontal line on the selected cross-sectional view

chemical composition (green) of Mn in the initial particles gradually changed to another composition (red) after 200 cycles.

In Fig. 4.14e, the measured XANES of the main composition (pristine and 200-cycled) is drawn together with a number of standard spectra of some Mn-based compounds, to follow the average change in the Mn valence state as a function of electrochemical cycling. Comparing the absorption spectra of the 200-cycled sample and the reference compounds, especially $LiMn_{1.5}Ni_{0.5}O_4$ and MnO_2, it is obvious that a spinel-like structure is produced when the initial LMR-NMC is cycled at higher voltage. It can be seen that the edge energy of Mn in the 200-cycled sample is lower than that of the pristine sample (6.5521 keV vs. 6.5538 keV), indicating that there is more reduced Mn in the cycle sample.

A spinel phase and reduced Mn appear in the cycle, because the electrochemical charge transfer in battery materials usually involves the insertion (and de-insertion) of ions, causing atoms to rearrange in the lattice. Moreover, due to atomic migration and/or diffusion, local phase segregation usually occurs [9]. These changes generally lead to the stabilization of the low-energy phase, followed by changes in particle size, shape, crystal grain and morphology [29, 30].

Except irreversible phase transition, one of the electrode degradation mechanisms, other degradation mechanisms of the electrode are also explored by researchers based on TXM-XANES technique. Xu et al. studied the particle

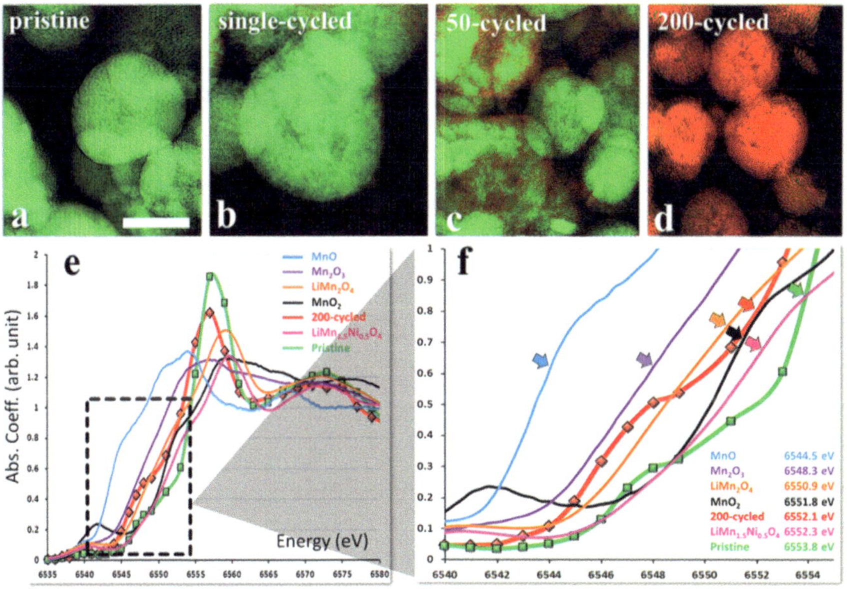

Fig. 4.14 TXM/XANES mappings of LMR-NMC particles. Panels **a–d** are 2D Mn valence maps, respectively, of pristine particle, single-cycled, 50-cycled, 200-cycled. Panels **e**, **f** show the X-ray near-edge spectra of the two main components as well as the standard spectra of some Mn-based compounds

evolution and electrochemical degradation process of $Li_2Ru_{0.5}Mn_{0.5}O_3$ [10]. Figure 4.15a–d shows the two-dimensional distribution of Mn oxidation states in particles under different cycles. For a more intuitive comparison, the overall Mn spectra are plotted in Fig. 4.15e. As shown in the enlarged illustration, after 20 cycles, the bulk spectra shift to lower energy by ~3 eV. They speculated that the change in the oxidation state of Mn in the cycle might be driven by the release of oxygen, which is one of the main reasons for the voltage drop. In order to obtain a deeper understanding of the reaction mechanism, Xu et al. conducted a three-dimensional imaging experiment to monitor the depth dependence of Mn oxidation state changes (Fig. 4.15f). As shown in the panel, Mn edge energy of pristine particles and 01-cycled particles has shifted, but it does not show depth dependence, indicating that Mn is uniformly reduced in the early stage of battery life. In the 10-cycled particles, it is obvious that the reduction rate of Mn on the surface of particles is faster (due to the release of surface oxygen), indicating that the release of oxygen starts from an earlier stage. Noteworthily, similar phenomenon has also been found in NMC-type lithium-rich electrodes [31].

In order to further understand the mechanism of oxygen release, Kan et al. studied the occurrence of oxygen release at different stages of delithiation [27]. It can be seen from Fig. 4.16 that after the gradual increase in the mole ratio of the oxidant (NO_2BF_4) to the original particles ($Li_xTa_{0.3}Mn_{0.4}O_2$), the overall color of map changes from blue to yellow (panel a), and the average edge energy also shifts to a higher value (panel b), indicating that the Mn in the particles is gradually

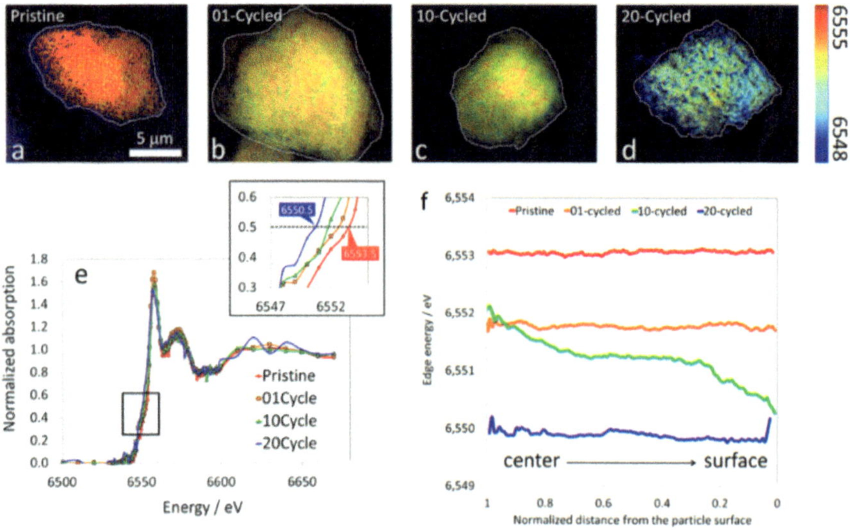

Fig. 4.15 TXM/XANES mappings of $Li_2Ru_{0.5}Mn_{0.5}O_3$ particles. Panels **a–d** are 2D Mn valence maps, respectively, of pristine particle, 01-cycled, 10-cycled, 20-cycled. Panel **e** shows the X-ray near-edge spectra and magnified images of samples under different cycles. Panel **f** shows the depth dependence of Mn oxidation state under different cycle periods

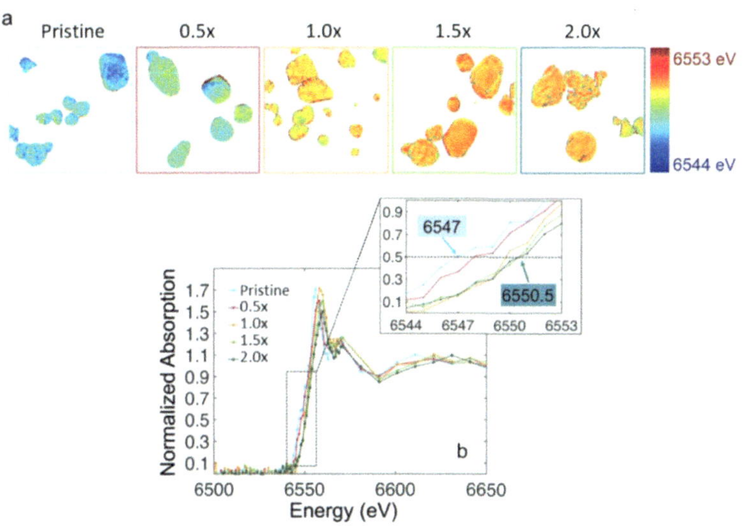

Fig. 4.16 **a** 2D Mn valence maps of $Li_xTa_{0.3}Mn_{0.4}O_2$ (L_xTMO) particles prepared using the specified mole ratio of oxidant to pristine particle. **b** Mn K-edges XANES spectra of particles at different degrees of oxidation

oxidized. When the NO_2BF_4 ratio exceeds 1 mol, the average edge energy increases very little, indicating that the contribution of Mn oxidation is saturated, and oxygen oxidation becomes the main charge compensation mechanism. This research result further explains the mechanism of oxygen release.

The reduction of transition metal cations caused by the above oxygen loss leads to the decrease of battery voltage with the cycle. In addition, when Xu et al. observed the cycle process of $LiCoO_2$ particles [32], they found from the chemical map that a larger chemical gradient appeared at the upper edge of the particles, indicating the nucleation of the inactive phase and indicating that the potential barrier in the particles hindered the free migration of Li-ions. Kan et al. also found that the dramatic change of Mn oxidation state on $Li_{1.3}Nb_{0.3}Mn_{0.4}O_2$ particles prevented the diffusion of Li-ions [33], indicating that the diffusion path of Li-ions has a certain relationship with the size of the regional chemical gradient in the particles.

D.2 High-Temperature Performances

Although Li-ion batteries are designed to work in a mild temperature window, extreme temperature peaks can be caused by many reasons, especially at the nanometer scale, where the complexity of structure and chemistry can cause very large local currents [11]. Studies have shown that when the parent phase becomes metastable due to temperature changes, solid-state phase changes generally occur in crystalline materials [34]. In the redox active materials, the phase transition is usually accompanied by a redox reaction, and the phase boundary can be determined as the interface between the oxidized and reduced domain [35]. Therefore, the local electronic properties determined on the spectra (e.g., the valence state of the elements involved in the redox reaction) can be used to visually characterize the phase transition and propagation in the material [34].

Nickel-rich NMC cathode materials are recent attractive candidate materials that can improve the energy density of next-generation Li-ion batteries. However, the practical application of these materials is hindered by poor thermal stability and the release of oxygen owing to structural decomposition. Therefore, research on the thermal stability of these materials becomes very necessary. This problem has attracted many research teams to carry out related in-depth research.

Because the electrolyte and the "deactivated" components (carbon and binder) play an important role in the local chemical reaction [36], compared with the samples prepared by the chemical delithiated method, the $Li_{0.5}Ni_{0.6}Mn_{0.2}Co_{0.2}O_2$(NMC-622) cycling in the real battery particles reacts more realistically to the influence of the active substances. Wei et al. disassembled the coin battery in the charged state (i.e., 4.5 V) in the glove box and collected the cathode for further characterization [11]. The top row in Fig. 4.17 shows the chemical evolution of NMC-622 secondary particles after in situ heating, and the bottom rows show the differential valence maps. It can be observed that after heating, Ni cations are reduced as a whole, but local Ni cation is oxidized. Wei et al. attributed the local oxidation of Ni to the thermally driven redistribution of Li-ions.

In this case, the valence state distribution of Ni can be explained how the NMC cathode particles undergo chemical transformation under thermal conditions, which provide information for the design of thermally stable electrode materials.

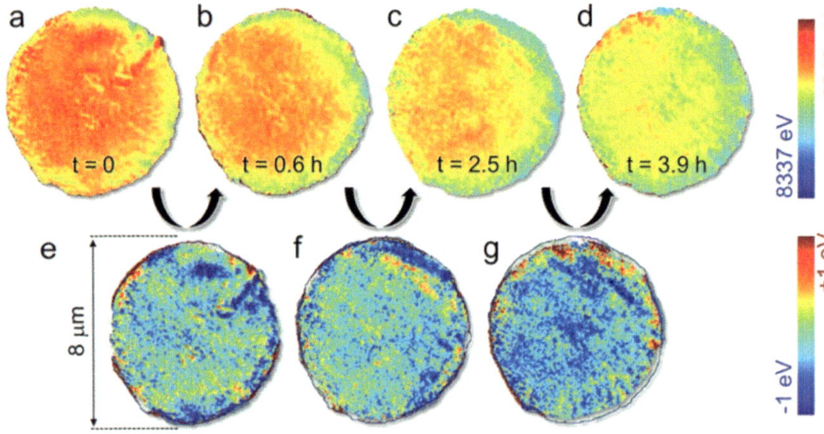

Fig. 4.17 Chemical evolution of $Li_{0.5}Ni_{0.6}Mn_{0.2}Co_{0.2}O_2$ secondary particles was characterized by TXM-XANES under thermal abuse conditions (380 °C). Panels **a–d** show the evolution of the two-dimensional distribution of the valence state of Ni when the particles are exposed to high temperatures. Panels **e–g** display the differential valence maps between adjacent panels from **a** to **d**

However, only the results of 2D imaging are not sufficient for researchers to analyze the cathode particles under heating. For this reason, 3D imaging was performed to obtain depth information.

In order to further understand the body-to-surface transition of NMC materials, Tian et al. performed XANES characterization of chemically delithiated NMC-622 particles [37]. This is explained by the fact that the interaction between the electrode particles and other components in the battery (such as carbon and binder) is very complicated. The phenomenon they observed was similar to those previously reported by Wei et al. [11]. In some cases (e.g., in a 50% delithiated sample), the surface transition metal would be oxidized after heating, although they would reduce overall.

Performing comparison between Fig. 4.18a, c, we can conclude that the surface of 50% delithiated sample is oxidized (red spots on the surface) after heating to 350 °C. In contrast, 100% delithiated particles are significantly reduced, according to Fig. 4.18d, f. However, all points of view indicate that the distribution of Ni oxidation states is heterogeneous. In order to eliminate the influence of heterogeneous distribution on data analysis and quantify the relationship between energy (i.e., Ni oxidation state) and depth during heating, the function of distance from the particle surface and average Ni K-edge energy is drawn in Fig. 4.18b. For 50% delithiated particle, the Ni oxidation state distribution from the surface to the whole is relatively uniform (Fig. 4.18b) before heating. But after heating, the surface oxidation state is relatively higher. Compared with 50% delithiated particles, the reduction of 100% delithiated particles is more uniform.

Detailed information of the TXM results of 50% delithiated NMC-622 particles before and after heating to 350 °C is shown in Fig. 4.19. From maps of Ni distribution difference before and after heating as shown in Fig. 4.19e, it can be observed that Ni is enriched in the inner and outer surface regions of the hollow NMC-622 particles. After the reaction, the apparent oxidation state gradient observed in the heated particles (Fig. 4.19b) may reflect the migration of lower valence nickel to the interior. The study results of Li et al. show that, for cathode materials rich in Li and Mn, the thermally driven redistribution of Li-ions may lead to a large amount of charge transfer behavior between oxygen anions and surface transition metal cations [38], which is consistent with the view of Wei et al. [11]. Therefore, Li-ions may also migrate during the heating process, and the transition metal oxidation state of the 50% delithiated particle changes locally. This study demonstrates the importance of combining surface and bulk sensitive, spatially resolved characterization techniques to reveal the importance of chemical changes for different complex surfaces and overall thermal behavior.

Since the content of Ni directly affects the energy density contained in the cathode of NMC particles, Alvarado et al. revealed the response of NMC-811 particles to high temperature [39]. Similarly, in order to eliminate the influence of carbon and binders, they used the electrode particles prepared by chemical delithiation.

Obviously, there is almost no change in the edge energy position of Ni from room temperature to 200 °C (Fig. 4.20a, c). However, a significant shift to low

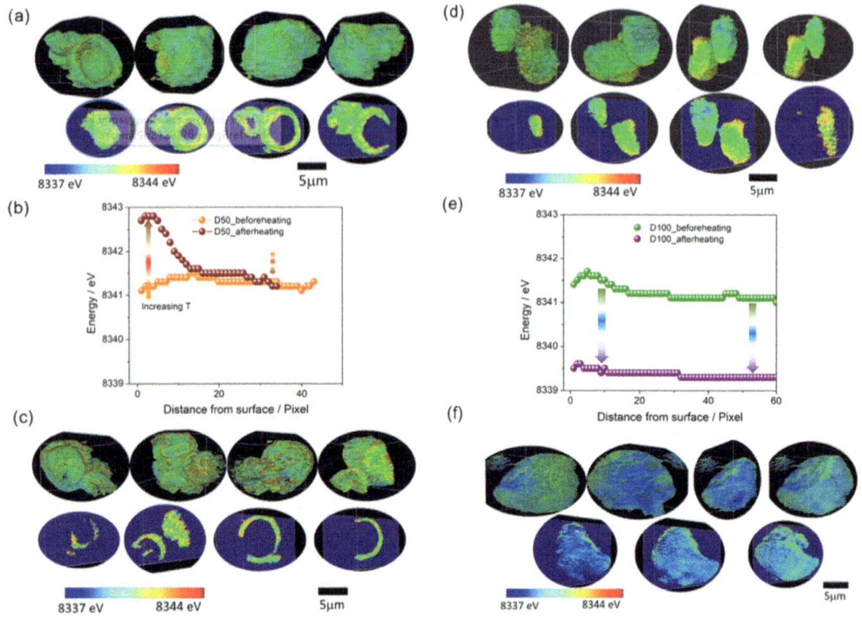

Fig. 4.18 TXM-XANES images of 50% delithiated and 100% delithiated of NMC-622 particles before heating **a**, **d** and after heating to 350 °C **c**, **f**. The blue represents the lower oxidation state, and the red represents the higher oxidation state. The relationship between the Ni K-edge energy and the distance from the particle surface of **b** 50% delithiated and **e** 100% delithiated particles before and after heating

energy will occur at 300 °C, especially for samples with 50% delithiated. In the heated sample, local differences in the oxidation state of Ni are particularly obvious (Fig. 4.20b, d). In particular, oxidized areas appeared on the sample surface at 120 °C, which is similar to the previous discovery by Wei et al. [11]. This further strengthens the thermally driven rearrangement theory of Li-ions in the secondary particles. Finally, Alvarado et al. compared dynamic thermal properties of NMC-622 and NMC-811 particles. The 50% and 75% of the chemically delithiated NMC-811 samples contain a phase mixture of H2 and H3, while the delithiated NMC-622 samples only contain a single phase [39], and due to its lower Ni content, it is more stable in heating as a whole. They found that the phase change of NMC-622 particles occurs at a higher temperature than NMC-811 particles, and the transition metal in NMC-811 particles is also more sensitive to heat. In other words, the main difference in thermal characteristics between NMC-622 and NMC-811 is that the initial temperature of the phase changes the behavior of the particle surface [39].

In short, the thermal behavior of Ni-rich NMC materials is extremely complex, involving structural reorganization, Li-ion migration, transition metal oxidation/ reduction, oxygen release and morphological changes, which have a major impact on battery safety, especially in the event of thermal runaway [39].

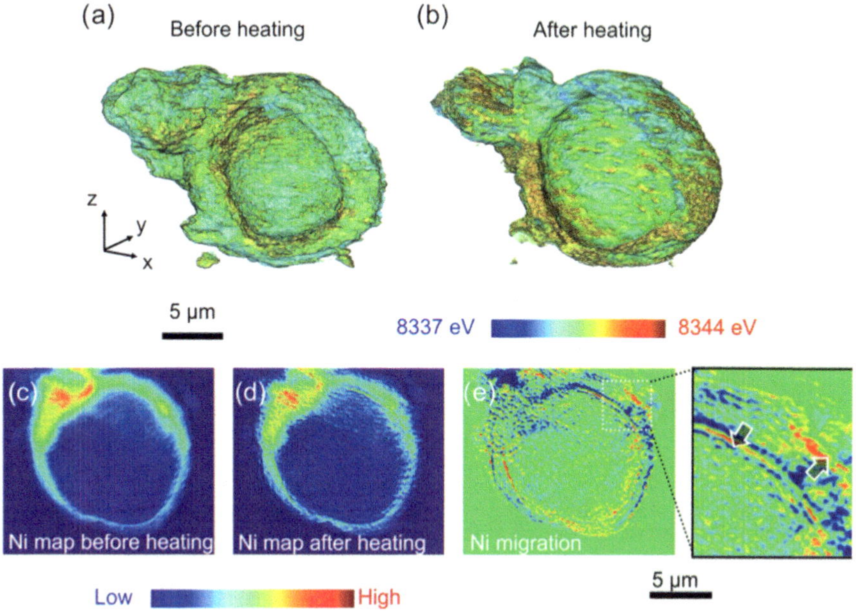

Fig. 4.19 3D TXM-XANES diagrams of NMC-622 particles with 50% delithiated **a** before and **b** after heating to 350 °C. Ni content distribution in the particles **c** before and **d** after heating. The map of Ni migration is shown in **e**, where the maps on the right are the enlarged part of the white box

4.4 Conclusions and Outlooks

In this chapter, we have clarified the principle of the TXM and reviewed its applications in energy materials, especially battery materials. From these researches, we can conclude the TXM is a powerful tool for morphological imaging, elemental distribution and valence change, which play a significant role in investigating the operation mechanism of battery materials.

With the development of hardware, such as the latest generation synchrotron radiation facilities, better performance grating monochromators and detectors, both the temporal resolution and spatial resolution of TXM can be improved. At the same time, the data volume of TXM will increase by several orders of magnitude. Therefore, software development like processing method of TXM data and quantitative analysis of processing result will become a hot topic. In the near future, more and more artificial intelligence (AI) technologies based on machine learning and deep learning will be used for TXM data processing and analysis. We believe the performance of batteries will be improved sharply based on sophisticated TXM technique.

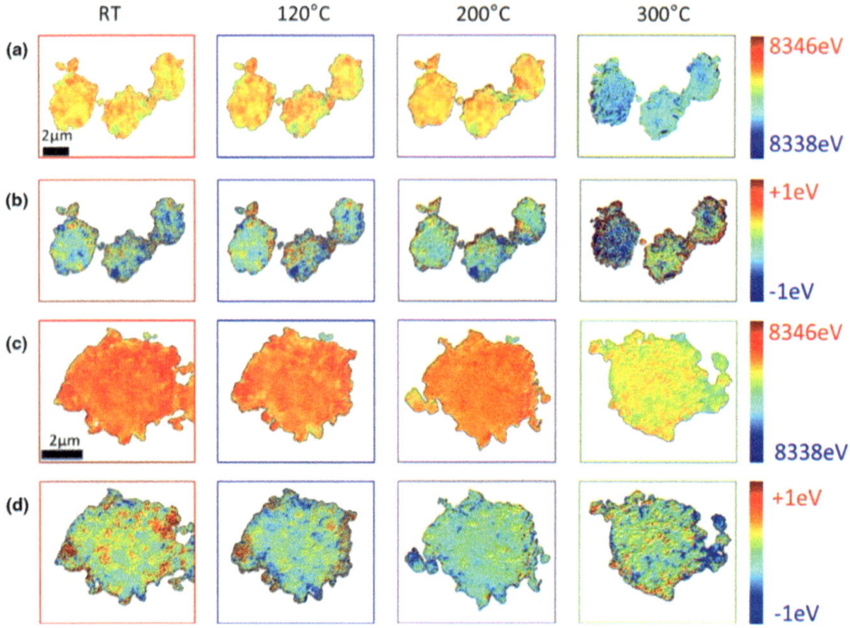

Fig. 4.20 2D Ni XANES maps of 50% delithiated NMC-811 particles varies with temperature **a** and relative edge energy differences **b**. 2D Ni XANES maps of 75% delithiated NMC-811 particles vary with temperature **c** and relative edge energy differences **d**

References

1. Chu, S., Majumdar, A.: Opportunities and challenges for a sustainable energy future. Nature **488**(7411), 294–303 (2012)
2. Zenyuk, I.V.: Bridging X-ray computed tomography and computational modeling for electrochemical energy-conversion and -storage. Curr. Opin. Electrochem. **13**, 78–85 (2019)
3. Lou, S., Yu, Z., Liu, Q., Wang, H., Chen, M., Wang, J.: Multi-scale imaging of solid-state battery interfaces: from atomic scale to macroscopic scale. Chem **6**(9), 2199–2218 (2020)
4. Wei, C.X., Xia, S.H., Huang, H., Mao, Y.W., Pianetta, P., Liu, Y.J.: Mesoscale battery science: the behavior of electrode particles caught on a multispectral X-ray camera. Acc. Chem. Res. **51**(10), 2484–2492 (2018)
5. Cao, C.T., Toney, M.F., Sham, S.K., Harder, R., Shearing, P.R., Xiao, X.H., Wang, J.J.: Emerging X-ray imaging technologies for energy materials. Mater. Today **34**, 132–147 (2020)
6. Tariq, F., Yufit, V., Kishimoto, M., Shearing, P.R., Menkin, S., Golodnitsky, D., Gelb, J., Peled, E., Brandon, N.P.: Three-dimensional high resolution X-ray imaging and quantification of lithium ion battery mesocarbon microbead anodes. J. Power Sources. **248**, 1014–1020 (2014)
7. Meirer, F., Cabana, J., Liu, Y.J., Mehta, A., Andrews, J.C., Pianetta, P.: Three-dimensional imaging of chemical phase transformations at the nanoscale with full-field transmission X-ray microscopy. J. Synchrotron Radiat. **18**, 773–781 (2011)
8. Wang, J.J., Chen-Wiegart, Y.C.K., Eng, C., Shen, Q., Wang, J.: Visualization of anisotropic-isotropic phase transformation dynamics in battery electrode particles. Nat. Commun. **7**, 7 (2016)

9. Yang, F.F., Liu, Y.J., Martha, S.K., Wu, Z.Y., Andrews, J.C., Ice, G.E., Pianetta, P., Nanda, J.: Nanoscale morphological and chemical changes of high voltage lithium manganese rich NMC composite cathodes with cycling. Nano Lett. **14**(8), 4334–4341 (2014)
10. Xu, Y.H., Hu, E.Y., Yang, F.F., Corbett, J., Sun, Z.H., Lyu, Y.C., Yu, X.Q., Liu, Y.J., Yang, X.Q., Li, H.: Structural integrity-searching the key factor to suppress the voltage fade of Li-rich layered cathode materials through 3D X-ray imaging and spectroscopy techniques. Nano Energy **28**, 164–171 (2016)
11. Wei, C.X., Zhang, Y., Lee, S.J., Mu, L.Q., Liu, J., Wang, C.X., Yang, Y., Doeff, M., Pianetta, P., Nordlund, D., Du, X.W., Tian, Y.C., Zhao, K.J., Lee, J.S., Lin, F., Liu, Y.J.: Thermally driven mesoscale chemomechanical interplay in $Li_{0.5}Ni_{0.6}Mn_{0.2}Co_{0.2}O_2$ cathode materials. J. Mater. Chem. A **6**(45), 23055–23061 (2018)
12. Shearing, P.R., Howard, L.E., Jorgensen, P.S., Brandon, N.P., Harris, S.J.: Characterization of the 3-dimensional microstructure of a graphite negative electrode from a Li-ion battery. Electrochem. Commun. **12**(3), 374–377 (2010)
13. Wang, J., Chen-Wiegart, Y.-C.K., Wang, J.: In situ three-dimensional synchrotron X-ray nanotomography of the (De)lithiation processes in tin anodes. Angewandte Chemie Int Edn **53**(17), 4460–4464 (2014)
14. Wang, J., Eng, C., Chen-Wiegart, Y.-C.K., Wang, J.: Probing three-dimensional sodiation-desodiation equilibrium in sodium-ion batteries by in situ hard X-ray nanotomography. Nat. Commun. **6** (2015)
15. Zhao, C.H., Wada, T., De Andrade, V., Gursoy, D., Kato, H., Chen-Wiegart, Y.C.K.: Imaging of 3D morphological evolution of nanoporous silicon anode in lithium ion battery by X-ray nano-tomography. Nano Energy **52**, 381–390 (2018)
16. Li, T.Y., Lim, C., Cui, Y., Zhou, X.W., Kang, H.X., Yan, B., Meyerson, M.L., Weeks, J.A., Liu, Q., Guo, F.M., Kou, R.H., Liu, Y.Z., De Andrade, V., De Carlo, F., Ren, Y., Sun, C.J., Mullins, C.B., Chen, L., Fu, Y.Z., Zhu, L.K.: In situ and operando investigation of the dynamic morphological and phase changes of a selenium-doped germanium electrode during (de)lithiation processes. J. Mater. Chem. A **8**(2), 750–759 (2020)
17. Nelson, J., Misra, S., Yang, Y., Jackson, A., Liu, Y.J., Wang, H.L., Dai, H.J., Andrews, J.C., Cui, Y., Toney, M.F.: In operando X-ray diffraction and transmission X-ray microscopy of lithium sulfur batteries. J. Am. Chem. Soc. **134**(14), 6337–6343 (2012)
18. Li, S., Jiang, Z., Han, J., Xu, Z., Wang, C., Huang, H., Yu, C., Lee, S.-J., Pianetta, P., Ohldag, H., Qiu, J., Lee, J.-S., Lin, F., Zhao, K., Liu, Y.: Mutual modulation between surface chemistry and bulk microstructure within secondary particles of nickel-rich layered oxides. Nat. Commun. **11**(1) (2020)
19. Cheng, J.H., Assegie, A.A., Huang, C.J., Lin, M.H., Tripathi, A.M., Wang, C.C., Tang, M.T., Song, Y.F., Su, W.N., Hwang, B.J.: Visualization of lithium plating and stripping via in operando transmission X-ray microscopy. J. Phys. Chem. C **121**(14), 7761–7766 (2017)
20. Foroozan, T., Sharifi-Asl, S., Shahbazian-Yassar, R.: Mechanistic understanding of Li dendrites growth by in-situ/operando imaging techniques. J. Power Sources **461**, 23 (2020)
21. Finegan, D.P., Cooper, S.J., Tjaden, B., Taiwo, O.O., Gelb, J., Hinds, G., Brett, D.J.L., Shearing, P.R.: Characterising the structural properties of polymer separators for lithium-ion batteries in 3D using phase contrast X-ray microscopy. J. Power Sources **333**, 184–192 (2016)
22. Grew, K.N., Chu, Y.S., Yi, J., Peracchio, A.A., Izzo, J.R., Jr., Hwu, Y., De Carlo, F., Chiu, W.K.S.: Nondestructive nanoscale 3D elemental mapping and analysis of a solid oxide fuel cell anode. J. Electrochem. Soc. **157**(6), B783–B792 (2010)
23. Guan, Y., Gong, Y., Li, W., Gelb, J., Zhang, L., Liu, G., Zhang, X., Song, X., Xia, C., Xiong, Y., Wang, H., Wu, Z., Tian, Y.: Quantitative analysis of micro structural and conductivity evolution of Ni-YSZ anodes during thermal cycling based on nano-computed tomography. J. Power Sources **196**(24), 10601–10605 (2011)
24. Rahman, M.M., Xu, Y.H., Cheng, H., Shi, Q.L., Kou, R.H., Mu, L.Q., Liu, Q., Xia, S.H., Xiao, X.H., Sun, C.J., Sokaras, D., Nordlund, D., Zheng, J.C., Liu, Y.J., Lin, F.: Empowering multicomponent cathode materials for sodium ion batteries by exploring three-dimensional compositional heterogeneities. Energy Environ. Sci. **11**(9), 2496–2508 (2018)

25. Wang, J., Chen-Wiegart, Y.-C.K., Wang, J.: In situ chemical mapping of a lithium-ion battery using full-field hard X-ray spectroscopic imaging. Chem. Commun. **49**(58), 6480–6482 (2013)
26. Li, L.S., Chen-Wiegart, Y.C.K., Wang, J.J., Gao, P., Ding, Q., Yu, Y.S., Wang, F., Cabana, J., Wang, J., Jin, S.: Visualization of electrochemically driven solid-state phase transformations using operando hard X-ray spectro-imaging. Nat. Commun. **6**, 8 (2015)
27. Kan, W.H., Wei, C.X., Chen, D.C., Bo, T., Wang, B.T., Zhang, Y., Tian, Y.C., Lee, J.S., Liu, Y.J., Chen, G.Y.: Evolution of local structural ordering and chemical distribution upon delithiation of a rock salt structured $Li_{1.3}Ta_{0.3}Mn_{0.4}O_2$ cathode. Adv. Funct. Mater. **29**(17), 11 (2019)
28. Zhang, F., Lou, S., Li, S., Yu, Z., Liu, Q., Dai, A., Cao, C., Toney, M.F., Ge, M., Xiao, X., Lee, W.-K., Yao, Y., Deng, J., Liu, T., Tang, Y., Yin, G., Lu, J., Su, D., Wang, J.: Surface regulation enables high stability of single-crystal lithium-ion cathodes at high voltage. Nat. Commun. **11**(1), 3050 (2020)
29. Lin, F., Markus, I.M., Nordlund, D., Weng, T.-C., Asta, M.D., Xin, H.L., Doeff, M.M.: Surface reconstruction and chemical evolution of stoichiometric layered cathode materials for lithium-ion batteries. Nat. Commun. **5**(1), 3529 (2014)
30. Lee, J., Urban, A., Li, X., Su, D., Hautier, G., Ceder, G.: Unlocking the potential of cation-disordered oxides for rechargeable lithium batteries. Science **343**(6170), 519–522 (2014)
31. Yabuuchi, N., Yoshii, K., Myung, S.-T., Nakai, I., Komaba, S.: Detailed studies of a high-capacity electrode material for rechargeable batteries, $Li_2MnO_3-LiCo_{1/3}Ni_{1/3}Mn_{1/3}O_2$. J. Am. Chem. Soc. **133**(12), 4404–4419 (2011)
32. Xu, Y.H., Hu, E.Y., Zhan, K., Wang, X.L., Borzenets, V., Sun, Z.H., Pianetta, P., Yu, X.Q., Liu, Y.J., Yang, X.Q., Li, H.: In situ visualization of state-of-charge heterogeneity within a $LiCoO_2$ particle that evolves upon cycling at different rates. Acs Energy Lett. **2**(5), 1240–1245 (2017)
33. Kan, W.H., Deng, B., Xu, Y.H., Shukla, A.K., Bo, T., Zhang, S., Liu, J., Pianetta, P., Wang, B.T., Liu, Y.J., Chen, G.Y.: Understanding the effect of local short-range ordering on lithium diffusion in $Li_{1.3}Nb_{0.3}Mn_{0.4}O_2$ single-crystal cathode. Chem **4**(9), 2108–2123 (2018)
34. Lei, L., He, D., Zou, Y., Zhang, W., Wang, Z., Jiang, M., Du, M.: Phase transitions of $LiAlO_2$ at high pressure and high temperature. J. Solid State Chem. **181**(8), 1810–1815 (2008)
35. Mu, L.Q., Yuan, Q.X., Tian, C.X., Wei, C.X., Zhang, K., Liu, J., Pianetta, P., Doeff, M.M., Liu, Y.J., Lin, F.: Propagation topography of redox phase transformations in heterogeneous layered oxide cathode materials. Nat. Commun. **9**, 8 (2018)
36. Tian, C., Xu, Y., Nordlund, D., Lin, F., Liu, J., Sun, Z., Liu, Y., Doeff, M.: Charge heterogeneity and surface chemistry in polycrystalline cathode materials. Joule **2**(3), 464–477 (2018)
37. Tian, C.X., Xu, Y.H., Kan, W.H., Sokaras, D., Nordlund, D., Shen, H., Chen, K., Liu, Y.J., Doeff, M.: Distinct surface and bulk thermal behaviors of $LiNi_{0.6}Mn_{0.2}Co_{0.2}O_2$ cathode materials as a function of state of charge. Acs Appl. Mater. Interfaces **12**(10), 11643–11656 (2020)
38. Li, S.F., Lee, S.J., Wang, X.L., Yang, W.L., Huang, H., Swetz, D.S., Doriese, W.B., O'Neil, G.C., Ullom, J.N., Titus, C.J., Irwin, K.D., Lee, H.K., Nordlund, D., Pianetta, P., Yu, C., Qiu, J.S., Yu, X.Q., Yang, X.Q., Hu, E.Y., Lee, J.S., Liu, Y.J.: Surface-to-bulk redox coupling through thermally driven Li redistribution in Li- and Mn-rich Layered cathode materials. J. Am. Chem. Soc. **141**(30), 12079–12086 (2019)
39. Alvarado, J., Wei, C.X., Nordlund, D., Kroll, T., Sokaras, D., Tian, Y.C., Liu, Y.J., Doeff, M. M.: Thermal stress-induced charge and structure heterogeneity in emerging cathode materials. Mater. Today **35**, 87–98 (2020)

Chapter 5
Coherent Diffractive Imaging and Its Application in Energy Materials and Devices Study

Xiaojing Huang

5.1 Introduction

Lenses are dominantly popular in various imaging systems, either as objective optics to form images directly or as focusing optics for scanning microscopy. While a lens-based imaging system is elegant and intuitive, the lens also limits achievable image quality. Alternative approaches seek complementary imaging methods without physical image-forming lenses, thus eliminating the associated limitations. Coherent diffractive imaging (CDI) emerged as one of the lensless imaging techniques, which can provide a diffraction-limited resolution with multiple contrasts.

5.2 Go Lensless

Understanding an objective lens's role in a conventional imaging system is essential to replace its functionality with a different approach. Considering a point light source is placed in front of an objective lens, the wavefront from the point source propagates through a distance z_1 to the entrance surface of the lens, and this propagation process introduces a quadratic Fresnel phase term $\exp[i(x^2 + y^2)/2z_1]$ [1]. The objective lens has a finite pupil aperture, which collects the signal inside. The lens profile (typically a convex shape for visible lights) is designed to create a quadratic phase term $\exp[-i(x^2 + y^2)/2f]$ that defines the focal length f. After penetrating the lens, the wavefront propagates z_2 further downstream to the image plane. Similarly, this propagation gives another quadratic phase term $\exp[i(x^2 + y^2)/2z_2]$. The image plane's location satisfies the lens law, $1/z_1 + 1/z_2 = 1/f$, where

X. Huang (✉)
National Synchrotron Light Source II, Brookhaven National Laboratory,
Upton, NY 11973, USA
e-mail: xjhuang@bnl.gov

J. Wang (ed.), *Advanced X-ray Imaging of Electrochemical Energy Materials and Devices*, https://doi.org/10.1007/978-981-16-5328-5_5

those mentioned above three quadratic phase terms cancel out. The remaining terms turn out to be a Fourier transform of the pupil function of the objective lens. For a typical lens with a round aperture, the image of the point source is not a point, but an Airy disk with a rounded central peak and concentric rings, as indicated in the middle row of Fig. 5.1. The Airy disk is also known as the point-spread function of the objective lens-based imaging system. In this imaging forming process, the function of the objective lens can be summarized in two folds: (1) the lens collects signal within its aperture and (2) the varying thickness of the lens brings a quadratic phase term to cancel the propagation phase terms and forms an image at the image plane.

For the first function, as the aperture of a lens has a finite size, it only collects signals within the aperture, while the high spatial frequency signal outside the aperture does not contribute to forming the image. This maximum spatial frequency is determined by the collectible maximum scattering angle, defined as the numerical aperture (NA) by the aperture size D and the focal length f as $D/(2f)$. The NA of the lens determines the achievable spatial resolution of the imaging system. We can use the formation of the Airy disk as an example to show this effect. Since the Airy disk is a Fourier transform of the lens aperture function, a larger aperture would generate a smaller Airy disk. Because an Airy disk is the image of one point source, if we place two point sources very close to each other, the corresponding two Airy disks will overlay together and be observed as an enlarged oval disk, where the two point sources cannot be resolved. When the separation between these two point sources

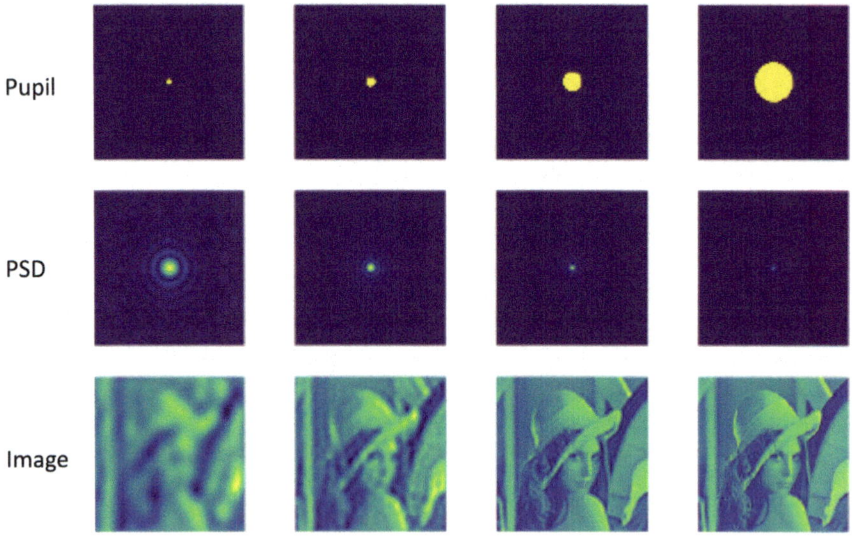

Fig. 5.1 Impact of the numerical aperture on the image quality. Top row, the pupil apertures with increasing sizes. Middle row, the corresponding point-spread function (PSF) for each aperture, which is an Airy disk. Bottom row, the obtained images with different apertures. A larger aperture gives smaller PSF and achieves higher spatial resolution

increases, the overlaid Airy disks will separate more. The spatial resolution of such an imaging system can be defined as the minimum separation between two point sources that can be resolved. Rayleigh criterion [2] provides an empirical threshold: when the peak of the second Airy disk locates at the first dark ring of the first Airy disk, the intensity at the center of the summation is about 73.5% of the peak, which is distinguishable to human eyes. The spatial resolution determined using the Rayleigh criterion can be expressed as $d = 0.61\lambda/\mathrm{NA}$. A larger NA is required to achieve a higher spatial resolution. Figure 5.1 shows the impact of NA on the obtained image quality. When a lens is used as a focusing optics, the focal size is directly determined by NA in the same manner, $0.61\lambda/\mathrm{NA}$, which also directly determines the scanning probe microscopy system's achievable spatial resolution.

For the second function of the objective lens, the lens's thickness profile introduces a quadratic phase term to cancel the other quadratic phase terms from propagations. If we remove the lens and setup a detector behind the sample to measure the propagated signal directly. The recorded pattern can be expressed as a Fourier transform of the product of the wavefront after sample and the quadratic phase term determined by the propagation distance. When the detector-to-sample distance is far enough to satisfy the Fraunhofer approximation, the quadratic phase term is negligible in the Fresnel propagation, which can be further simplified by a Fourier transform. Thus, the measured propagation data can be inverse Fourier transferred numerically to form the desired real-space image. In principle, the imaging form process can be completed with the absence of the objective lens. Consequently, we can also remove the resolution limitation imposed by the lens's finite aperture and possible artifacts introduced by aberrations of the lens.

5.3 Phase Retrieval

Removing the objective lens from the imaging system removes the drawbacks of the lens and also takes out some advantages. One such advantage is that the lens collects both magnitude and phase information of the scattered signal and forms an image with complex-valued contrast. When measuring the image with a photon-sensitive detector, only the magnitude of the complex-valued image can be recorded, which gives a real-space image with the absorption contrast. The phase contrast has to be converted to measurable magnitude variation using methods such as the Zernike phase contrast mechanism. When recording the scattering signal directly with the objective lens removed, we face the same challenge: only the scattering intensity (magnitude square) can be measured, while the phase part of the complex-valued scattering field is lost during the measurement process. This is known as the "missing phase" problem [3].

The unmeasured phase actually plays a critical role in the image-forming process. For instance, as shown in Fig. 5.2, we start with a picture of the BNL logo and apply a Fourier transform to convert it into the scattering pattern in Fourier space. This complex-valued scattering pattern has magnitude and phase parts. If we only take the magnitude part and inverse Fourier transfer back to real space, the obtained image does not show the initial BNL logo features. If we inverse Fourier transfer only the phase part instead, we can clearly recognize the BNL logo on the obtained image. This example suggests that the scattering signal phase encodes more critical information of the initial real-space image. Unfortunately, the phases cannot be measured as easily as the magnitudes.

The lost phase problem was encountered in many scientific research scenarios. A similar case can be found in crystallography, where only the diffraction intensity is measured. Fourier transform of the diffraction intensities does not give the real-space atomic map, but its autocorrelation, known as Patterson maps, gives interatomic spacings [4]. Another example can be found in stellar speckle interferometry, where the recorded image is blurred as the initial object convoluted with the point-spread function due to the atmospheric turbulence [5]. The object's Fourier intensity can be obtained by decoupling the turbulence term measured without the object in the imaging system.

The main task for CDI technique is to recover the unmeasured phase information and thus reconstruct the desired real-space image with diffraction-limited resolution. David Sayre firstly pointed out that it is possible to recover the lost phase from the Fourier intensity measurement [6]. This idea was inspired by the Nyquist–Shannon sampling theorem, which states that a band-limited continuous signal can be completely specified when its Fourier transform is measured with a sampling rate that doubles its highest spatial frequency [7]. Figure 5.3 illustrates the impact of the sampling rate on signal recovery. Here, the reconstruction target is a step function with 200 pixels, and the 20 pixels in the center are one while all other pixels are zero. The Fourier amplitude of this step function is a sinc function with a band limit of 0.1. The Nyquist–Shannon sampling theorem suggests that a sampling rate of 0.2 is required to fully recover the initial step function, which means sampling at least one out of every 5 pixels. Figure 5.3 shows that with a low sampling rate at measuring one out of every 6 and 7 pixels, the recovered functions deviate from the

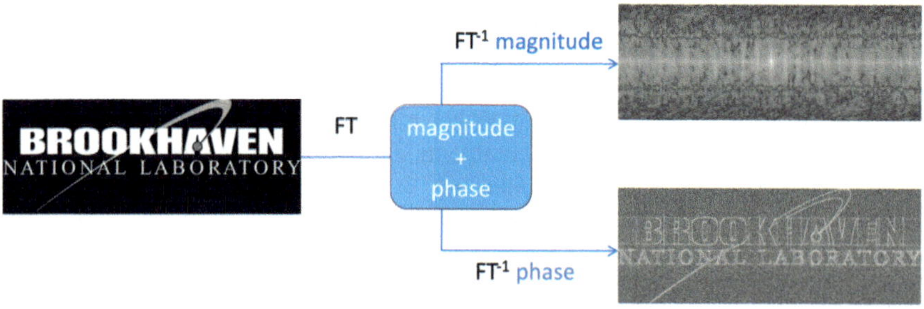

Fig. 5.2 Recovering real-space image from Fourier components. The Fourier transform of a real-space image, the BNL logo, gives a complex-valued pattern in reciprocal space. Inverse Fourier transform only the magnitude part gives a real-space image with no recognizable features for the BNL logo. Inverse Fourier transform only the phase part provides an image representing the logo. It suggests more critical information is encoded in the Fourier phases

target step function while sampling every 5 pixels starts to recover the exact step function. For a CDI measurement, the smallest feature in the sample defines the highest spatial frequency generated by the sample. In other words, the diffraction pattern (Fourier intensity) is band-limited. If one can sample the diffraction pattern using a detector with sufficiently fine pixels, the sample function can be accurately recovered.

An intuitive way to understanding the sample requirement for CDI is to match the number of known and unknown variables in this numerical imaging process. Consider a sample image with $N \times M$ pixels, and each pixel value is a complex number representing the absorption and phase contrasts. As a result, $2 \times M \times N$ variables represent the sample, which needs to be reconstructed. A detector records $N \times M$ pixels of Fourier intensities giving $N \times M$ known variables in the measurement. To make the phase-retrieval problem solvable, one has to reduce the number of unknown variables by at least 50%. The sample has to be compact and confined in a region occupies less than $N \times M/2$ pixels inside the field of view. In practice, the area that occupied by the sample is defined as the support, which is widely used as a constraint in real-space. Since the smallest Fourier space speckle is determined by the largest real-space feature, the support, the Fourier space's sampling condition, is expressed as the smallest speckle has to spread to at least 2 detector pixels. It should note that the oversampling ratio of two

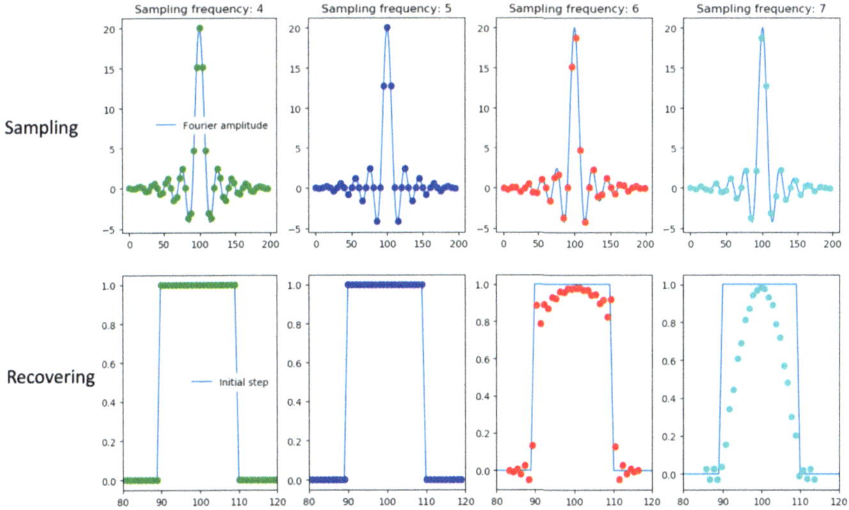

Fig. 5.3 An illustration of Nyquist–Shannon sampling theorem. Consider a 1D step function with 200 pixels in total and the central 20 pixels nonzero, and its Fourier transform is a sinc function with the band-limited to 0.1. The Nyquist–Shannon sampling theorem suggests that a sampling rate higher than 0.2 (one measurement out of every 5 pixels) is needed to recover the step function fully. Sampling rates equal or above the Nyquist–Shannon sampling rate (one out of every 4 and 5 pixels) recover the step function accurately. In comparison, sampling rates coarser than the Nyquist–Shannon sampling rate (one out of every 6 and 7 pixels) fail to recover the exact step function

ideally applies to 1D, 2D, and 3D cases, instead of sampling each dimension by two times [8]. However, a higher oversampling ratio is usually preferable to improve the phase-retrieval process's convergence and robustness, especially when the dataset has unmeasured areas, such as pixels blocked by the central beamstop or pixels with counts below the noise level [9].

Generally speaking, the phase-retrieval process's goal is to find an object function whose Fourier intensity matches with the measurement. The reconstruction ambiguity raises when more than one object functions give the same scattering intensity. For instance, the Fourier transform property suggests that the object function with an arbitrary translation, a multiplicative phase offset, or its central reflected and complex conjugated twin give exactly the same Fourier intensity. The linear combinations of all these ambiguities are all allowable.

In the reconstruction process, an accurate support function can effectively exclude the translational ambiguity and the twin if the shape of the object is not central symmetric. Another type of ambiguity comes from the propagation property of the wavefront. Considering the wavefront exits the object and propagates to the detector plane, every wavefront along the propagation path gives the same Fourier intensity pattern recorded on the detector, known as the propagation uncertainty. Adjusting the size and shape of the support constraint effectively determines which propagation plane the reconstruction is performed. In other words, a tight support constraint breaks the propagation uncertainty to a single plane [10]. Other real-space constraints such as the positivity constraint [11] and the phase constraint [12] eliminate the propagation uncertainty similarly.

A good initial guess of the support and an effective refinement strategy during iterations are critical for the convergence speed and the final reconstruction quality. One way to initiate the support is to use the autocorrelation function, which is the Fourier transform of the measured far-field scattering intensity and gives the object's shape with a doubled dimension. This property has been used as the Patterson maps in crystallography, and it was introduced to the CDI reconstruction [13]. The presence of noise and other imperfections in real experimental datasets makes it very challenging to obtain very accurate supports using the autocorrelation function. The shrinkwrap method was proposed to automatically and progressively refine the support function as the iterative reconstruction gradually converges [14]. As shown in Fig. 5.4, the shrinkwrap method starts with a rough estimation of the support first. The obtained reconstruction image is then smoothened by convoluting with a Gaussian function, where an intensity threshold redefines the support function for the following iterations. This refinement process can be repeated with adjustable iteration intervals. The width of the blurring Gaussian function and the threshold setting control how aggressively the refinement process is conducted.

One of the most generic and basic phase-retrieval algorithms is known as error reduction (ER), which alternatively imposes the support constraint in real space and the measured Fourier magnitude (modulus constraint) in reciprocal space. The left panel of Fig. 5.5 shows the ER algorithm's schematic, which starts by assigning a random phase map to the measured Fourier magnitude, then inverse Fourier transfers them into real space, where the pixel values outside the support area are set to zero. We denote the operation of applying the support constraint as $P_{\mathrm{s}}(\psi)$, where ψ is the sample function. The modified real-space image is then Fourier transferred back to reciprocal space, while replacing the Fourier magnitude with the measurement and keeping the calculated phases. We denote the operation of applying the modulus constraint as $P_{\mathrm{m}}(\psi)$. This process iterates until it converges to a solution, which satisfies both the support and the modulus constraints. To quantify the convergence of the iterative reconstruction engine, one can define the error metrics as the difference in ψ before and after P_{s} and P_{m} operation [15],

$$E^2_{\mathrm{s,m}} = \Sigma\left|P_{\mathrm{s,m}}(\psi) - \psi\right|^2 \tag{5.1}$$

The gradients of these two error metrics can be calculated as:

$$_{\psi}E^2_{\mathrm{s,m}} = -2\left[P_{\mathrm{s,m}} - I\right]\psi \tag{5.2}$$

The operations of applying constraints can be expressed by the error gradients as:

$$P_{\mathrm{s,m}}(\psi) = \psi - 0.5\nabla_{\psi}E^2_{\mathrm{s,m}} \tag{5.3}$$

which suggests that the ER algorithm updates the object function following the steepest descent direction, so it ensures the error metrics' decrement.

The object function ψ can be considered as a vector in a high-dimension space, and the support and Fourier modulus constraints define two sets of points in the

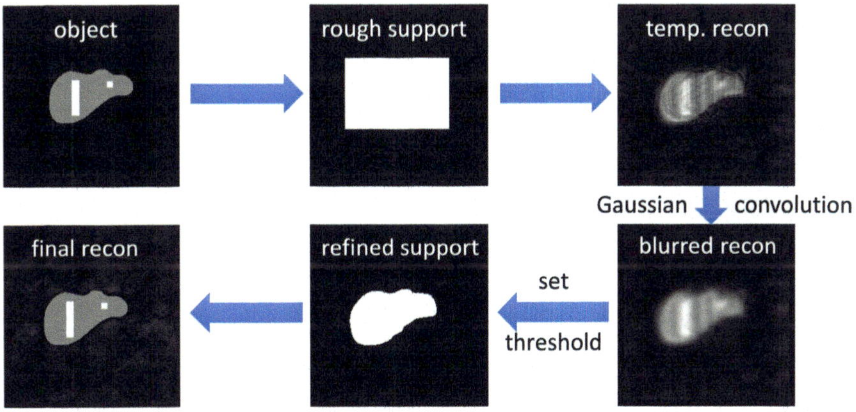

Fig. 5.4 Shirnkwrap method for refining the support constraint. A preliminary reconstruction is conducted with a rough estimation of the support. The obtained image magnitude is then convolved with a Gaussian function to smooth out sharp features and boundaries. An updated support is obtained by applying a threshold to the blurry magnitude image. The support refinement process can be iteratively conducted if necessary. A smaller width of the Gaussian function and a higher mask threshold determines a more aggressive refinement process

same space [15]. Enforcing the support constraint P_s is equivalent to project the ψ vector onto a set defined by the support, as shown in Fig. 5.5a. Applying the modulus constraint P_m is equivalent to project the ψ vector to a circle with its radius defined by the measured Fourier magnitude, as shown in Fig. 5.5b. The intersectional region of these two sets represents the allowable solutions to the phase-retrieval problem. Because of the ER algorithm's steepest descend nature, it can be trapped into local minimums and fails to find the true solution at the global minimum. This phenomenon is known as stagnation, as shown in Fig. 5.5c. The ER algorithm is modified to various versions to avoid the stagnation problem. One of the most popular variations is the hybrid input–output (HIO) algorithm, which relaxes the support using negative feedback instead of setting outside pixels directly to zero [16]. Using the projection concept, we can see the HIO algorithm takes a spiral trajectory, capable of avoiding or moving out of the local minimums, as shown in Fig. 5.5d. A comprehensive review of the performance of phase-retrieval algorithms can be found in [15].

The phase-retrieval process can be generally considered as minimizing a defined error metrics following the gradient direction. This problem is also well suited for nonlinear optimization algorithms with various gradient computation and direction searching methods [17, 18]. Besides the algorithm family based on the oversampling concept and alternatively enforcing constraints, the compressive sensing algorithm was proposed to utilize the sparsity of the measured dataset [19], which can reconstruct images with a sampling rate beyond the Nyquist–Shannon sampling

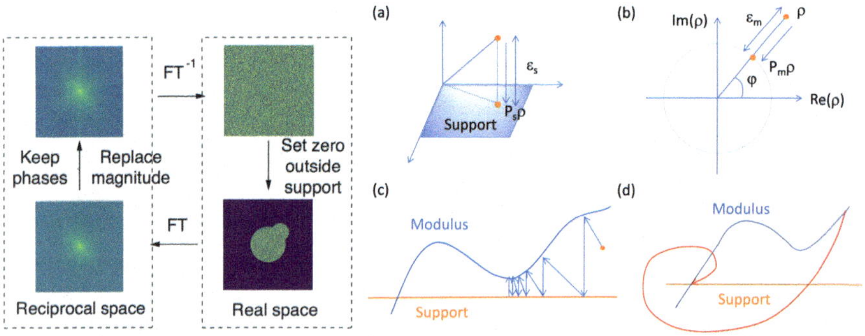

Fig. 5.5 Phase-retrieval algorithms. Left panel, the schematic of the error reduction algorithm. A random phase map is assigned to the measured Fourier magnitude, which is then inverse Fourier transferred to real space. The support constraint is applied to the real-space image by setting pixels outside the support to zero. The modified real-space image is Fourier transferred to reciprocal space. The modulus constraint is applied by replacing the magnitude with the measurement and keeping the calculated phase. This process iterates until it converges to a solution. Right panel: **a** the support constraint projects an image vector onto a set defined by the support; **b** the modulus constraint projects a Fourier vector onto a circle with its radius defined by the measured Fourier magnitude; **c** the alternative projection process in the ER algorithm can be trapped in local minimums; **d** modified algorithms such as HIO is able to avoid local minimums and converge to the global minimum

theorem. This algorithm relies on prior knowledge about the data sparsity on a certain mathematical basis, which may not always be available [20].

5.4 Contrast and Image Quality

The object image is visualized through specific contrast mechanisms. In visible light and X-ray regimes, the absorption and phase contrasts are generated through the interaction between the photons and the object materials. The propagation of a plane wave in the vacuum can be described using an oscillating function $\psi_0 e^{ikz}$, where ψ_0 is the initial oscillating magnitude, k is the wave number ($2\pi/\lambda$), and z is the propagation distance. When the wave penetrates through a material, the propagation function changes to $\psi_0 e^{inkt}$, where n is the refractive index of the material for the wavelength of the incident wave and t is the sample thickness. The refractive index is defined to represent how the material modifies the wavefront: the oscillating magnitude is reduced as an absorption effect, and the oscillation phase will be retarded because the wavelength expands inside the material compared with the vacuum. The refractive index is denoted as a complex-valued number as $n = 1 - \delta + i\beta$. The wave function in materials can then break into three terms:

$$\psi_0 e^{inkt} = \psi_0 e^{ikt} e^{-i\delta kt} e^{-\beta kt} \tag{5.4}$$

where $\psi_0 e^{ikt}$ is the vacuum propagation, $e^{-i\delta kt}$ and $e^{-\beta kt}$ represent the phase shift and the magnitude decay. The last two terms provide the phase and absorption contrasts. Omitting the vacuum propagation phase term, one can express the transmission function of the object ψ as $Ae^{i\varphi}$, where A denotes for the absorbed magnitude $\psi_0 e^{-\beta kt}$ and φ is the phase shift $-\delta kt$. The object is thus described as the complex-valued function, which is the function that the phase-retrieval reconstruction algorithms aim to recover. We can see that the CDI method simultaneously provides quantitative images with both absorption and phase contrasts through phase retrieval, which are directly related to the refractive index of the sample.

In CDI measurements, without an objective lens between the sample and the detector, the numerical aperture of this lensless imaging system is defined by the maximum scattering angle collected by the detector centered at zero spatial frequency. Similar to the NA definition of an objective lens, the numerical aperture of the CDI imaging system is $\text{NA} = N\Delta/(2z)$, where N is the number of pixels on the detector grid, Δ is the detector pixel size, and z is the sample-to-detector distance, as shown in Fig. 5.6a. The maximum spatial frequency associated with NA is $q_{max} = \text{NA}/\lambda = N\Delta/(2\lambda z)$. The real-space pixel size of the reconstructed image is $\Delta r = 1/(2q_{max}) = \lambda/(2\text{NA}) = \lambda z/(N\Delta)$. Intuitively, to achieve a finer spatial resolution, the NA has to be increased by enlarging the detector size with more and larger pixels, placing the detector closer to the sample, or using shorter wavelength illumination. These options need to be evaluated in realistic experiments: increasing the detector pixel size and moving the detector closer to the sample will debilitate the oversampling condition; and a higher energy illumination with a shorter wavelength will impact the contrast mechanism. Another useful note is that the field of view can be calculated by the reconstruction pixel size and pixel numbers as FoV = $\Delta rN = \lambda z/\Delta$. To satisfying the oversampling condition, the sample dimension should be less than 0.5 FoV. On the other hand, if the object has a diameter of D, the associated speckle size is $\lambda z/D$, which should be at least two times larger than the detector pixel size Δ. Thus, the detector-to-sample distance z has to satisfy $z > 2D\Delta/\lambda$.

When pursuing large NA for achieving a fine spatial resolution, the depth of field also shrinks with enlarged NA as $A\lambda$/NA2, where A is a scaling factor. The Huygens principle suggests that the scattering signal spreads out as a spherical wavefront from the source. In reciprocal space, it is known as the Ewald's sphere, with a radius of $1/\lambda$. In experiments, a flat detector is used to measure the spherical wavefront, which inevitably introduces a distortion. This distortion progressively deteriorates from the scattering center to the high spatial frequency region. As shown in Fig. 5.6b, the maximum frequency component perpendicular to the detector surface is $q_z = \lambda/\left[2\left(q_x^2 + q_y^2\right)\right]$, where q_x and q_y are the maximum spatial frequency on the detector plane, and they are determined by the NA of the imaging system as NA = $q_{x,y}\lambda$. Similar to the reconstruction pixel size in the lateral plane $\Delta x = \lambda/2$NA, the q_z component corresponds to a pixel size in the thickness direction $\Delta z = 1/2q_z = \lambda$/NA2. In the published literature, this pixel size along the thickness direction [21] or half of the pixel size [22] was chosen to define the depth of field, which corresponds the scaling factor A of 1 and 0.5, respectively. If the Rayleigh

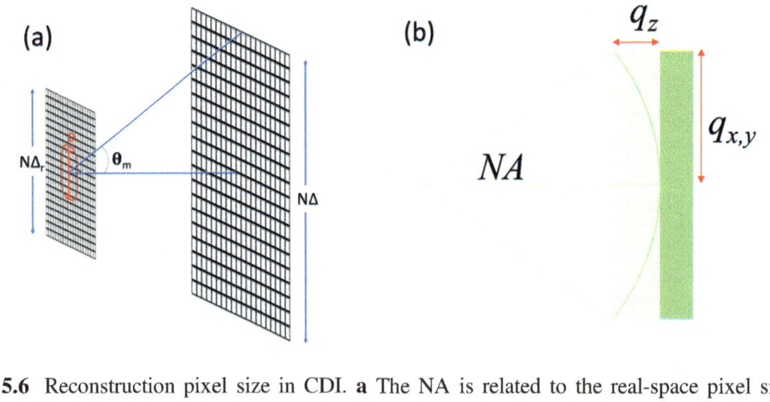

Fig. 5.6 Reconstruction pixel size in CDI. **a** The NA is related to the real-space pixel size as NA = $\lambda/(2\Delta r)$. The NA can be also calculated from the maximum scattering angle recorded by the detector as NA = $N\Delta/z$. Thus, the reconstruction pixel size in real-space can be calculated as $\Delta r = \lambda z/(N\Delta)$. **b** Recording the scattering signal on the Ewald's sphere using a flat detector introduces a distortion in the direction norm to the detector surface, $q_z = \lambda/\left[2\left(q_x^2 + q_y^2\right)\right]$. This distortion is more severe toward the high spatial frequency region, and it determines the depth of field and the maximum sample thickness

resolution ($0.61\lambda/NA$) along the thickness direction is chosen to define the depth of field, which gives a scaling factor of 1.22 [23, 24]. Analogous to transmission electron microscopy (TEM) applications, the sample thickness is required to be thinner than the depth of field without sacrificing achievable resolution.

With a detector capable of recording a high-angle scattering signal, on the other hand, the illuminated sample has to scatter enough photons at corresponding spatial frequencies to make a meaningful measurement with a sufficiently high signal-to-noise ratio. Theoretical modeling shows that the scattering power dramatically decays along with the spatial frequency with third or fourth power [25, 26]. This effect marks CDI as a photon-hungry method. More incident photon flux and longer exposure time are needed to measure high spatial frequency signals for achieving higher resolution. The inquiry of intense photon illumination brings the radiation damage problem as a side effect. To quantify the radiation intensity, one can define the radiation dose as the absorbed energy per sample mass with an SI unit of gray (Gy), which equals 1 J of energy absorbed by 1 kg of material. Taking the square of the wave function penetrating a material with a thickness t, one can obtain the transmitted intensity as $I(t) = I_0 e^{-\mu t}$, where μ is the attenuation length $\mu = 2k\beta$. The radiation dose depositing on the sample can be calculated as $D = \bar{n} E_{photon} \mu / \rho$, where $\bar{n}$ is the photon flux density, Ephoton is the photon energy, and ρ is the sample density. Theoretical modeling and experimental data show that fine features will be destroyed by the X-ray radiation first, and larger features will be damaged with the increment of radiation dose. For biological

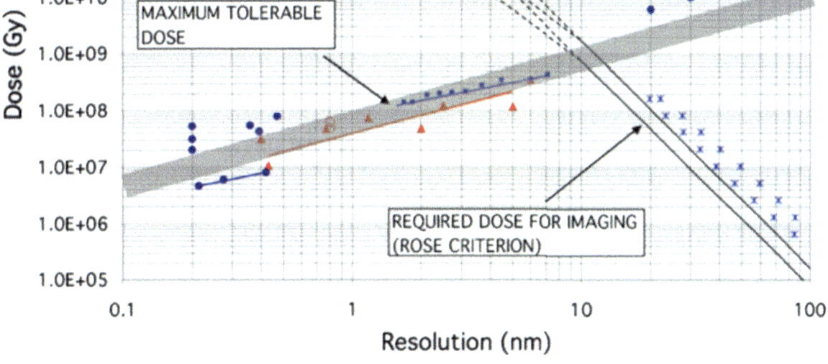

Fig. 5.7 Dose requirement and dose tolerance for biological imaging. More radiation dose is required to achieve higher spatial resolution, but finer structures are less tolerant to radiation dose. These two competing effects set a limit for achievable resolution for biological specimens at 10 nm. The figure is adapted from [27] Fig. 3

samples, the battle of the radiation dose on reaching finer resolution and damaging larger features sets an achievable resolution limit around 10 nm [27], as shown in Fig. 5.7. Materials sciences sample systems are expected to be radiation harder, but the radiation damage effect should be kept in mind for X-ray imaging experiments, especially when aiming for high spatial resolutions.

Another consequence of pursuing high radiation dose for measuring weak scattering signal at a high spatial frequencies is that the detector's dynamic range has to be sufficiently large to collect the full scattering intensity from the intense center to the dim outer ring. A typical pixel array detector with single-photon counting capability has a dynamic range of six decades in the hard X-ray regime. The state-of-the-art detector has a dynamic range of nine decades [28]. In the soft X-ray regime, the dynamic range for the actual photon detection is usually smaller due to analog-to-digital conversion. In practice, the detector dynamic range can be extended by accumulating multiple exposures and blocking or attenuating the intense scattering center with an opaque or a semi-transparent beamstop.

5.5 Coherence

A coherence illumination allows the constructive interference and the encoding of the phase information into measurable scattering intensity as the square of amplitude summation, making it possible to recover the unmeasured phase through scattering intensity alone. As a result, a coherent incident beam is required for a CDI experiment. For the third generation of synchrotron radiation sources, the coherent portion of photons must be selected to illuminate the specimen. The transverse or spatial coherence can describe the coherence property in the lateral plane and the longitudinal or temporal coherence in the beam propagation direction.

The transverse coherence can be improved by using a small aperture as a spatial coherence filter. Consider an aperture with a diameter of d and the observation plane located z away from the aperture, as shown in the left panel of Fig. 5.8. The transverse coherence length can be estimated by defining a point t_{coh} away from the center on the observation plane, and the optical path difference of an opposite point pair on the aperture boundary is $\lambda/2$, where these two lights will be in opposite phase and cancel each other. This transverse coherence length can be calculated as $t_{\mathrm{coh}} = \lambda z/d$. The transverse coherence length of the incident illumination has to be larger than the entire reconstruction field of view [29], which is the sample size multiplying the oversampling ratio. To increase the transverse coherence length, one can reduce the aperture size or move the observation plane further away from the aperture, however, both of which are expected to reduce the incident photon flux to the sample.

When the incident beam is not entirely monochromatic but with various wavelengths distributed in a bandwidth $\Delta\lambda$, it is longitudinally incoherent. As shown in the right panel of Fig. 5.8, the longitudinal coherence length can be estimated from two wavefronts starting with the same phase and traveling a distance

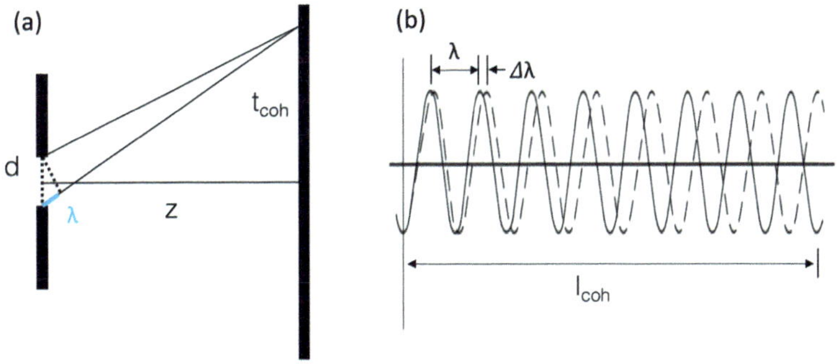

Fig. 5.8 Transverse (spatial) and longitudinal (temporal) coherence lengths estimated as good correlations in wavefront phases. **a** Consider two lights emitting from the opposite sides of an aperture d reach an observation plane placed z downstream and centered with the aperture. The transverse coherence length can be estimated as the two lights arrive at a point with a lateral distance t_{coh} from the observation plane center with an optical path difference of λ, and thus, they have the opposite phases. Under this condition, $t_{\mathrm{coh}} = \lambda z/d$. **b** Considering a wavefront ensemble with a central wavelength λ and a bandwidth $\Delta\lambda$, by traveling a distance l_{coh}, the phases of λ and $\lambda + \Delta\lambda$ waves will be offset by π. Under this condition, $l_{\mathrm{coh}} = \lambda 2/(2\Delta\lambda)$

l_{coh} to create a phase offset of π, where these two wavefronts will have opposite phases. The longitudinal coherence length l_{coh} can be calculated as $l_{\mathrm{coh}} = \lambda 2/(2\Delta\lambda)$. For the maximum spatial frequency q_m captured at the detector's edge with $N \times N$ pixels, the corresponding scattering angle is $\theta = N\Delta/(2z) = q_m\lambda$. The wavelength bandwidth changes this scattering angle by $\Delta\theta = q_m\Delta\lambda$. The longitudinal coherence condition can be estimated by limiting this angular change no larger than half of the solid angle defined by a detector pixel [21], $\Delta/(2z)$, which gives $\Delta\lambda/\lambda < 1/N$.

In practice, relaxing the coherence condition significantly increases the photon flux, thus improving the throughput and achievable resolution. The Fourier intensity is convolved with the incident beam's mutual coherence function for the partial transverse coherence case. It has been demonstrated that the coherence function can be reconstructed simultaneously with the object function by adding an iterative deconvolution step in the phase-retrieval process [30]. A successful reconstruction was obtained for the partial longitudinal coherence case by implementing the intensity spectrum of a polychromatic beam into the reconstruction engine and modeling the measured Fourier intensity as the summation from all wavelengths [31]. The capability to deal with partial coherence heavily depends on the dataset quality and the sampling condition. A study showed that it is possible to relax both transverse and longitudinal coherence length by factor 2 without sacrificing reconstruction quality [32].

5.6 Resolution Estimation

Although the diffraction-limited achievable resolution can be estimated using the imaging system's NA, the evaluation of the obtained spatial resolution of an image is not straightforward and can be sample dependent.

If there is a sharp edge present in the sample, the line profile across the sharp boundary can be considered as a convolution of the step function with the point-spread function of the imaging system, which can fit with an error function or a Gaussian function to its derivative. The obtained full-width-at-half-maximum (FWHM) can be used to estimate of the spatial resolution representing the impact from the PSF.

The reconstruction quality can be evaluated by checking the consistency of two independent reconstructions with random initiate guesses. Considering the reliably reconstructed features can be reproducible with different starts, while the imperfection in data and the reconstruction inaccuracy due to constraint inconsistency will introduce fluctuations in the recovered image, the image quality can be quantified as an overall signal-to-noise ratio (SNR) calculated from the correlation coefficient r of two independent reconstructions.

as $\mathrm{SNR} = p1/(1-r)$,

and $r = \langle (I_1 - \langle I_1 \rangle)(I_2 - \langle I_2 \rangle)^* \rangle / \sqrt{\left(\left\langle (I_1 - \langle I_1 \rangle)^2 \right\rangle \left\langle (I_2 - \langle I_2 \rangle)^2 \right\rangle \right)}$ [33].

The resolution can also be estimated from how faithfully the image is recovered in reciprocal space by computing the ratio between the Fourier magnitude calculated from the reconstructed image and the measured Fourier magnitude. Plotting this ratio at all spatial frequencies gives the phase-retrieval transform function (PRTF) [21, 34], analogous to the modulus transform function. The image resolution can be estimated by the frequency where the PRTF drops below 0.5.

The reconstruction consistency in reciprocal space can also be evaluated using Fourier shell correlation (FSC). This method requires two reconstructed images from two independent measurements or two halves split from a full dataset and calculate their correlation in Fourier space at different spatial frequency shells. The reconstruction repeatability is estimated by the intersection between the FSC curve and the threshold curve defined by the noise level [35]. Two noise levels are typically used: the stringent 1-bit criterion requires the full dataset and half datasets to give SNR values of 1 and 0.5, respectively, while the 0.5-bit criterion requires SNRs of 0.4142 and 0.2071, respectively.

5.7 Fresnel CDI

Conducting CDI measurements with planar illuminations simplifies the interaction between the beam and the sample. However, this setup introduces reconstruction ambiguities, and the convergence of the iterative engine is highly sensitive to the data quality, the constraint accuracy, and the initial guess. Targeting these problems, it has been pointed out that using an illumination with finite curvature created by focusing X-ray optics can effectively eliminate ambiguities and provide a unique solution [36]. This method is referred to the Fresnel CDI. Instead of using a plane wave, Fresnel CDI uses a focused beam and places the sample at a de-focal plane. Thus, the incident beam on the sample captures a phase curvature through a Fresnel propagation from the focus, which is the key to remove the reconstruction ambiguities. A simulation study suggests the optimal reconstruction quality is achieved with a Fresnel number of 5 [37]. The Fresnel number can be calculated as FN = $a2/(\lambda z)$, where a is the characteristic feature size, λ is the incident beam wavelength, and z is the sample-to-detector distance. The recorded far-field scattering pattern in Fresnel CDI can be classified into two regions: (1) the holographic area in the center diverged from the focused beam creates an in-line hologram, which can be back-propagated to the sample plane to recover the object image with a spatial resolution limited by the NA of the focusing optics. (2) The scattered fringes and speckles outside the central cone capture high spatial frequency signals, which can be used in the iterative phase-retrieval process to deliver an image with refined spatial resolution [38].

Fresnel CDI reconstruction utilizes the same algorithms as CDI, but the focused beam profile has to be pre-characterized [39]. The support constraint is required to satisfy the oversampling criterion and eliminate propagation uncertainty. However, Fresnel CDI can be applied to the image object with an extended dimension through a mosaic scan scheme. Diffraction data at each exposed location gives a 2D image, and images from adjacent positions can be stitched together to create an image with an extended field of view [40, 41].

5.8 3D Imaging

A 2D diffraction dataset reconstructs a projection image of the object through the iterative phase-retrieval process. A 3D imaging can be obtained by adapting the tomography approach by rotating the object along an axis perpendicular to the incident beam direction, recording diffraction patterns at each rotation angle, and recovering 2D projections using CDI, and then reconstructing a 3D image using tomography algorithms. However, the reconstruction ambiguities in CDI make it challenging to follow this recipe directly. For instance, the translational ambiguity introduces an extra burden in aligning 2D projections, and the arbitrary phase offset and twin image bring inconsistency for 3D reconstruction. Since the CDI measurement is performed in the Fourier space, the diffraction datasets at all projection angles can be mapped to a 3D Fourier volume through a rotation and proper interpolation [42]. A critical step to interpolate a 2D diffraction frame into the 3D data volume is to map the flat 2D data frame onto the Ewald's sphere. Here, we follow the expressions described in [21]. An incident beam along the z axis defines kin = $1/\lambda\hat{z}$, and a pixel $(nx\Delta,\ ny\Delta)$ on a detector in the $(x,\ y)$ plane defines

$k_{out} = 1/\lambda(n_x\Delta\hat{x} + n_y\Delta\hat{y} + z\hat{z})/\sqrt{(n_x^2\Delta^2 + n_y^2\Delta^2 + z^2)}$. The wave-vector transfer $q = k_{out} - k_{in}$ gives the three components on a Cartesian grid in the Fourier space:

$$q_{x,y} = \frac{1}{\lambda}\frac{n_{x,y}\Delta}{\sqrt{(n_x^2\Delta^2 + n_y^2\Delta^2 + z^2)}}, \quad q_z = \frac{1}{\lambda}\left(\frac{z}{\sqrt{n_x^2\Delta^2 + n_y^2\Delta^2 + z^2}} - 1\right) \tag{5.5}$$

The relative rotation between diffraction frames collected at different angles can be processed by a 3D rotation matrix. The assembled 3D data volume in Fourier space can then be reconstructed using the same CDI algorithms with 3D Fourier transform to directly recover a 3D image. For a 3D object with confined dimensions, the overall oversampling ratio can be more generous than 2D cases. As a result, it facilitates the tolerance on typical sampling issues in tomography, such as limited angular range, missing angular wedges, and coarse angular steps under the Crowther criterion [43]. Recent algorithm developments show that the diffraction patterns from a randomly oriented sample or multiple identical copies of the same structure can be recognized and assembled in Fourier space, which can then be used for directly 3D reconstructions [44].

5.9 Bragg CDI

The CDI technique can be applied to crystalline specimens. When illuminating X-rays onto a cluster of crystals with random orientations, it creates Debye–Scherrer rings at characteristic diffraction angles. The locations of the powder rings are determined by the Bragg law, $2d\sin\theta = \lambda$, where d is the corresponding lattice d-spacing. If the number of illuminated crystals is gradually reduced, the continuous powder rings become sparse and eventually break into isolated diffraction peaks. If the incident beam is coherent, the diffraction fringes and speckles determined by the shape, size, and facets of the crystal will form in the vicinity of the Bragg peak. When the diffraction speckles are adequately sampled, they can be fed into a CDI reconstruction engine and convert into real-space images [45].

The transverse coherence length requirement for the Bragg CDI experiment keeps the same as for the transmission case, where the transverse coherence length should be larger than the reconstruction field of view. The longitudinal coherence length for Bragg CDI, instead of directly related to sample thickness, requires larger than the maximum optical path length difference (OPLD) of all the diffracted rays inside the sample. Varying the OPLD by picking different Bragg peaks from the same crystal was used to tune the fringe visibilities, which can then be used to estimate the incident illumination's longitudinal coherence length [47]. The partial coherence condition blurs the measured diffraction pattern as a convolution with the mutual coherence function. This impact can be effectively removed by decoupling the coherence function in 3D using iterative blind deconvolution algorithms [48], such as the Lucy–Richardson method [49, 50].

To excite a certain Bragg peak, one can manipulate the crystalline sample with both translational and rotational degrees of freedom. The detector is motorized on a diffractometer arm or alternative options (such as a robotic arm) to catch the target Bragg peak. Unlike the isolation requirement in the transmission CDI experiment, the crystal sample is not necessary to be physically isolated, because the excitations condition and the diffraction peaks will be separated if adjacent crystals with a different lattice structure or the same structure with different orientations [51]. In a typical Bragg CDI experiment, as shown in the left panel of Fig. 5.9, the detector is oriented and fixed at the target Brag peak with a momentum transform vector Q, and the crystal sample is rocked along an axis with a large component perpendicular to the Q direction, which is more efficient to move the 3D diffraction pattern across the detector plane. At each rocking angle, the detector records one slice of the 3D diffraction pattern. A typical rocking angle for a Bragg CDI measurement is about 1 degree, which is another advantage compared with 3D imaging transmission mode, where the sample needs to be rotated over 180 degrees angular range. Stacking the recorded diffraction frames together assembles to the 3D data volume in Fourier space, which can be reconstructed to give 3D real-space images.

Analogous to the contrast mechanism in the transmission geometry, the diffraction intensity in Bragg CDI can be expressed as $I(q) = |\int\rho(r)s(r)e^{iQu}e^{iqr}dr|^2$, where ρ is the electron density, s is the shape function, Q is the momentum transform vector associated, u is the displacement vector in the crystal, and q is the spatial frequency. The magnitude part of the reconstructed image $\rho(r)s(r)$ represents the electron density and the shape of the measured diffraction volume. Because Bragg CDI only measures the excited crystalline structure, an area with zero magnitudes in the reconstruction does not necessarily mean this area has zero electron density as avoid, while it is also possible that the area has a different lattice structure or a different orientation [52, 53]. The phase part of the reconstructed image $eiQ{\cdot}r$ manifests the component of the displacement field u projected on the

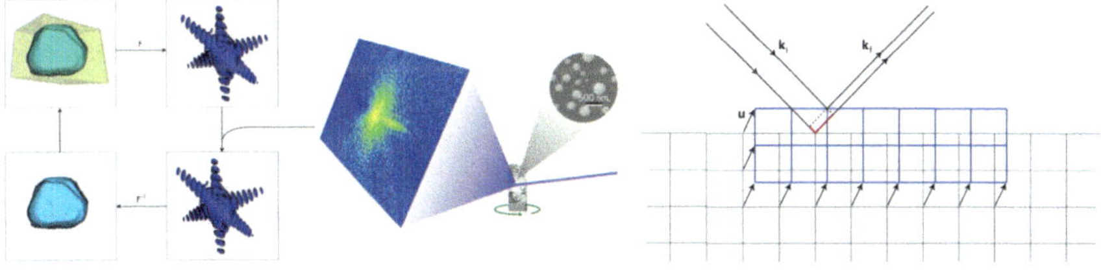

Fig. 5.9 Schematic for Bragg CDI. Left panel, the 3D Bragg diffraction signal is recorded using a pixel array detector slice-by-slice by rocking the crystal across its Bragg peak. The stacked 3D Fourier magnitude can be fed into phase-retrieval algorithms and converted into real-space image. Right panel, the magnitude of the reconstructed image represents the shape and size of the diffraction volume. The phase is determined by the projection of the displacement field u onto the momentum transform vector Q direction. The figures are adapted from [46] Box 1 and Box 2

Q vector [46], as shown in the right panel of Fig. 5.9. Because Bragg reconstructs the displacement field rendered in phases, the displacement sensitivity can reach pm level [54], while the spatial resolution of the reconstructed image is typically 10 s of nm.

One Bragg CDI measurement gives one component of the displacement field. Obtaining all three Cartesian components requires at least three measurements at non-planar Bragg peaks [55, 56]. With the fully resolved displacement field, the full nine-component strain tensor can be calculated as $\varepsilon_{ij} = 1/2(\delta\mu_j/\delta x_i + \delta\mu_i/\delta x_j)$, where x_i is the orthogonal axis.

Recent algorithm developments adopt machine-learning networks to solve the Bragg CDI phase-retrieval problem as a complementary approach to the computation-heavy and time-consuming iterative reconstruction algorithms. The designed deep-neural-networks trained from synthetic datasets can provide real-time and accurate reconstructions [57, 58].

5.10 Applications in Electrochemical Energy Materials and Devices

Because the CDI technique relies on the stringent sampling condition to make the phase-retrieval process possible, the requirement of the sample environment is strict. For instance, the isolated sample is expected in an empty or uniform background, which does not create parasitic diffraction. Such a requirement makes it challenging to conduct in-situ and in-operando experiments such as the electrochemical sample systems. Fresnel CDI relaxes the requirement on sample dimension with the assistant of the phase curvature in the incident beam. It has been applied to image an integrated circuit with extended size and about 20 nm spatial resolution [40]. This approach is expected to be more favorable for imaging functional materials in their working conditions. A newly developed approach referred as in-situ CDI combines the holography principle with CDI, which uses a dual-pinhole aperture with a sample only placed at one of the pinholes. The empty pinhole provides a static reference signal, which improves convergence robustness and removes ambiguities [59]. With the field of view confined in the aperture and improved reconstruction performance, this method can provide high-quality images in-situ of a dynamic sample system. An emerging trend of the X-ray microscopy method development clearly shows that the combination of CDI with the scanning probe technique unifies both advantages of achieving diffraction-limited resolution and extended dimension and removes the requirement of the support constraint by using the overlapping measurements from adjacent scan spots. This method is known as ptychography, which quickly becomes popular in the X-ray microscopy community and has been proven to be more convenient for in-situ and in-operando investigations [60]. The ptychography method is described in this Chap. 6.

On the other hand, since Bragg CDI uniquely reconstructs the 3D displacement field and is relatively less sensitive to the surrounding environment outside the diffraction volume, Bragg CDI is still considered as a powerful tool to study morphology and strain evolutions in functioning sample systems. For example, the 3D internal strain evolution of a zeolite crystal was measured in-situ during a calcination process using Bragg CDI [61]. An abnormal transient triangular-shaped strain was observed before the full calcination, as shown in the left panel of Fig. 5.10. Bragg CDI reconstruction is sensitive to this unique strain pattern caused by small traces of residual organic impurities inside the zeolite crystal, which introduces a core–shell distribution of the crystal lattice and impacts the adsorption and diffusion of molecules through zeolite pores. A zeolite crystal's strain response to the catalytic nitrogen oxide deoxygenation process was investigated by measuring a continuous series of Bragg diffraction patterns with XFEL pulses during the propene absorption [62]. This deoxygenation reaction plays a critical role in purifying vehicle exhausts. The reconstructed image frames reveal a time-resolved structural deformation due to the inhomogeneous distribution of the active cations.

Bragg CDI has also been applied to study the strain distribution of Li-ion battery cathode nanoparticles. The nanoscale strain mapping in $LiNi_{0.5}Mn_{1.5}O_4$ cathode nanoparticles was measured ex-situ and in-situ (in a modified coil cell) using Bragg CDI [64], and observed compressive/tensile strains up to 0.4%. The strain map of a $LiNi_{0.5}Mn_{1.5}O_4$ nanoparticle during the first discharge process at a C/2 rate under operando condition was visualized using Bragg CDI [65]. The discharge-driven evolution of the inhomogeneous, coherency strain/striping between coexistent phases and the elastic energy landscape was quantitatively visualized in 3D. The topological defect dynamics of a $LiNi_{0.5}Mn_{1.5}O_4$ was visualized in-operando during charge [66]. The reconstructed strain maps revealed 3D edge dislocation lines as singularities in the recovered displacement field, which were observed to be stable at room and evolves as a function of charging toward the particle boundary. As shown in the right panel of Fig. 5.10, the Bragg CDI imaging of the nucleation and dynamics of dislocations in a $Li_{1.2}Ni_{0.133}Mn_{0.533}Co_{0.133}O_2$ nanoparticle captured the formation of the edge dislocation during lithium extraction and monitored its growth to a dislocation network during charge [63]. A significantly higher rate of dislocation formation was observed compared with classical materials, which associates with higher strain energy and acts as the origin of the voltage fade.

Bragg CDI will undoubtedly find wide applications on electrochemical energy materials and devices with the relaxed requirement on sample environments and its inherent selectivity on the diffraction volume. With the emerging of the next generation of coherent X-ray sources such as the diffraction-limited storage ring and XFEL, the coherent imaging methods are expected to keep playing an important role in boosting energy conversion and storage-related researches.

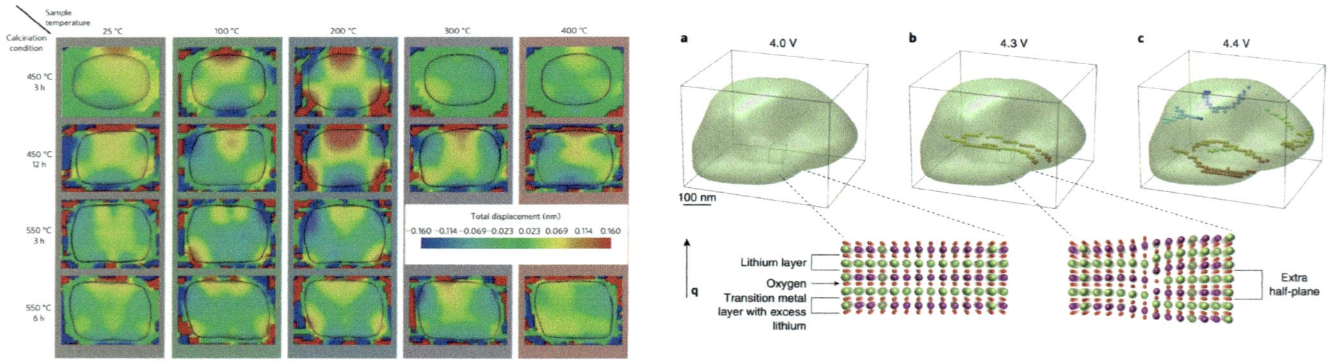

Fig. 5.10 Bragg CDI applications on electrochemical energy materials. Left panel, the strain evolution inside a zeolite crystal was revealed by Bragg CDI during a calcination process. An abnormal triangular strain pattern introduced by residual organic impurities appears at 200°. This picture is adapted from [61]. Right panel, the formation and evolution of a dislocation network inside a lithium-rich layered oxide crystal during a charging process. The edge dislocation initiates during lithium extraction, grows to a dislocation network, and moves toward the crystal boundary. This picture is adapted from [63]

References

1. Goodman, J.: Introduction to Fourier Optics, 3rd edn. Roberts and Company (2005)
2. Rayleigh, L.: Investigations in optics, with special reference to the spectroscope. Philos. Mag. **8**(49), 261–274 (1879)
3. Duane, W.: The calculation of the X-ray diffracting power at points in a crystal. Proc. Natl. Acad. Sci. USA **11**(8), 489–493 (1925)
4. Patterson, A.: A fourier series method for the determination of the components of interatomic distances in crystals. Phys. Rev. **46**, 372–376 (1934)
5. Labeyrie, A.: Attainment of diffraction limited resolution in large telescopes by fourier analysing speckle patterns in star images. Astron. Astrophys. **6**, 85–87 (1970)
6. Sayre, D.: Some implications of a theorem due to shannon. Acta Crystallogr. A **5**, 843 (1952)
7. Shannon, C.: Communication in the presence of noise. Proc. Inst. Radio Eng. **37**(1), 10–21 (1949)
8. Miao, J., Sayre, D., Chapman, H.N.: Phase retrieval from the magnitude of the Fourier transforms of non-periodic objects. J. Opt. Soc. Am. A **15**(6), 1662–1669 (1998)
9. Thibault, P., Elser, V., Jacobsen, C., Shapiro, D., Sayre, D.: Reconstruction of a yeast cell from X-ray diffraction data. Acta Crystallographica A **62**, 248–261 (2006)
10. Spence, J.C.H., Weierstall, U., Howells, M.: Phase recovery and lensless imaging by iterative methods in optical, X-ray and electron diffraction. Phil. Trans. Royal Soc. Lon. A. **360**, 875–895 (2002)
11. Luke, D.: Relaxed averaged alternating reflections for diffraction imaging. Inverse Prob. **21**, 37–50 (2005)
12. Huang, X., Harder, R., Xiong, G., Shi, X., Robinson, I.: Propagation uniqueness in three-dimensional coherent diffractive imaging. Phys. Rev. B **83**, 224109 (2011)
13. Crimmins, T., Fienup, J., Thelen, B.: Improved bounds on object support from autocorrelation support and application to phase retrieval. J. Opt. Soc. Am. A **7**(1), 3–13 (1990)
14. Marchesini, S., He, H., Chapman, H., Hau-Riege, S., Noy, A., Howells, M., Weierstall, U., Spence, J.: X-ray image reconstruction from a diffraction pattern alone. Phys. Rev. B **68**(14), 140101 (2003)
15. Marchesini, S.: A unified evaluation of iterative projection algorithms for phase retrieval. Rev. Sci. Instrum. **78**, 011301 (2007)
16. Fienup, J.: Phase retrieval algorithms: a comparison. Appl. Opt. **21**(12), 2758–2769 (1982)
17. Guizar-Sicairos, M., Fienup, J.: Phase retrieval with transverse translation diversity: a nonlinear optimization approach. Opt. Express **16**(10), 7264–7278 (2008)
18. Thibault, P., Guizar-Sicairos, M.: Maximum-likelihood refinement for coherent diffractive imaging. New J. Phys. **14**, 063004 (2012)
19. Candes, E., Romberg, J., Tao, T.: Robust uncertainty principles: exact signal reconstruction from highly incomplete Fourier information. IEEE Trans. Inf. Theor. **52**(2), 489–509 (2006)
20. Szameit, A., Shechtman, Y., Osherovich, E., Bullkich, E., Sidorenko, P., Dana, H., Steiner, S., Gazit, S., Kley, E., Cohen-Hyams, T., Shoham, S., Zibulevsky, M., Yavneh, I., Eldar, Y., Cohen, O., Segev, M.: Sparsity-based single-shot subwavelength coherent diffractive imaging. Nat. Mater. **11**, 455–459 (2012)
21. Chapman, H., Barty, A., Marchesini, S., Noy, A., Hau-Riege, S., Cui, C., Howells, M., Rosen, R., He, H., Spence, J., Weierstall, U., Beetz, T., Jacobsen, C., Shapiro, D.: High resolution ab initio three-dimensional X-ray diffraction microscopy. J. Opt. Soc. Am. A **23**, 1179–1200 (2006)
22. Rodenburg, J., Bat, R.: The theory of super-resolution electron microscopy via Wigner-distribution deconvolution. Phil. Trans. R. Soc. Lond. A **339**, 521–553 (1992)
23. Born, M., Wolf, E.: Principles of Optics, 7th edn. Cambridge University (1999)
24. Jacobsen, C.: Relaxation of the crowther criterion in multislice tomography. Opt. Lett. **43**(19), 4811–4814 (2018)

25. Marchesini, S., Chapman, H., Hau-Riege, S., London, R., Szoke, A., He, H., Howells, M., Padmore, H., Rosen, R., Spence, J., Weierstall, U.: Coherent X-ray diffractive imaging: applications and limitations. Opt. Express **11**(19), 2344–2353 (2003)
26. Shen, Q., Bazarovc, I., Thibault, P.: Diffractive imaging of nonperiodic materials with future coherent X-ray sources. J. Synchrotron Rad. **11**, 432–438 (2004)
27. Howells, M., Beetz, T., Chapman, H., Cui, C., Holton, J., Jacobsen, C., Kirz, J., Lima, E., Marchesini, S., Miao, H., Sayre, D., Shapiro, D., Spence, J., Starodub, D.: An assessment of the resolution limitation due to radiation-damage in X-ray diffraction microscopy. J. Electron Spectrosc. Relat. Phenom. **170**, 4–12 (2009)
28. Casanas, A., Warshamanage, R., Finke, A., Panepucci, E., Olieric, V., Noll, A., Tampe, R., Brandstetter, S., Forster, A., Mueller, M., Schulze-Briese, C., Bunk, O., Wang, M.: EIGER detector: application in macromolecular crystallography. Acta Cryst. **D72**, 1036–1048 (2016)
29. Spence, J., Weierstall, U., Howells, M.: Coherence and sampling requirements for diffractive imaging. Ultramicroscopy **101**, 149–152 (2004)
30. Clark, J., Peele, A.: Simultaneous sample and spatial coherence characterisation using diffractive imaging. Appl. Phys. Lett. **99**, 154103 (2011)
31. Abbey, B., Whitehead, L., Quiney, H., Vine, D., Cadenazzi, G., Henderson, C., Nugent, K., Balaur, E., Putkunz, C., Peele, A., Williams, G., McNulty, I.: Lensless imaging using broadband X-ray sources. Nat. Photonics **5**, 420–424 (2011)
32. Chen, B., Abbey, B., Dilanian, R., Balaur, E., van Riessen, G., Junker, M., Tran, C., Jones, M., Peele, A., McNulty, I., Vine, D., Putkunz, C., Quiney, H., Nugent, K.: Diffraction imaging: the limits of partial coherence. Phys. Rev. B **86**, 235401 (2012)
33. Huang, X., Miao, H., Steinbrener, J., Nelson, J., Stewart, A., Shapiro, D., Jacobsen, C.: Signal to noise considerations in diffraction and conventional microscopy. Opt. Express **17**(16), 13541–13553 (2009)
34. Shapiro, D., Thibault, P., Beetz, T., Elser, V., Howells, M., Jacobsen, C., Kirz, J., Lima, E., Miao, H., Neiman, A.M., Sayre, D.: Biological imaging by soft X-ray diffraction microscopy. Proc. Nat. Acad. Sci. **102**(43), 15343–15346 (2005)
35. Heel, M., Schatz, M.: Fourier shell correlation threshold criteria. J. Struct. Biol. **151**, 250–262 (2005)
36. Nugent, K., Peele, A., Quiney, H., Chapman, H.: Diffraction with wavefront curvature: a path to unique phase recovery. Acta Crystallogr. A **61**, 373–381 (2005)
37. Quiney, H., Nugent, K., Peele, A.: Iterative image reconstruction algorithms using wave-front intensity and phase variation. Opt. Lett. **30**(13), 1628–1640 (2005)
38. Williams, G., Quiney, H., Dhal, B., Tran, C., Nugent, K., Peele, A., Paterson, D., de Jonge, M.: Fresnel coherent diffractive imaging. Phys. Rev. Lett. **97**(2), 025506 (2006)
39. Quiney, H., Peele, A., Cai, Z., Paterson, D., Nugent, K.: Diffractive imaging of highly focused X-ray fields. Nat. Phys. **2**, 101–104 (2006)
40. Abbey, B., Williams, G., Pfeifer, M., Clark, J., Putkunz, C., Torrance, A., McNulty, I., Levin, T., Peele, A., Nugent, K.: Quantitative coherent diffractive imaging of an integrated circuit at a spatial resolution of 20 nm. Appl. Phys. Lett. **93**, 214101 (2008)
41. Abbey, B., Nugent, K., Williams, G., Clark, J., Peele, A., Pferfer, M., de Jonge, M., McNulty, I.: Keyhole coherent diffractive imaging. Nat. Phys. **4**, 394–398 (2008)
42. Miao, J., Ishikawa, T., Johnson, B., Anderson, E.H., Lai, B., Hodgson, K.O.: High resolution 3D X-ray diffraction microscopy. Phys. Rev. Lett. **89**(8), 088303 (2002)
43. Nishino, Y., Takahashi, Y., Imamoto, N., Ishikawa, T., Maeshima, K.: Threedimensional visualization of a human chromosome using coherent X-ray diffraction. Phys. Rev. Lett. **102**, 018101 (2009)
44. Saldin, D., Poon, H., Bogan, M., Marchesini, S., Shapiro, D., Kirian, R., Weierstall, U., Spence, J.: New light on disordered ensembles: ab initio structure determination of one particle from scattering fluctuations of many copies. Phys. Rev. Lett. **106**, 115501 (2011)
45. Pfeifer, M., Williams, G., Vartanyants, I., Harder, R., Robinson, I.: Threedimensional mapping of a deformation field inside a nanocrystal. Nature **442**, 63–66 (2006)

46. Robinson, I., Harder, R.: Coherent X-ray diffraction imaging of strain at the nanoscale. Nat. Mater. **8**, 291–298 (2009)
47. Leake, S., Newton, M., Harder, R., Robinson, I.: Longitudinal coherence function in X-ray imaging of crystals. Opt. Express **17**(18), 15853–15859 (2009)
48. Clark, J., Huang, X., Harder, R., Robinson, I.: High-resolution three-dimensional partially coherent diffraction imaging. Nat. Commun. **3**, 933 (2012)
49. Richardson, W.: Bayesian-based iterative method of image restoration. J. Opt. Soc. Am. **62**, 55–59 (1972)
50. Lucy, L.: Iterative technique for rectification of observed distributions. Astron. J. **79**, 745–754 (1974)
51. Yau, A., Cha, W., Kanan, M., Stephenson, G., Ulvestad, A.: Bragg coherent diffractive imaging of single-grain defect dynamics in polycrystalline films. Science **356**, 739–742 (2017)
52. Ulvestad, A., Clark, J., Harder, R., Robinson, I., Shpyrko, O.: 3d imaging of twin domain defects in gold nanoparticles. Nano Lett. **15**, 4066–4070 (2015)
53. Huang, X., Yang, W., Harder, R., Sun, Y., Lu, M., Chu, Y., Robinson, I., Mao, H.: Deformation twinning of a silver nanocrystal under high pressure. Nano Lett. **15**, 7644–7649 (2015)
54. Clark, J., Beitra, L., Xiong, G., Higginbotham, A., Fritz, D., Lemke, H., Zhu, D., Chollet, M., Williams, G., Messerschmidt, M., Abbey, B., Harder, R., Korsunsky, A., Wark, J., Robinson, I.K.: Ultrafast three-dimensional imaging of lattice dynamics in individual gold nanocrystals. Science **341**, 56–59 (2013)
55. Newton, M., Leake, S., Harder, R., Robinson, I.: Three-dimensional imaging of strain in a single ZnO nanorod. Nat. Mater. **9**, 120–124 (2010)
56. Hofmann, F., Tarleton, E., Harder, R., Phillips, N., Ma, P., Clark, J., Robinson, I., Abbey, B., Liu, W., Beck, C.: 3d lattice distortions and defect structures in ionimplanted nano-crystals. Sci. Rep. **7**, 45993 (2017)
57. Cherukara, M., Nashed, Y., Harder, R.: Real-time coherent diffraction inversion using deep generative networks. Sci. Rep. **8**, 16520 (2018)
58. Wu, L., Juhas, P., Yoo, S., Robinson, I.: Complex imaging of phase domains by deep neural networks. IUCrJ **8** (2021)
59. Lo, Y., Zhao, L., Gallagher-Jones, M., Rana, A., Lodico, J., Xiao, W., Regan, B., Miao, J.: In situ coherent diffractive imaging. Nat. Commun. **9**, 1826 (2018)
60. Pfeifer, F.: X-ray ptychography. Nat. Photonics **12**, 9–17 (2017)
61. Cha, W., Jeong, N., Song, S., Park, H., Pham, T., Harder, R., Lim, B., Xiong, G., Ahn, D., McNulty, I., Kim, J., Yoon, K., Robinson, I., Kim, H.: Core-shell strain structure of zeolite microcrystals. Nat. Mater. **12**, 729–734 (2013)
62. Kang, J., Carnis, J., Kim, D., Chung, M., Kim, J., Yun, K., An, G., Cha, W., Harder, R., Song, S., Sikorski, M., Robert, A., Thanh, N., Lee, H., Choi, Y., Huang, X., Chu, Y., Clark, J., Song, M., Yoon, K., Robinson, I., Kim, H.: Time-resolved in situ visualization of the structural response of zeolites during catalysis. Nat. Commun. **11**, 5901 (2020)
63. Singer, A., Zhang, M., Hy, S., Cela, D., Fang, C., Wynn, T., Qiu, B., Xia, Y., Liu, Z., Ulvestad, A., Hua, N., Wingert, J., Liu, H., Sprung, M., Zozulya, A., Maxey, E., Harder, R., Meng, Y., Shpyrko, O.: Nucleation of dislocations and their dynamics in layered oxide cathode materials during battery charging. Nat. Energy **3**, 641–647 (2018)
64. Ulvestad, A., Cho, H., Harder, R., Kim, J., Dietze, S., Fohtung, E., Meng, Y., Shpyrko, O.: Nanoscale strain mapping in battery nanostructures. Appl. Phys. Lett. **104**, 073108 (2014)
65. Ulvestad, A., Singer, A., Cho, H., Clark, J., Harder, R., Maser, J., Meng, Y., Shpyrko, O.: Single particle nanomechanics in operando batteries via lensless strain mapping. Nano Lett. **14**, 5123–5127 (2014)
66. Ulvestad, A., Singer, A., Clark, J., Cho, H., Kim, J., Harder, R., Maser, J., Meng, Y., Shpyrko, O.: Topological defect dynamics in operando battery nanoparticles. Science **348**, 1344–1347 (2015)

Chapter 6
Synchrotron Radiation Based X-ray Fluorescence Imaging

Biao Deng and Xiaolu Ju

Knowing the inhomogeneous distribution of chemical elements in matter is meaningful in both scientific research and industrial production. X-ray fluorescence (XRF) imaging is a powerful technique for nondestructive investigation of chemical element distribution in the samples.

6.1 X-ray Fluorescence Principle

Fluorescent X-rays are created when irradiated X-rays force inner-shell electrons of the constituent atoms to an outer shell and outer shell electrons promptly move to inner shells to fill the vacancies. The released energy is emitted in the form of radiation and produces X-ray fluorescence, whose energy is equal to the energy difference between two energy levels. Therefore, the energy or wavelength of X-ray fluorescence is characteristic and has one-to-one correspondence with elements. Figure 6.1 shows how fluorescent X-rays are produced. XRF analysis is a method that uses character critic X-rays (fluorescent X-rays) generated. X-ray fluorescence analysis instruments can be largely categorized into wavelength-dispersive X-ray spectroscopy (WDX) and energy-dispersive X-ray spectroscopy (EDX). The detector in EDX has superior energy resolution and requires no dispersion system, which allows the instrument to be smaller in size. Energy-dispersive X-ray fluorescence is

B. Deng (✉) · X. Ju
Shanghai Synchrotron Radiation Facility, Shanghai Advanced Research Institute, Chinese Academy of Sciences, Shanghai, China
e-mail: dengbiao@zjlab.org.cn

B. Deng · X. Ju
Shanghai Institute of Applied Physics, Chinese Academy of Sciences, Shanghai, China

B. Deng · X. Ju
University of Chinese Academy of Sciences, Beijing, China

J. Wang (ed.), *Advanced X-ray Imaging of Electrochemical Energy Materials and Devices*, https://doi.org/10.1007/978-981-16-5328-5_6

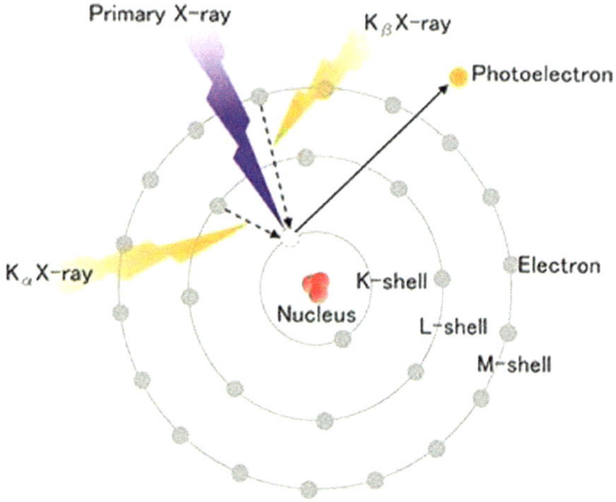

Fig. 6.1 Fluorescent X-ray generation

used widely in X-ray fluorescence analysis. And XRF is used to obtain both qualitative and quantitative information about elemental distribution in the sample.

Qualitative Analysis: The fluorescence X-ray of different elements have their own specific energy, so the composition of elements can be determined according to the energy of fluorescence X-rays. The Qualitative Analysis of fluorescent X-ray analysis uses this regularity (Moseley's law). Computer software qualitatively determines the composition of the sample. Qualitative analysis can be accurately performed by displaying the marker and observing the intensity comparison of multiple characteristic X-rays.

Quantitative Analysis: The basis of quantitative analysis by X-ray fluorescence spectroscopy is that the fluorescence X-ray intensity of the element is directly proportional to the content of the element in the sample. There are two basic methods of quantitative analysis with fluorescent X-rays. The first method is to create a calibration curve. This method involves measuring several samples of known element concentration and finding the relationship between the intensity of the measured element's fluorescent X-rays and the concentration. This relationship allows you to obtain the element concentration of an unknown sample from its fluorescent X-ray intensity. The other method is known as the fundamental parameter method of theoretical calculation method. This method allows us to theoretically derive the intensity of the fluorescent X-rays if the type and properties of all elements that compose a sample are known. The composition of the unknown sample can be extrapolated by the fluorescent X-ray intensities of each element.

6.2 Synchrotron Radiation Based X-ray Fluorescence Imaging Methodology Development

Synchrotron radiation based X-ray fluorescence (SR-XRF) is developed on the basis of traditional X-ray fluorescence. It uses synchrotron radiation source instead of X-ray tube as excitation source. Synchrotron radiation source has the characteristics of high intensity and high collimation. The XRF excited by synchrotron radiation source has higher sensitivity and spatial resolution than that excited by traditional X-ray tube or electron beam. Synchrotron radiation source may be the most ideal excitation source for X-ray fluorescence analysis. SR-XRF has been applied to a wide range because of synchrotron radiation high-sensitivity X-ray fluorescence imaging technologies development, including micro/nano-beam scanning X-ray fluorescence mapping, confocal X-ray fluorescence imaging, full-field X-ray fluorescence imaging, X-ray fluorescence computer tomography (XFCT), and combining X-ray fluorescence imaging with other methods using synchrotron radiation. Some development on X-ray fluorescence imaging methodology will be introduced in this section.

6.2.1 Micro/Nano-Beam Scanning X-ray Fluorescence Microscopy

Micro/Nano-beam X-ray fluorescence mapping is a qualitative and quantitative analysis method, based on the energy and intensity of the X-ray emitted from the micro area on the surface of the sample with extremely fine focused X-ray. At present, this method is frequently referred to as scanning X-ray fluorescence microscopy (SXFM). The spatial resolution of SXFM depends on the size of the focused X-ray beam. This hard X-ray micro/nano probe offers advanced X-ray microanalysis capabilities delivering unique elemental, chemical, structural, and morphological information about the sample at the micrometer scale of spatial resolution, with improvement to 1–10 nm resolution expected in the near future.

Figure 6.2 is Schematic of a micro/nano-beam scanning X-ray fluorescence microscopy. The workflow of scanning X-ray fluorescence microscopy is as follows: (1) firstly, the beam is focused on the sample by focusing elements such as K-B optics, capillary, zone plate, etc. (2) Step-scan sample through focused X-ray beam. (3) A fluorescent detector was used to record the full XRF spectrum of each scanning point. (4) The distribution and content of elements in the sample were obtained by analyzing the XRF spectrum.

Scanning X-ray fluorescence microscopy has proven to be a powerful tool to obtain two-dimensional (2D) elemental distributions for sectioned samples. However, samples with varying thickness will complicate image analysis. In addition, physical sectioning often leads to significant damage and artifacts, and it is extremely difficult to get continuous axial information as well. To evaluate the capabilities and limitations of SXFM for obtaining depth-dependent compositional and 3D information, confocal X-ray fluorescence microscope and X-ray fluorescence computed tomography was constructed.

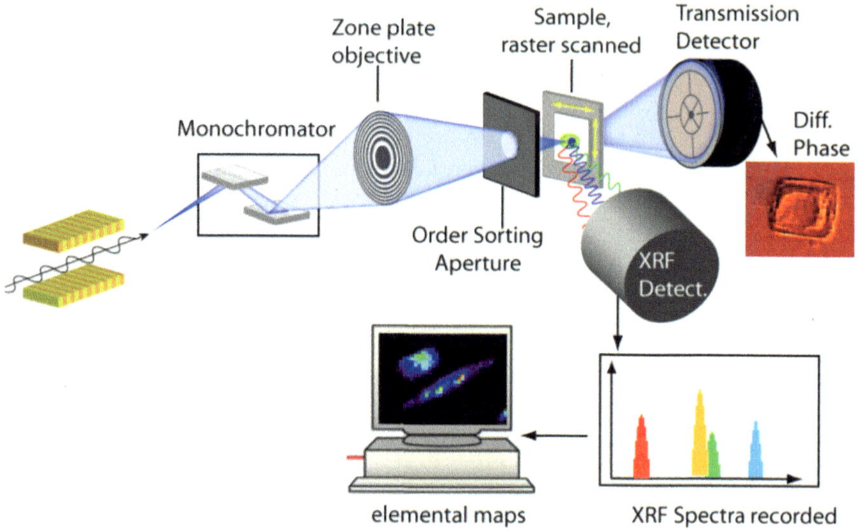

Fig. 6.2 Schematic of a micro/nano-beam scanning X-ray fluorescence microscopy [1]

6.2.2 Confocal X-ray Fluorescence Microscope

Confocal X-ray fluorescence uses an optic on both the excitation source for focusing and the detector for collimating fluorescence X-ray. The excitation optic focuses a microbeam onto the specimen. The detection optic collects fluorescent X-rays from the sample. Specifically, elemental concentrations are measured within the small probe volume ("confocal volume") defined by the intersection of the output focal spot of the excitation optic and the input focal spot of the collection optic. The detector can only detect the X-ray fluorescence signal from the confocal volume. When the confocal spot moves in the sample, the three-dimensional X-ray fluorescence information inside the sample can be obtained. A schematic layout of a typical confocal XRF is given in Fig. 6.3. This volume depends on the size of the exciting microbeam and the acceptance of the detection optic element. The characterized confocal setup provides good absolute detection limits which are comparable to those achievable in the regular XRF mode of operation. Confocal XRF can be used in such applications as materials sciences, particulate characterization, nanotechnology, and many others.

6.2.3 X-ray Fluorescence Computed Tomography

As nondestructive 3D imaging technique, X-ray fluorescence computed tomography (XFCT) is capable of reconstructing the distribution of elements inside the

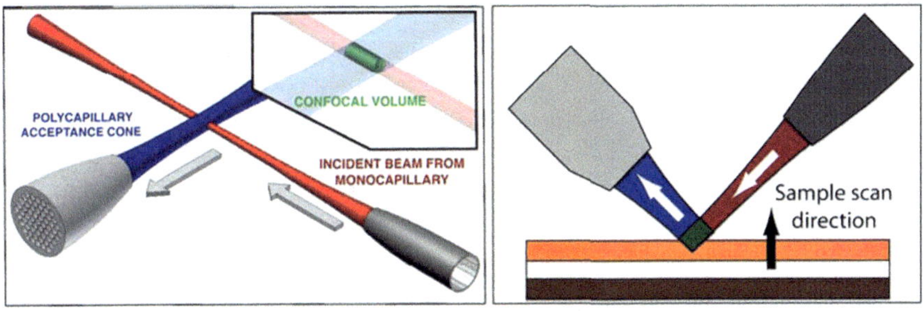

Fig. 6.3 Schematic layout of a typical confocal XRF [2] (Left) Three-dimensional schematic view of the optical arrangement for the confocal XRF. (Right) Side-view indicating the sample motion when performing a depth scan

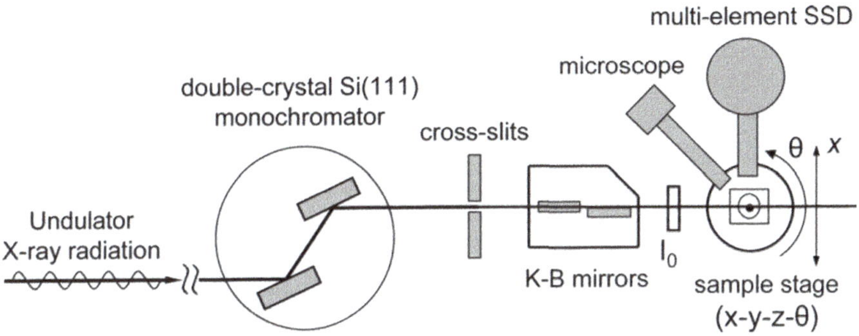

Fig. 6.4 Schematic of X-ray fluorescence computed tomography setup [3]

sample without physical sectioning. In synchrotron-based XFCT, the sample is irradiated by a high intensity pencil-beam X-ray and the incident X-rays interacting through the photoelectric effect will generate XRF inside the sample. By illuminating the object at different translation positions and along different projection angles, XRF data recorded by the spectral detector to reconstruct the distribution of elements using CT reconstruction algorithm. Figure 6.4 is the Schematic of X-ray fluorescence computed tomography setup.

X-ray fluorescence CT is an organic combination of XRF and CT. So reconstruction algorithm is very important for XFCT. The ordered subsets expectation maximization reconstruction algorithm (OSEM) is introduced to X-ray fluorescence computed tomography at SSRF. The results indicate that OSEM is more accurate than the filtered back-projection reconstruction algorithm, and it can efficiently suppress the deterioration of image quality within a large range of angular sampling intervals. And the absorption-corrected ordered subsets expectation maximization (AC-OSEM) algorithm also had been developed for the application of biomedical samples to solve self-absorption [4, 5]. Based on these algorithms, the experimental efficiency and imaging quality of fluorescent CT can be improved.

With respect to 3D microanalysis, XFCT will be compared to confocal X-ray fluorescence imaging. In case of the latter, instead of using computed tomography techniques, direct local information from an arbitrary microscopic volume-element within the sample can be obtained (limited in depth by signal self-absorption) by employing a polycapillary half-lens in front of the energy-dispersive detector. This optical element restricts the XRF-spectra to be detected only from the microscopic intersection volume of the incoming microbeam and the coinciding focus of the polycapillary. XFCT and confocal XRF have their own advantages and disadvantages, which need to be flexibly selected for different experimental samples.

6.2.4 Full-Field X-ray Fluorescence Microscopy

In typical micro/nano-XRF studies, the sample is scanned step by step with a micro/nano beam of synchrotron X-rays, which takes an enormous amount of acquisition time to acquire a 2D/3D image. In order to overcome this difficulty, some new data acquisition modes were proposed to accelerate the image acquisition. Geng Fua et al. have proposed a novel imaging geometry for XFCT studies with a position- and energy-sensitive detector collimated by one or more pinholes [6]. Sunaguchi et al. propose an XFCT using a pinhole methodology which utilizes an X-ray CCD (charge coupled device) camera with no energy resolution [7]. And a new data acquisition mode has been proposed to accelerate the image acquisition using full-field X-ray fluorescence computed tomography (FF-XFCT) at SSRF [8]. The FF-XFCT setup consists of a pinhole collimator and an X-ray imaging detector with no energy resolution. FF-XFCT is based on the same geometric principle as in pinhole single-photon emission computerized tomography (SPECT). Energy tuning can be easily and precisely accomplished with the use of a monochromator at synchrotron beam lines over wide energy ranges. Using FF-XFCT system with subtracting imaging, 3D elemental distributions for elements having absorption edges within the working energy range can be retrieved. The results demonstrate FF-XFCT is a sensitive, inexpensive, and simple approach for effective 3D element imaging (Fig. 6.5).

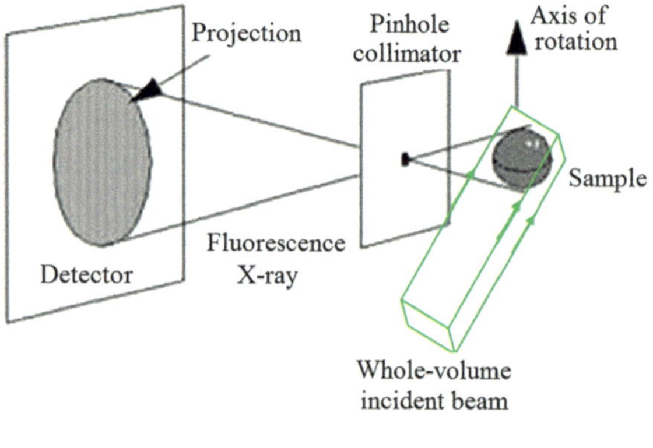

Fig. 6.5 Sketch of the FF-XFCT with pinhole collimator [8]

6.3 X-ray Fluorescence Imaging Applications

In this section, we introduced the typical applications of X-ray fluorescence imaging in materials science, biology, and others.

6.3.1 Materials Science

6.3.1.1 Trace Element Doping for $LiCoO_2$ Cathode

In order to improve the energy density of lithium-ion batteries, cation doping is a widely used method to modify the positive electrode of lithium-ion batteries. However, a deep understanding of its internal mechanism is still elusive. The segregation of trace Doped Ti in $LiCoO_2$ cathode was characterized and analyzed by means of nano-XRF imaging, nano resolution X-ray spectroscopy, scanning nano diffraction, and in-situ XRD. Figure 6.6 is the distribution Co and Ti in an isolated primary particle by XRF imaging and the Ti-rich particle surface can be clearly visualized. The defects of layered structure were revealed, and the cycle stability was improved by inhibiting the unwanted phase transition, which was in a passionate state. The results show that due to the low solubility of Ti in the LCO lattice, micro doping of Ti can change the microstructure of single particle, thus changing the defect properties inside and between particles.

6.3.1.2 Mg and Al Doping for Cathode Materials

$LiCoO_2$ is the one of main cathode materials for Li ion batteries because of its high volumetric energy density, which can be further improved by high-voltage charging. However, the structural instability of $LiCoO_2$ in the state of deep lithium removal and related safety problems hinder the practical application of high-voltage charging. In order to study the effect of trace element doping on the performance of

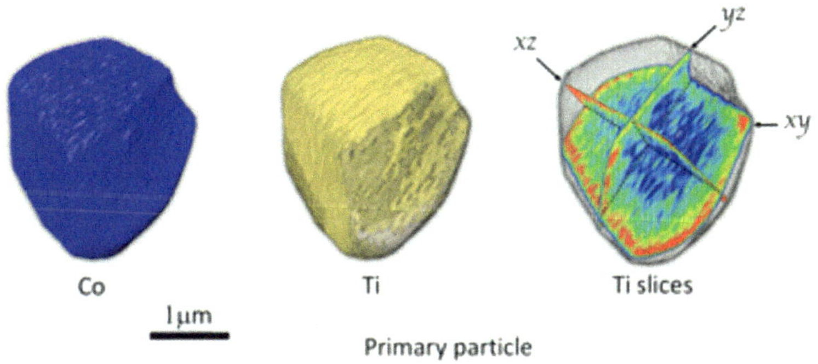

Fig. 6.6 The distribution Co and Ti in an isolated primary particle by nano-XRF imaging [9]

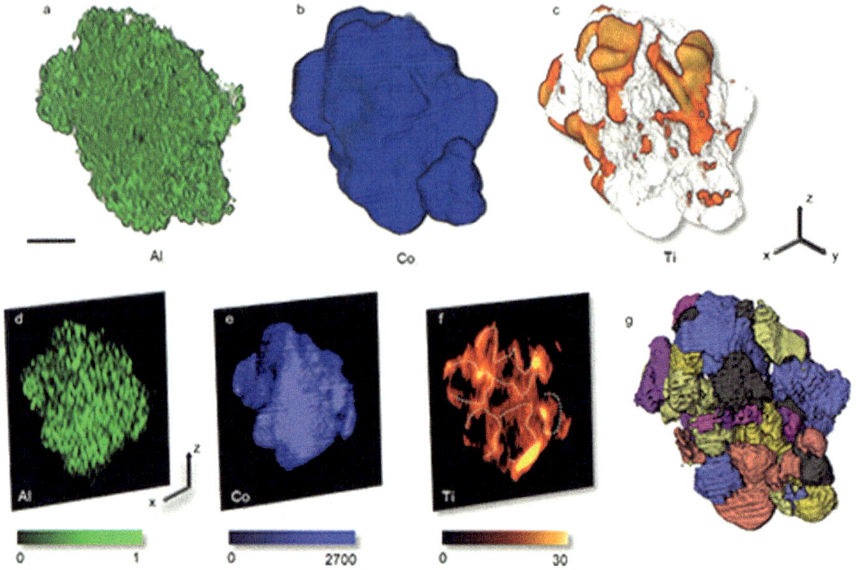

Fig. 6.7 3D X-ray tomography reconstruction and element distribution in TMA-LCO [10]

lithium-ion batteries, the doping of Mg and Al into $LiCoO_2$ lattice was imaged by X-ray fluorescence imaging and spectroscopy. Figure 6.7 is 3D X-ray tomography reconstruction and element distribution in TMA-LCO. The results show that even trace Ti can segregate significantly at the grain boundary and surface, stabilize the surface oxygen under high pressure and change the microstructure of the particles. By CO doping with trace Ti–mg–Al, $LiCoO_2$ can be obtained stable cycle at 4.6 V (relative to Li/Li^+).

6.3.1.3 Single Crystal Ni-Base Superalloy

Single crystal Nickel base superalloy is one of the basic materials developed to meet the specific requirements of turbine blades. However, because the quality of castings is very sensitive to production parameters and post production heat treatment, efforts need to be made to more accurately control the fine details of the macro and micro structures of single crystal blades. XRF imaging is used to study the segregation of refractory elements during solidification of single crystal castings. Figure 6.8 is tungsten and rhenium distribution in Single crystal Nickel base superalloy. The high concentration of element is marked with the dark color and the small with the light one. The strong segregation of refractory elements like tungsten or rhenium to the dendrite core is observed.

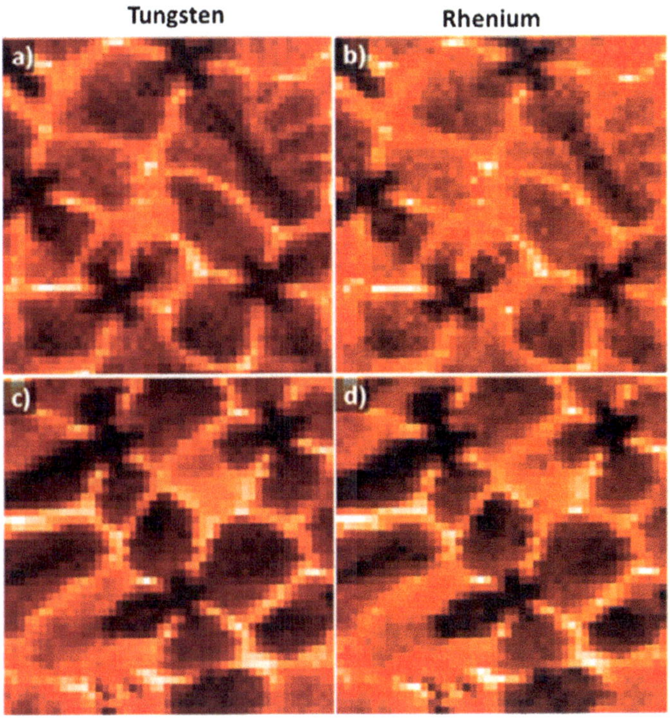

Fig. 6.8 Tungsten and rhenium distribution during solidification of single crystal Nickel base superalloy [11]

6.3.1.4 Lithium-ion Battery Particles (LixFePO_4)

The combination of nano-XRF and nano XANES provides a new way to study the chemical composition of materials with high sensitivity and resolution. The new technique has been used to study secondary phases in lithium iron phosphate (LFP) particles of lithium-ion batteries. Figure 6.9 is the Chemical imaging of LFP. This image shows how different species are distributed with the nanoscale. Single iron phosphide nanoparticles were observed in the original LFP, while some (DE) lithiated particles showed iron phosphide Nanonetworks. These findings reveal the contradictory reports about the morphology of iron phosphide in the literature.

6.3.1.5 Catalyst Particles

Fluid catalytic cracking (FCC) catalyst is a kind of important gasoline production chemical catalyst. FCC particles are composed of activated alumina and zeolite phases, which are precracked and cracked respectively. With the aging of the

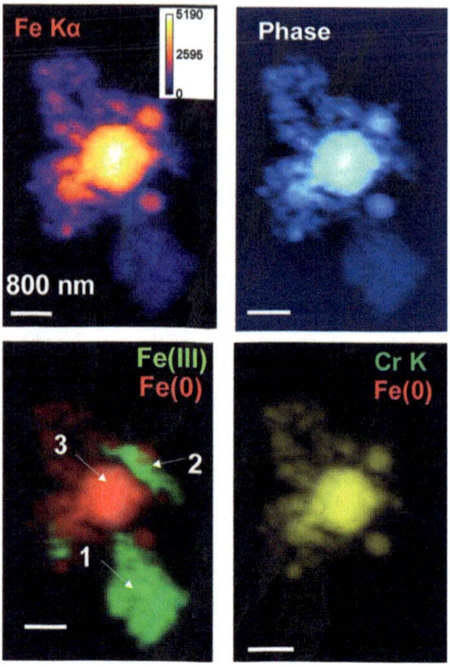

Fig. 6.9 Fe fluorescence image (XRF), phase image (ptychography), Fe chemical image (XANES), and Cr/Fe map in Lithium-ion battery particle [12]

catalyst, the severe physical and chemical conditions lead to the decline of the overall cracking capacity of FCC catalyst particles. The Investigation of the deposition and distribution of Fe, Ni, and V is essential for understanding FCC deactivation and ultimately designing more robust cracking catalysts. The deposition of metal poisons in single, intact, and commercial deactivated FCC particles at different life stages was studied by microprobe X-ray fluorescence tomography. Figure 6.10 is Fe, Ni, V, and Ca distribution in ECAT1(younger) and ECAT2 (older) FCC catalyst particles. By detecting the correlation of metal poisons, it is found that Fe, Ni, and Ca are highly correlated, especially on the surface of the catalyst particles, which form a shell around the catalyst particles. This indicates that the three elements have similar depositional mechanism.

6.3.1.6 Structure, Composition, and Accessibility of a Single Catalyst Particle

The fluid catalytic cracking (FCC) catalyst is designed as a multi-component, graded porous particle with a diameter of 50–100micro, and consists of a highly active phase (zeolite) embedded in a matrix composed of an active component (alumina) and an inactive part made of silica and clay. In the process of catalytic cracking, the residue containing a large number of heavy hydrocarbon molecules after distillation of crude oil is converted into lighter and more valuable fractions.

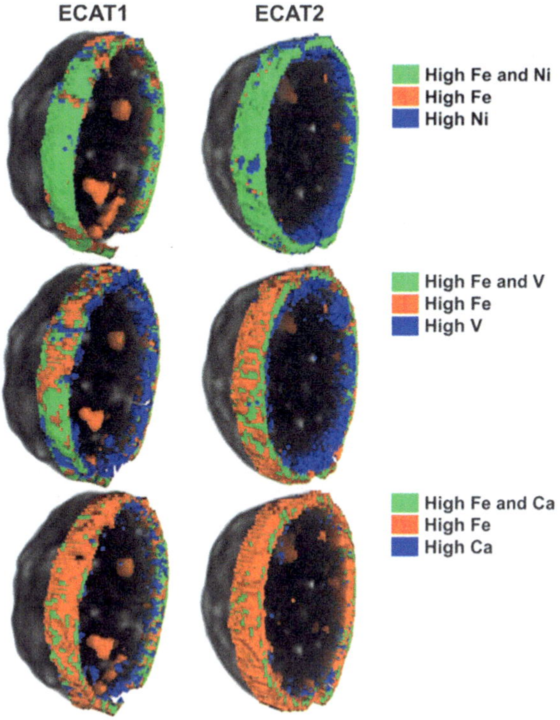

Fig. 6.10 Fe, Ni, V, and Ca distribution in ECAT1 (younger) and ECAT2(older) FCC catalyst particles [13]

In order to understand how hierarchical functional materials work, analytical tools that can reveal small structural and chemical details in large sample size are needed. To obtain a complete picture of pore network of a single catalyst particle and related catalyst accessibility, a correlative 3D micro-spectroscopic approach combining the methods of nano-TXM and multiple-element micro X-ray fluorescence (m-XRF) tomography has been developed to study the distribution of accumulated metals in the whole catalyst particle from a three-dimensional perspective. Figure 6.11 is micro-XRF and nano-TXM tomography of an individual FCC catalyst particle. The results clearly show that Fe, Ni, and Ca are preferentially accumulated in/on the catalyst particle surface, while V was found heterogeneously distributed throughout the fluid catalytic cracking (FCC) catalysts particle.

6.3.1.7 Phase Separation in Single $In_xGa_{1-x}N$ Nanowires

Using nanoscale materials and heterostructures can achieve unique performance and improved device functions. The features of these complex structures require a combination of several techniques, preferably with high spatial resolution, to avoid hiding the average value of relevant local features. The composition and the short- and long-range structural order of single $In_xGa_{1-x}N$ nanowires were studied by

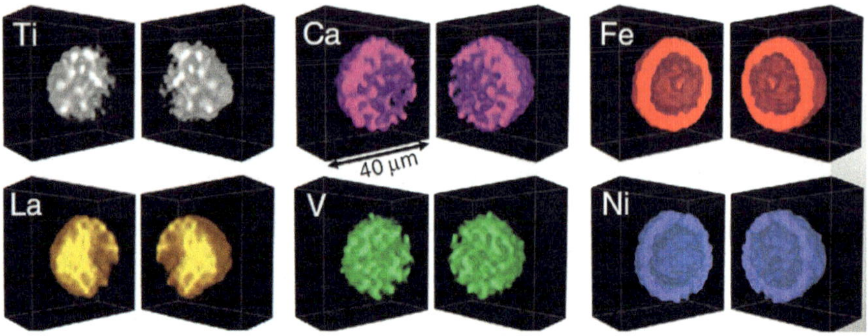

Fig. 6.11 3D distributions of multiple elements at an individual FCC catalyst particle [14]

Fig. 6.12 SEM image (top), In and Ga distributions in nanowires [15]

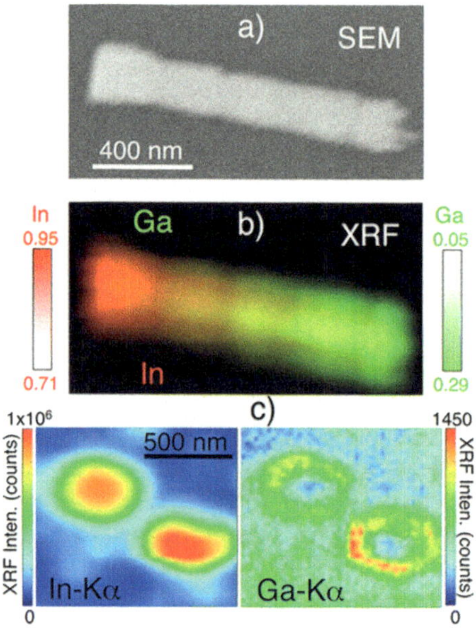

using hard nano-XRF. Figure 6.12 shows the heterogeneity of the axial and radial elements and Ga accumulates in the bottom and outer regions of the nanowires. The methodology presented here may be contribute to a better understanding of the underlying growth concepts of many nanostructures in materials science.

6.3.1.8 Highly Heterogeneous Cementitious Materials

Cementitious materials are commonly used for the safe disposal of hazardous and radioactive wastehig Hardened cement paste (HCP) is an important part of engineering barrier. HCP is composed of a very uneven mineral assemblage, and its

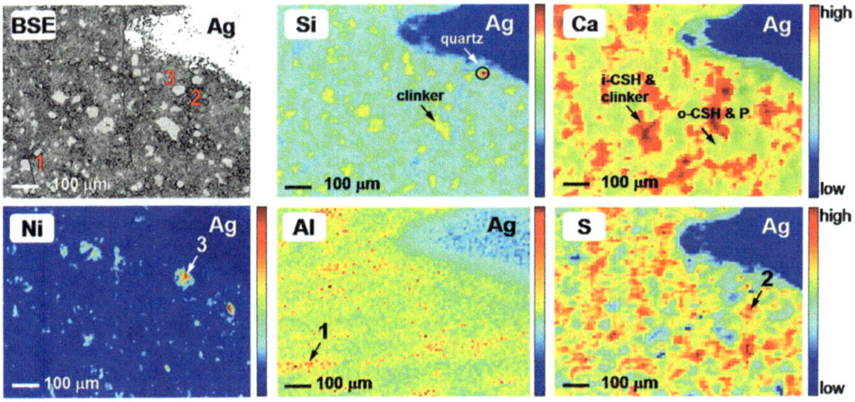

Fig. 6.13 BSE image and micro-XRF elemental distribution maps of Si, Ca, Ni, Al, and S in a Ni enriched hydrated cement matrix [16]

particle size ranges from nanometer to micron. Ignoring the micro heterogeneity of this complex matrix may lead to misinterpretation of the immobilization mechanism inferred from macroscopic studies. The element, chemical, and structural information of spatial decomposition is often the key to explain the binding mechanism in cement matrix and provides new insights into the chemical reactivity of cement-based systems. Synchrotron-based micro X-ray fluorescence (XRF) combined with scanning electron microscopy-based energy-dispersive microanalysis (EDS) has been used to determine the elemental distribution of contaminants and of chemical elements inherent to the cement matrix in hardened cement paste. Figure 6.13 is BSE image and micro-XRF elemental distribution maps of Si, Ca, Ni, Al, and S in a Ni enriched hydrated cement matrix.

6.3.1.9 Zinc Electrodeposits Multi-Element X-ray Movie Imaging

Full-field X-ray movie has always been regarded as a promising tool to explore and provide insight into chemical reactions. Multi-element sensitivity X-ray movies can further clarify the behavior differences of various elements and help to study the interaction between them. Many researchers wish to establish their own setup for visualizing chemical diffusion in various reactions. The growth process of zinc dendrites during electro-deposition was recorded by multi-element X-ray movie imaging in the BL-14B Photon Factory of KEK, Japan. In this work, a spatial resolution of 15 μm was achieved. In the X-ray movie, a movie frame acquisition time of 2 min and a spatial resolution of 50 μm were simultaneously achieved. Figure 6.14 is some key frames of the movie to observation of growing zinc electrodeposits.

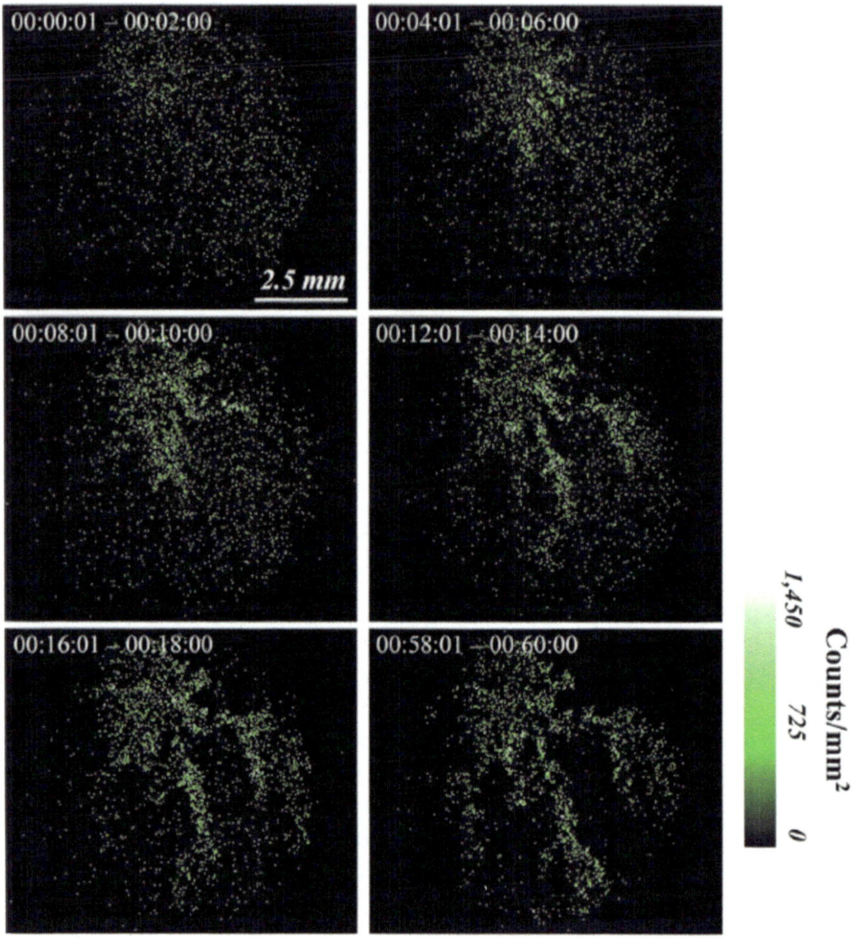

Fig. 6.14 Some key frames of the movie to observation of growing zinc electrodeposits [17]

6.3.2 *Biology*

6.3.2.1 Cyclotella Meneghiniana

The location of elements in cells affects the accumulation of these elements in the food chain. In order to understand a series of key problems, it is necessary to accurately describe the distribution of elements in cells. The elemental composition of phytoplankton, especially those that are easy to sink from surface water, determines the amount of carbon that can be fixed in the ocean to a certain extent. X-ray fluorescence tomography has been applied to subcellular imaging of a whole cell of the freshwater diatom species *Cyclotella meneghiniana* because of their distinctive

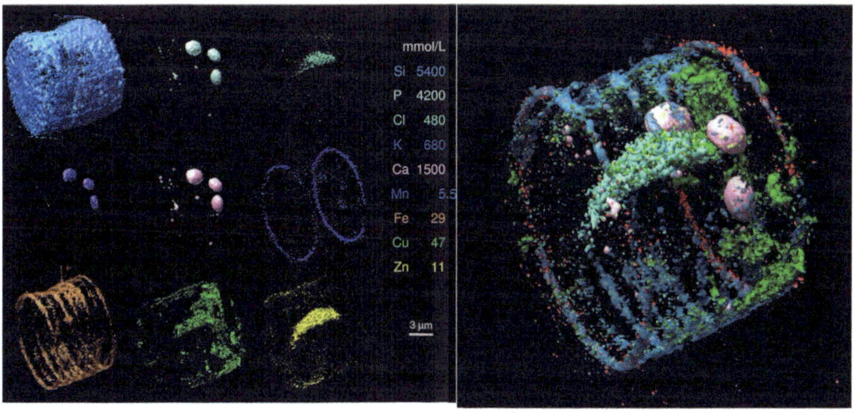

Fig. 6.15 3D renderings of elemental distributions in *Cyclotella meneghiniana* [18]

3D structure. Figure 6.15 is 3D renderings of elemental distributions in *C. meneghiniana*. These images faithfully reproduce the cell structure and may clarify the role of metals in diatom biology and of diatoms in the global elemental cycles.

6.3.2.2 Pteris Vittata Fronds

Pteris vittata fronts is the object of many studies because of its extreme arsenic hyperaccumulation. In order to better understand the biotransformation pathway of arsenic in this plant, it is necessary to know the chemical speciation and distribution across cell types within *P. vittata* fronds. Using a combination of planar and confocal μ-X-ray fluorescence imaging and fluorescence computed μ-tomography to get the information of the arsenic chemical speciation and distribution across. Figure 6.16 is the planar μXRF maps of Ca, K, As, and Compton Scatter of a live *P. vittata* pinnules. The distributions of As and K are again similar although the distribution of K extends into the lamina. It would be better to understand the arsenic biotransformation pathways in this unusual fern.

6.3.2.3 Green Microalgae

Coccomyxa actinabiotis is a newly discovered unicellular microalgae. It has strong anti radiation ability and can enrich a large number of radionuclides and toxic metals. The distribution and combination of Fe, Zn, P, Mn, Mo, K, s, Cl, and Cu in microalgae subcellular were studied by synchrotron radiation nano X ray fluorescence spectroscopy. Figure 6.17 are multi-elemental distribution and abundance in micro-alga exposed to10^{-5} mol$L^{-1}$of cobalt. Iron and zinc were well separated and distributed in complementary manner. Iron was mainly located inside chloroplast,

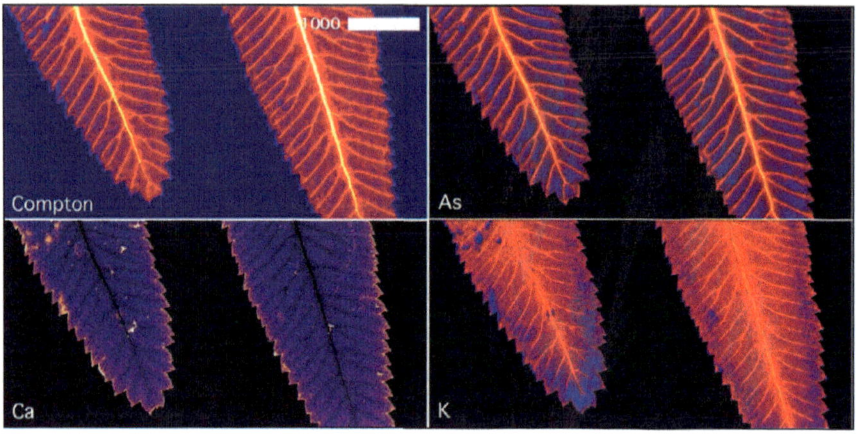

Fig. 6.16 Planar μXRF maps of Ca, K, As, and Compton Scatter of a live *Pteris vittata* pinnules [19]. Relative concentrations are depicted by shading with brighter colors equating to higher concentrations ('heatmap'). Scale bars denote μm

while zinc was located outside chloroplast. This study assesses the association and exclusion of elements in the subcellular microalgae chamber and helps to understand the mechanisms of silver and cobalt accumulation and detoxification.

6.3.2.4 Fire Scars in Tree-Ring

Fire scars in tree-ring are mainly formed in the place where forest fire occurs. It is usually used to study the history of forest fire frequency and various exogenous climate factors that may control such events. In order to understand the relationship between element changes and fire scars and undamaged contemporary growth, to provide new information about the specific element behavior of this species, and to contribute to a broader study of the history of chemical fire in trees, XRF was used to study elemental intensities of the scarred and un-scarred vectors of *Pseudotsuga macrocarpa* (Vasey) Mayr (bigcone Douglas-fir) tree-rings from a site in the Los Padres National Forest, Southern California. Figure 6.18 is multi-element intensity maps of the sample based XRF mapping. There was a high correlation between the element distribution and wood anatomical structure. The content of Ca was higher in the traumatic resin tube, and Sr was higher in the resin tube. The possibility of establishing elemental fingerprints of forest fire in undamaged growth rings provides a basis for future research.

6.3.2.5 Daphnia Magna

The freshwater crustacean *Daphnia magna* is a common model organism for studying the toxic mechanism of metals. Micro X-ray fluorescence allows visualizing the trace level metal distributions within a specimen in an essentially

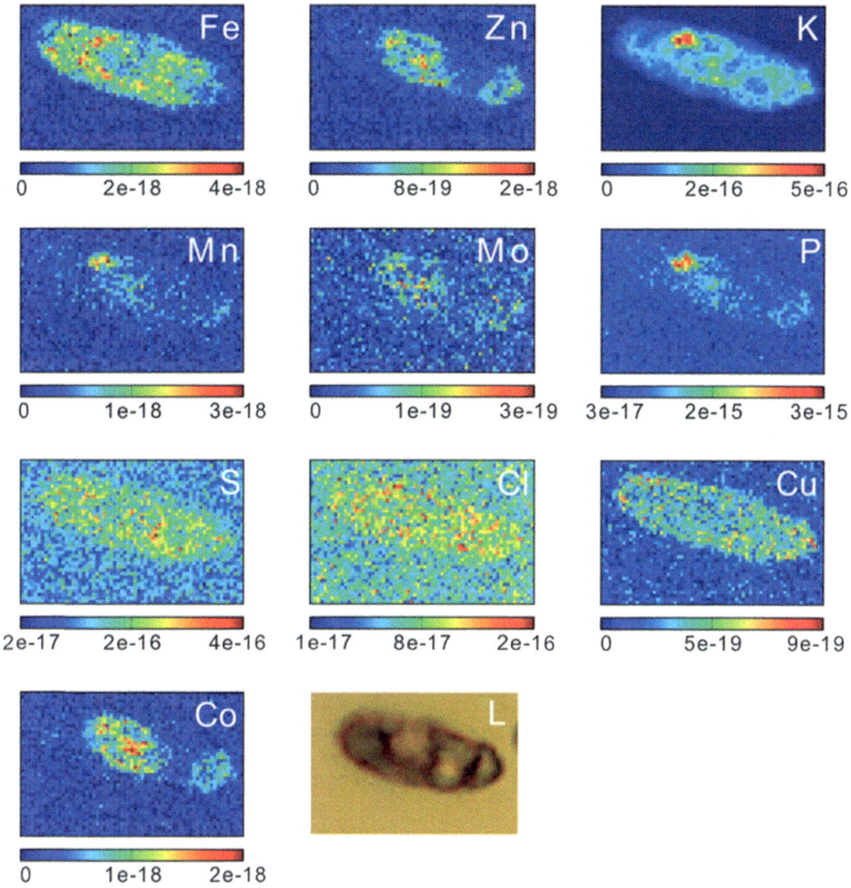

Fig. 6.17 Multi-elemental distribution and abundance in micro-alga exposed to 10^{-5} $molL^{-1}$of cobalt [20]

nondestructive manner. Biological samples pose challenges in the—XRF due to their high water content requiring dehydration procedures which may alter their chemical composition. Achieved the use of a cryostream instrument at a microfocus beamline enabling micro-XRF cryomapping and cryotomography in a dual detector arrangement on biological specimens. It was used on *D. magna*, a model organism to study the environmental impact of metals, to make a comparison between the elemental distributions within a chemically fixed and a cryogenically frozen *D. magna*. And X-ray fluorescence cryotomography coupling with laboratory absorption micro-CT is used to investigate the tissue-specific elemental distributions within this model organism. Figure 6.19 is the metal distributions (obtained by micro-SR-XRF combined with micro-CT. It can very clearly be illustrated that the exoskeleton contains Ca, the gut region contains Zn, and Fe hot spots coincide with the gill tissue.

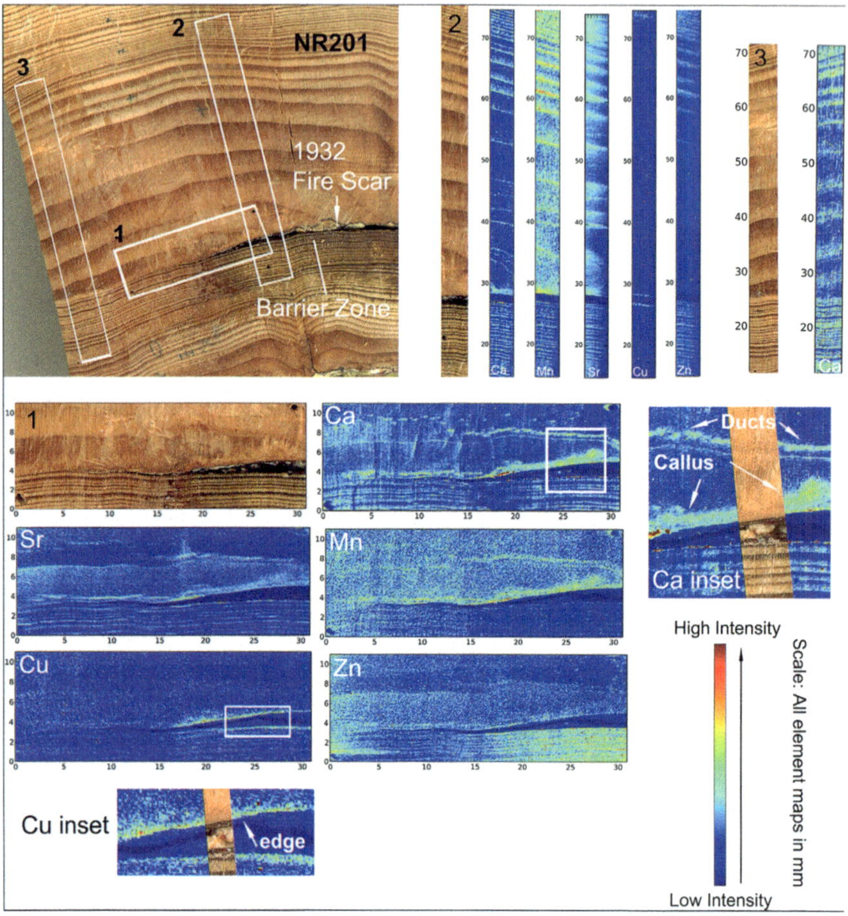

Fig. 6.18 Element intensity maps of the sample based XRF mapping [21]

6.3.2.6 Single Bacteria

Trace elements are necessary for biochemical reactions and are the structural components of cells. To truly understand the important role of trace elements, especially metal elements, in cell metabolism, it is necessary to image and quantitatively analyze the distribution of elements in the native cellular environment. XRF nanotomography has been used to image the distribution of trace elements in intact biological cells in three dimensions, while synchronous ptychography technology can be used for the co localization of trace elements and subcellular structure. Figure 6.20 is Zn, Au distribution, and the structure of the cells. The two-dimensional zinc distribution is shown in a series of two cells lined end-to-end. Again, the distribution of zinc is uneven within the cells and varies from cell to cell.

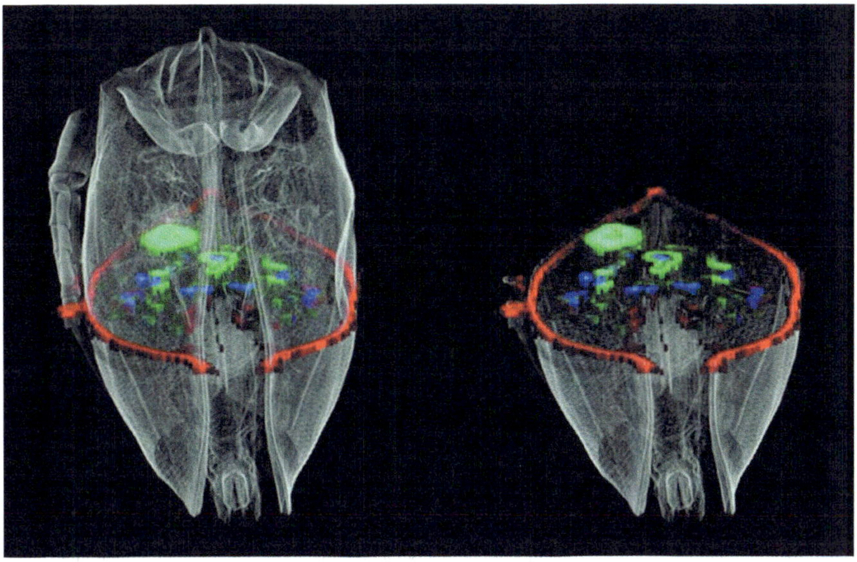

Fig. 6.19 The metal distributions (Ca(Red), Zn(green), and Fe(blue)) in *Daphnia magna* [22]

This work demonstrates that simultaneous two-dimensional ptychography and XRF nanotomography can be performed with a sub-15 nm beam size on unfrozen biological cells to co-localize elemental distribution and nanostructure simultaneously.

6.3.2.7 Daphnid

In addition to conventional scanning X-ray fluorescence imaging, Full-Field Micro X-ray Fluorescence (FF-μXRF) imaging technology does not need to scan the sample in space. The FF-μXRF system in combination with polycapillary optics provides a spatially resolved elemental image of the sample. This can for example be used to image biological samples. The biological model organism (*D. magna*) was imaged using FF-XRF imaging. Figure 6.21 shows the elemental distributions of Ca, Fe, and Zn in *D. magna* specimens. The internal organs and overall structure of the insect are clearly visible, as the FF-μXRF system allows wet, small, unpolished specimens to be analyzed with trace level sensitivity and high position resolution.

6.3.2.8 Tumor

Element mapping and visible fluorescence imaging techniques are often used to detect the distribution of platinum chemotherapeutic drugs in biological systems. Element mapping methods usually require strict sample preparation, which may

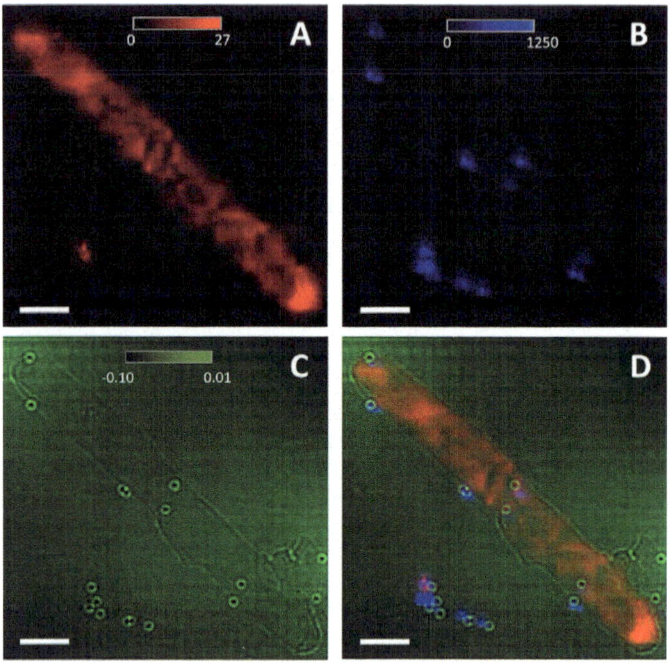

Fig. 6.20 **a** Zn and **b** Au distribution in cell **c** Phase image based ptychography reconstructed **d** Overlay of the structure and elements images of the cells. Scale bar is 1 μm [23]

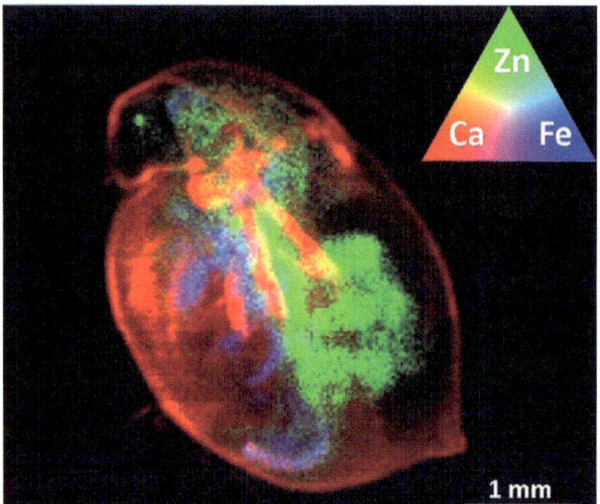

Fig. 6.21 Elementals distributions in a *Daphnia magna* [24]

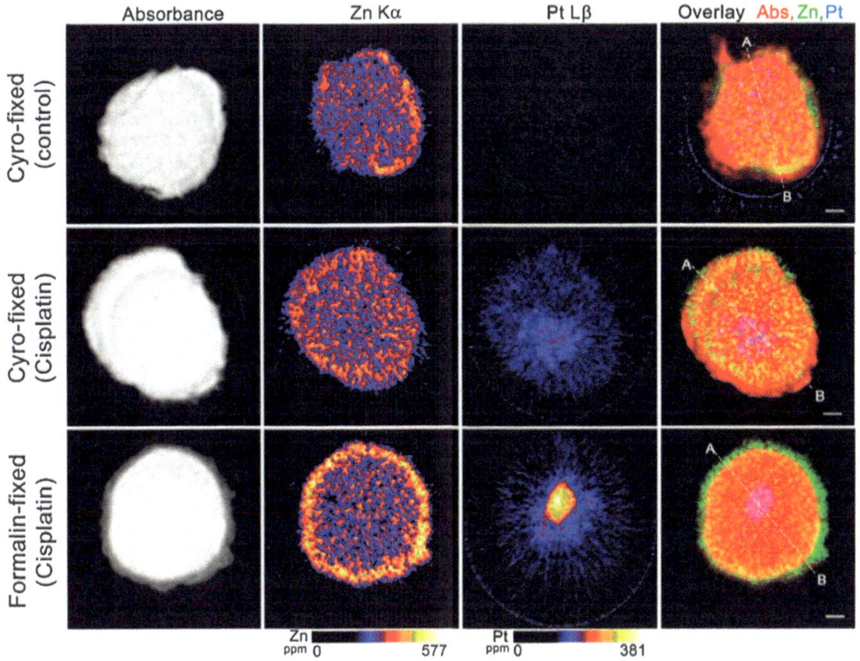

Fig. 6.22 Zn and Pt distributions in DLD-1 cancer cell spheroids using XRF CT [25]

change the chemical distribution. In-situ visible fluorescence studies need to label platinum components, which may be confused by ligand loss. In order to accurately detect the biodistribution of platinum compounds in the model tumor microenvironment, the spheres were frozen and fixed, and the element distribution in the "virtual" section of the complete sphere was obtained by XRF tomography at low temperature, and the element distribution in the sphere without physical section, chemical fixation or freeze-drying was obtained. Figure 6.22 is elemental distributions of representative cisplatin treated DLD-1 cancer cell spheroids. The results showed that cisplatin could easily penetrate the DLD-1 sphere and accumulate in the central hypoxia and necrosis area of the sphere. In addition, formalin fixation has been shown to cause significant changes in element distribution and concentration, especially in the case of platinum and zinc. This effect was not observed in frozen fixed and frozen sectioned samples. The platinum distribution and visible fluorescence image showed a similar diffusion trend, which supported the conclusion that the charge on the compound could slow down cell uptake and enhance tumor infiltration.

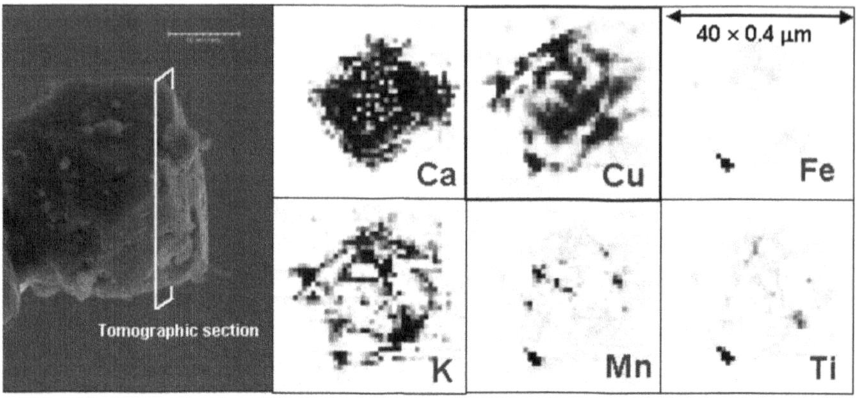

Fig. 6.23 Elementals maps on a Zeolite particle based XRF microtomography [26]

6.3.3 Others

6.3.3.1 Environmental Science

Zeolite, as a cation exchanger, is widely used in the purification of heavy metal contaminated soil to reduce the environmental hazards caused by organic matter and inorganic alien organisms. In order to study the formation process of zeolite and its effect on the mobility and availability of metals in Artificially Polluted agricultural soil with high concentration of Cu treated by fused fly ash. The spatial distribution and morphology of Cu on newly formed zeolites were determined by Micro-XRF tomography and XANES. Figure 6.23 is elementals maps on a Zeolite particle. Cu is unevenly distributed inside and outside zeolite particles. The distribution of Ca is more uniform, which can be used to describe the periphery of particles on the chromatographic plane. In the center of the cross section, a condensed copper rich nucleus can be seen clearly. There are other layered and clustered copper rich structures inside and outside the molecular sieve particles. Compared with the more uniform distribution of Ca, the non-uniform distribution of K can be attributed to the existence of K-rich phases in the interior and surface of the studied particles.

6.3.3.2 Paintings

Cadmium sulfide (CdS) has good durability and has been used as a pigment for painting. However, in the past decade, people have observed the physical performance of photo degradation of Cadmium yellow (CdS) pigment. In order to understand the synthesis and degradation pathway of pigments in complex painting, it is necessary to identify and locate the original paint components and degradation

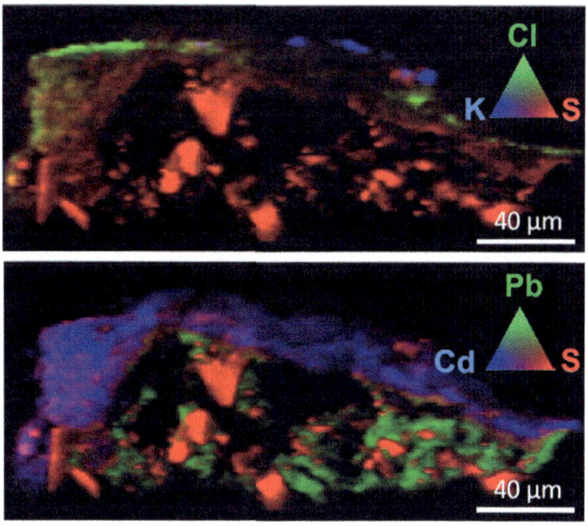

Fig. 6.24 Elemental maps of Cl, K, S, Cd, and Pb in the Joy of Life [27]

materials. The chemical and physical alterations of cadmium yellow (CdS) paints in Henri Matisse's The Joy of Life were systematically studied by XRF and other techniques. Figure 6.24 is elemental maps of Cl, K, S, Cd, and Pb in the part of The Joy of Life. X-ray fluorescence analysis showed that the white rich substrate was a mixture of coarse particles containing barium and sulfur, dispersed in the lead white matrix. This degradation seems to be limited to a few microns of the uppermost part of the cadmium yellow pigment and is related to the browning of the original yellow pigment. Through XRF and other methods, we can locate and identify the alteration products in Henry Matisse's works. This information is necessary to formulate one or multiple mechanisms for degradation of Matisse's paints from this period, and thus ensure proper environmental conditions for the storage and the display of his works.

6.3.3.3 Biomedical

Nanoparticles (NPS) are widely used in biomedical imaging as molecular contrast agents and have been widely studied as carriers of targeted drug delivery and treatment. In small animal applications, targeted nanoparticles based on gold and other materials have been used as contrast agents to replace traditional X-ray absorption imaging contrast agents. X-ray fluorescence (XRF) from targeted nanoparticles provides a path to functional/molecular biomedical imaging in living rodents. The method is used on mice for 3D tumor imaging based passive targeting molybdenum NPs. The Mo NPs used in the experiments are synthesized by wet-chemistry. NPs were injected 24 h before scanning and "low dose" XRF mode was used. Figure 6.25 is NP bio distribution mice in full-body 3D reconstruction based XRF and CT. The majority of the injected NPs accumulate in the lungs and

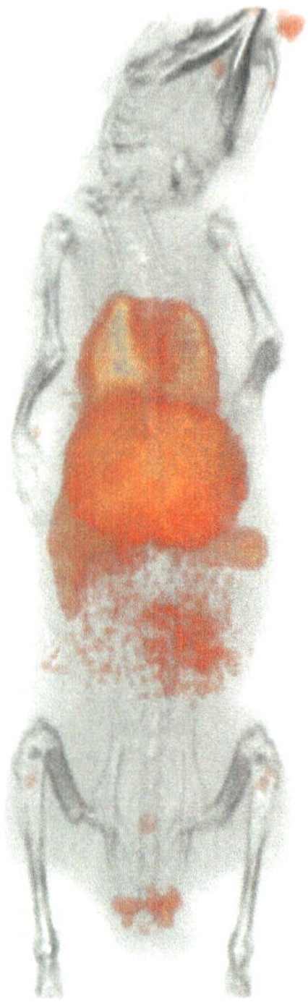

Fig. 6.25 NP distribution in full-body of mice reconstruction based XRF and CT [28]

liver but smaller amounts are also found in the spleen, abdominal area, and in nose sinuses. The experimental results show that the 3D imaging of nanoparticles for in vivo experiments based on XRF CT can be carried out with high spatial resolution, under exposure time, nanoparticle dose, and radiation dose suitable.

6.4 Conclusion and Outlook

X-ray fluorescence analysis is a well-established analytical technique with a long research history. Many applications have been reported in various fields. Recent advances in XRF imaging have been realized by the development of the diffraction limited synchrotron radiation source, new X-ray optics (nano focusing), and X-ray detectors ("color detector", et al.), leading to improved spatial resolution, imaging speed, and minimum determination limit. Combining X-ray fluorescence imaging with other methods using synchrotron radiation, enabling X-ray fluorescence imaging to be widely used in the near future.

References

1. Hong, Y., Gleber, S., O'Halloran, T., et al.: Alignment of low-dose X-ray fluorescence tomography images using differential phase contrast. J. Synchrotron Rad. **21**, 229–234 (2014)
2. Woll, A., Mass, J., Bisulca, C., et al.: Development of confocal X-ray fluorescence (XRF) microscopy at the cornell high energy synchrotron source. Appl. Phys. A **83**, 235–238 (2006)
3. Qiu, J., Deng, B., Yang, Q., Yan, F., Li, A., Yu, X.: Internal elemental imaging by scanning X-ray fluorescence microtomography at the hard X-ray microprobe beamline of the SSRF: preliminary experimental results. Nucl. Instrum. Meth. Phys. Res. B. **269**, 2662 (2011)
4. Yang, Q., Deng, B., Lv, W.W., et al.: Fast and accurate X-ray fluorescence CT imaging by ordered-subsets expectation maximization algorithm. J. Synchrotron Radiat. **19**, 210–215 (2012)
5. Yang, Q., Deng, B., Du, G., Xie, H., Zhou, G., Xiao, T., Xu, H.: X-ray fluorescence computed tomography with absorption correction for biomedical samples. X-Ray Spectrom. **43**, 278–285 (2014)
6. Fua, G., Meng, L.: Experimental demonstration of novel imaging geometries for x-ray fluorescence computed tomography. Med. Phys. **40**(6), 061903 (2013)
7. Scharf, O., Ihle, S., Ordavo, I., et al.: Compact pnCCD-based x-ray camera with high spatial and energy resolution: a color x-ray camera. Anal. Chem. **83**, 2532–2538 (2011)
8. Deng, B., Du, G., Zhou, G., Wang, Y., Ren, Y., Chen, R., Sun, P., Xie, H., Xiao, T.: 3D elemental sensitive imaging by full-field XFCT. Analyst **140**, 3521–3525 (2015)
9. Hong, Y., Huang, X., Wei, C., Wang, J., Zhang, J., Yan, H., Chu, Y., Pianetta, P., Xiao, R., Yu, X., Liu, Y., Li, H.: Hierarchical defect engineering for $LiCoO_2$ through low-solubility trace element doping. Chem **6**(10), 2759–2769 (2020)
10. Zhang, J.N., Li, Q., Ouyang, C., Yu, X., Ge, M., Huang, X., Hu, E., Ma, C., Li, S., Xiao, R., Yang, W., Chu, Y., Liu, Y., Yu, H., Yang, X., Huang, X., Chen, L., Li, H.: Trace doping of multiple elements enables stable battery cycling of $LiCoO_2$ at 4.6 V. Nat. Energy **4**(7), 594–603 (2019)
11. Matuszewski, K., Matysiak, H., Jaroszewicz, J., et al.: Influence of bridgman process conditions on microstructure and porosity of single crystal Ni-base superalloy CMSX-4. Cast Metals. **27**(6), 329–335 (2014)
12. Pattammattel, A., Tappero, R., Ge, M., Chu, Y.S., Huang, X., Gao, Y., Yan, H.: High-sensitivity nanoscale chemical imaging with hard x-ray nano-XANES. Sci. Adv. **6** (37), eabb3615 (2020)
13. Kalirai, S., Boesenberg, U., Falkenberg, G., et al.: X-ray fluorescence tomography of aged fluid-catalytic-cracking catalyst particles reveals insight into metal deposition processes. Chemcatchem **7**(22) (2015)
14. Liu, Y., Meirer, F., Krest, C.M., Webb, S., Weckhuysen, B.: Relating structure, composition and accessibility of a single catalyst particle with correlative 3-dimensional micro-spectroscopy. Nat. Commun. **7**, 12634 (2016)
15. Segura-Ruiz, J., Martínez-Criado, G., Denker, C., et al.: Phase separation in single $In_xGa_{1-x}N$ nanowires revealed through a hard X-ray synchrotron nanoprobe. Nano Lett. **14**(3), 1300–1305 (2014)
16. Vespa, M., Wieland, E., Dähn, R., Grolimund, D., Scheidegger, A.M.: Determination of the elemental distribution and chemical speciation in highly heterogeneous cementitious materials using synchrotron-based micro-spectroscopic techniques. Cement Concr. Res. (2007)
17. Zhao, W., Sakurai, K.: Multi-element X-ray movie imaging with a visible-light CMOS camera. J. Synchrotron Radiat. **26**(1), 230–233 (2019)
18. de Jonge, M.D., et al.: Quantitative 3D elemental microtomography of *Cyclotella meneghiniana* at 400-nm resolution. Proc. Natl. Acad. Sci. USA. **107**(36), 15676–15680 (2010)

19. Ent, A., Jonge, M., Spiers, K., et al.: Confocal volumetric XRF and fluorescence computed tomography reveals arsenic three-dimensional distribution within intact Pteris vittata Fronds. Environ. Sci. Technol. **54**(2) (2019)
20. Leonardo, T., Farhi, E., Boisson, A., et al.: Determination of elemental distribution in green microalgae using synchrotron radiation nano X-ray fluorescence (SR-nXRF) and electron microscopy techniques—subcellular localization and quantitative imaging of silver and cobalt uptake by *Coccomyxa actinabiotis*. Metallomics **6**(2), 316–329 (2014)
21. Pearson, C.L., Dale, D., et al.: An investigation of fire scars in *Pseudotsuga macrocarpa* by scanning X-ray fluorescence microscopy. Forest Ecol. Manag. **262**(7), 1258–1264 (2011)
22. Samber, B., Vanblaere, S., Evens, R., et al.: Dual detection X-ray fluorescence cryotomography and mapping on the model organism *Daphnia magna*. Powder Diffr. **25**(2), 169–174 (2010)
23. Victor, T., Easthon, L., Ge, M., et al.: X-ray fluorescence nanotomography of single bacteria with a sub-15 nm beam. Sci. Rep. **8**(1) (2018)
24. Samber, B., Scharf, O., Buzanich, G., et al.: Three-dimensional X-ray fluorescence imaging modes for biological specimens using a full-field energy dispersive CCD camera. Anal. At. Spectrom. **34**, 2083 (2019)
25. Zhang, J., Bryce, N., Lanzirotti, A., Chen, C., et al.: Getting to the core of platinum drug bio-distributions: the penetration of anti-cancer platinum complexes into spheroid tumour models. Metallomics **4**, 1209–1217 (2012)
26. Terzano, R., Spagnuolo, M., Medici, L., et al.: Copper stabilization by zeolite synthesis in polluted soils treated with coal fly ash. Environ. Sci. Technol. **39**(16), 6280–6287 (2005)
27. Pouyet, E., Cotte, M., Fayard, B., et al.: 2D X-ray and FTIR micro-analysis of the degradation of cadmium yellow pigment in paintings of Henri Matisse. Appl. Phys. A **121**(3), 967–980 (2015)
28. Jakob, L., Carmen, V., William, V., et al.: High-spatial-resolution X-ray fluorescence tomography with spectrally matched nanoparticles. **63**(16), 164001 (2018)

Chapter 7
Applications of Soft X-ray Spectromicroscopy in Energy Research from Materials to Batteries

Jigang Zhou and Jian Wang

7.1 Introduction

Modern energy storage systems, particularly lithium-ion batteries (LIBs), supercapacitors and fuel cells, are being implemented very rapidly in applications ranged from portable consumer electronics to large-scale electric vehicles and energy storage facilities. Their performance is critically related to the energy materials and electrode fabrication technologies. A detailed structural and chemical understanding of energy materials and the interplay among assembled electrochemical components and systems is key to performance enhancement and development of the next-generation energy storage systems.

Synchrotron-based soft X-ray spectromicroscopy generally consists of scanning transmission X-ray microscopy (STXM), scanning photoelectron microscopy (SPEM), X-ray fluorescence microscopy (XFM), full field transmission X-ray microscopy (TXM) and X-ray photoemission electron microscopy (X-PEEM) [1–4]. Figure 7.1 illustrates the schematics of these techniques or experimental approaches. Due to the popularity and relevance to energy materials science and research, this chapter will only focus on STXM and X-PEEM. In addition, the technical aspects and energy research applications will be primarily illustrated from the Ambient-STXM, Cryo-STXM and X-PEEM available at the soft X-ray spectromicroscopy (SM) beamline at the Canadian Light Source (CLS). These soft X-ray microscopes are well suited to the requirements for analysis of energy materials and systems in regard to characterizations of morphology/structure, identification and quantitative distributions of chemical components in 2D/3D and in bulk/surface, molecular orientation and polarization, etc., using X-ray Absorption

J. Zhou · J. Wang (✉)
Canadian Light Source Inc., University of Saskatchewan, Saskatoon, SK S7T 0P1, Canada
e-mail: jian.wang@lightsource.ca

J. Zhou
e-mail: jigang.zhou@lightsource.ca

J. Wang (ed.), *Advanced X-ray Imaging of Electrochemical Energy Materials and Devices*, https://doi.org/10.1007/978-981-16-5328-5_7

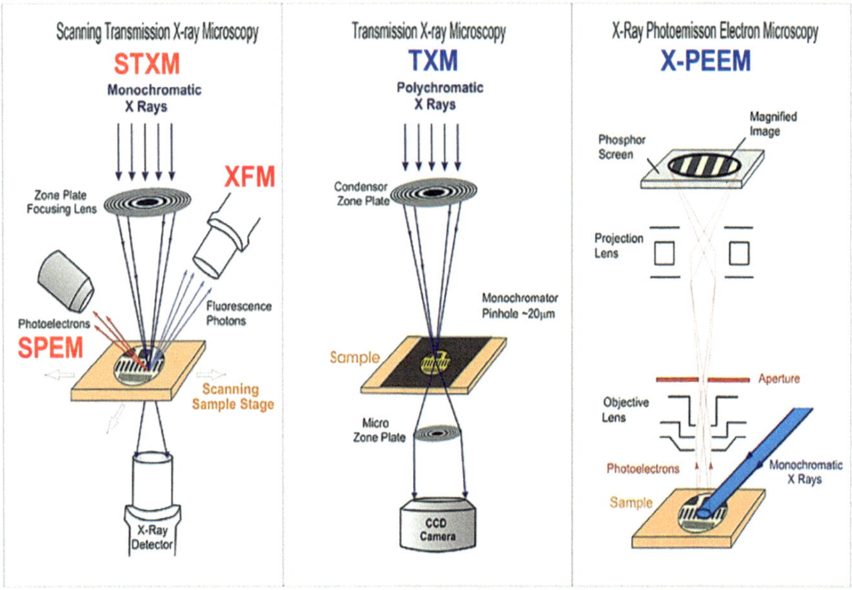

Fig. 7.1 Schematics of the general types or experimental approaches of soft X-ray spectromicroscopy, i.e., STXM, SPEM, XFM, TXM, and X-PEEM. Reprinted from Ref. [3], with permission from John Wiley and Sons

Near Edge Structure (XANES) based spectromicroscopy. This chapter is mainly organized into three parts: first is the instrumentation details of STXM and X-PEEM; then selected applications in energy materials studies will be presented; finally, future developments of STXM and X-PEEM will be discussed, particularly with respect to the next-generation diffraction-limited synchrotron sources (DLSR).

7.2 Instrumentation

7.2.1 STXM

7.2.1.1 Conventional STXM

Synchrotron soft X-ray-based scanning transmission X-ray microscopy (STXM) using zone plate optics for focusing was pioneered at the National Synchrotron Light Source (NSLS, USA) by Sayre, Kirz and their co-workers in 1980s [5–7]. By early 1990s spatially resolved XANES spectroscopy from local sample regions identified by STXM imaging had been achieved [8], which was eventually advanced to the image sequence-based spectromicroscopy in early 2000 [9]. On the other hand, efforts were taken to improve the stability and precision of STXM by the laser interferometry [10], and eventually a sophisticated modern interferometer-controlled STXM was developed in 2001 at the Advanced Light Source (ALS, USA), i.e., Beamline 5.3.2.2 polymer STXM [11]. Since then, in the following two decades many state-of-the-art interferometer-controlled STXMs have been developed in-house or even provided by commercial companies with custom design, and installed at different synchrotron light source facilities all over the world [1, 2, 12]. Currently, there are a considerable number of STXM microscopes operating or under construction or commissioning around the world, as listed in Table 7.1. These STXMs are either focused on the conventional high-throughput spectromicroscopy measurements including the growing applications of soft X-ray fluorescence (XRF), or developed as an advanced imaging tool for high-resolution 2D and 3D chemical imaging. A steady growth of STXM deployment in synchrotrons is highly anticipated, particularly for the next-generation diffraction-limited synchrotron sources with superb performance on beam coherence and brilliance.

In STXM as illustrated in Fig. 7.1, a spatially and temporally coherent beam, i.e., monochromatic X-rays obtained from an undulator source followed by a monochromator, is employed in order to achieve the diffraction-limited spatial resolution. Then, a Fresnel zone plate (ZP) is used to focus the X-ray beam into a fine spot, typically 30–50 nm in spot size depending on the specification of the zone plate, and the spot is focused and landed onto a soft X-ray penetrated sample. Transmitted X-ray intensity is detected by a phosphor scintillator coupled photomultiplier tube (PMT) or a silicon photodiode-based solid detector, while the sample is mechanically raster scanned at the focal point or the ZP is scanned within a very limited spatial range for samples not suitable for scanning. More details of STXM principle, instrumentation, experimental methods, and applications can be found in the literature [1–4, 11–14]. A typical STXM facility could be illustrated by the 10ID-1 Spectromicroscopy (SM) Beamline of the Canadian Light Source [15]. Figure 2a shows a photograph of the in-line CLS Ambient-STXM and Cryo-STXM, and Fig. 2b shows a close-up inside view of the Ambient-STXM with the transmission detection as the fundamental detection method. The CLS Ambient-STXM provides high quality 2D and 3D chemical and morphological imaging together with the spatially resolved XANES spectroscopy from 130 to about 2700 eV with $\sim$30 nm spatial resolution in the conventional mode for a wide range of samples from the fields of materials sciences, environmental/earth sciences, and life sciences [1, 2, 16]. Dry and wet samples can both be measured in a helium environment or in vacuum from rough to $\sim 10^{-6}$ Torr. Sample manipulating devices for azimuthal and polar sample rotation [17, 18], as well as sample in situ devices, including electro- and permanent magnets with magnetic fields up to 5 k Gauss [19], sample mild cooling and heating (5–80 °C) devices [20], controlled humidity cells [21], and electrochemical devices [22], are all available. These in-house made devices are still very primitive mainly satisfying certain types or levels of experiments, and need further improvement or modernization using the sophisticated electromechanical, computer, and software systems like the emerging state-of-the-art in situ STXM devices in the literature [23, 24]. For thick samples, transmission detection in the soft X-ray energy range sometimes could be very

Table 7.1 List of the existing operating and commissioning STXM facilities around the world

Instrument	Facility	Country	X-ray Source	X-ray Energy Range (eV)	Status	Remarks
STXM	ALS BL5.3.2.1	USA	BM	250–750	Operating since 2002; new STXM 2018	In-house custom design
STXM	ALS BL5.3.2.2	USA	BM	250–2500	Operating since 2011	In-house custom design, ptychography
STXM	ALS BL7.0.1 COSMIC	USA	EPU	250–2500	Operating since 2018	In-house custom design, ptychography
STXM	ALS BL11.0.2	USA	EPU	100–2000	Operating since 2003; new STXM 2015	In-house custom design, ptychography
STXM	BESSY II MAXYMUS	Germany	EPU	250–1500	Operating since 2009	Bruker, UHV
STXM	CLS 10ID-1 SM Ambient-STXM	Canada	EPU	130–2700	Operating since 2007	In-house custom design, ptychography, XRF
STXM	CLS 10ID-1 SM Cryo-STXM	Canada	EPU	130–2700	Operating since 2017	In-house custom design, Cryo-STXM, ptychography
STXM	Diamond I08	UK	EPU	250–4200	Operating since 2015	Bruker, XRF
(S)TXM	Elettra TwinMic	Italy	Und-L	250–2000	Operating since 2002	STXM/TXM
STXM	MAX IV SoftiMAX	Sweden	EPU	275–2500	Commissioning	In-house custom design, XRF, ptychography
STXM	PF BL19A	Japan	EPU	200–1500	Operating since 2013	Atto-cube, FPGA
STXM	PLS BL10A1 SXN	South Korea	EPU	100–2000	Operating since 2014	Bruker
STXM	SLS PolLux	Switzerland	BM	250–1600	Operating since 2006	Bruker, TEY
STXM	Soleil HERMES	France	EPU	70–2500	Operating since 2017	Bruker, ptychography
STXM	SSRF 08U	China	EPU	200–2000	Operating since 2011; new STXM commissioning	Xradia; new STXM based on CLS Cryo-STXM
STXM	TPS 27A	Taiwan, China	EPU	90–2500	Commissioning	In-house custom design, ptychography
STXM	UVSOR BL4U	Japan	Und-L	50–800	Operating since 2013	Bruker

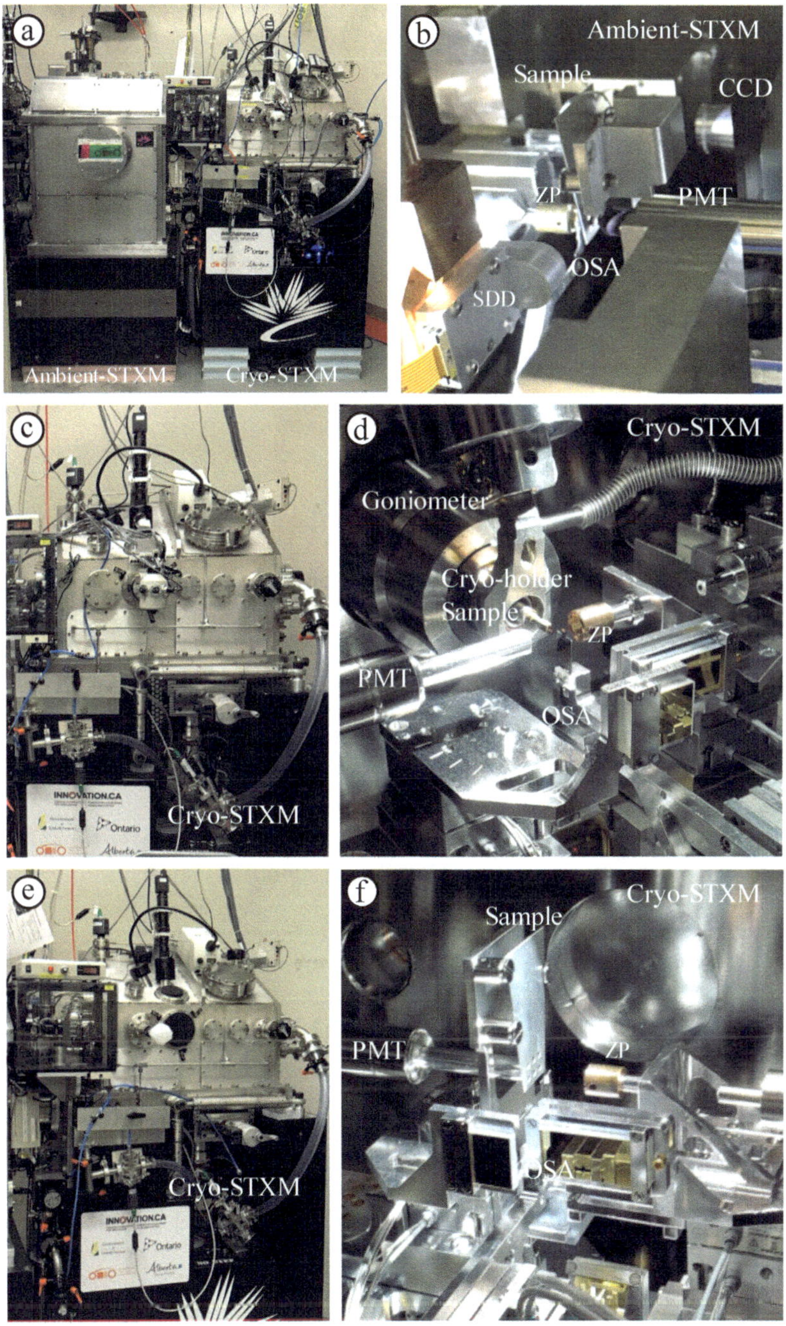
a
Ambient-STXM
Cryo-STXM
b
Ambient-STXM
Sample
CCD
ZP
PMT
OSA
SDD
c
Cryo-STXM
d
Cryo-STXM
Goniometer
Cryo-holder
Sample
ZP
PMT
OSA
e
Cryo-STXM
f
Cryo-STXM
Sample
PMT
ZP
OSA

◂**Fig. 7.2** **a** Photograph of the in-line CLS Ambient-STXM and Cryo-STXM; **b** close-up inside view of the Ambient-STXM; **c** photograph of the Cryo-STXM in the Cryo-Tomography mode; **d** close-up inside view of the Cryo-STXM in the Cryo-Tomography mode; **e** photograph of the Cryo-STXM in the conventional mode; **d** close-up inside view of the Cryo-STXM in the conventional mode. Adapted from Ref. [16] with permission from Cambridge University Press

challenging or even impossible, for instance micrometer-sized lithium-ion battery materials and electrode assemblies. To circumvent this problem, advanced STXM imaging approaches, using energy dispersive XRF recorded with a silicon drift detector (SDD), shown in Fig. 2b, and total electron yield (TEY) detection through sample photocurrent were implemented in the CLS Ambient-STXM [25].

7.2.1.2 Cryo-STXM

Another important development for STXM is the extreme sample environment, particularly the cryogenic condition. This experimental approach significantly reduces structural and chemical changes, mostly in terms of mass loss, for radiation-sensitive specimens, such as biological materials, organic molecular compounds and polymers. It is the best option for radiation-sensitive materials to conduct soft X-ray tomography and coherent diffractive imaging (CDI) measurements at near liquid nitrogen (LN2) temperatures in order to have enhanced detection sensitivity and reduced Debye-Waller factors and artifacts for scattering [26, 27]. In addition, many chemical and physical effects or phenomena, such as temperature-dependent metal-insulator transitions, magnetic phase transitions and superconductivity [28, 29], are only achievable at cryogenic temperatures. Therefore, it is highly desirable to develop cryogenic sample environment for STXM and other soft X-ray microscopes, e.g., the full field transmission X-ray microscope (TXM).

The first Cryo-STXM was developed at the NSLS, using an in-vacuum electron microscope cryo-specimen holder coupled to coarse stages for large movement, a piezo-driven flexure stage for fine scanning, and a manual actuator for specimen rotation for tomography [30, 31]. Later, the same group designed an apparatus that was comprised of a cryo-holder in a goniometer mounted on a vacuum chamber, and installed it for cryo-tomography and diffraction microscopy at the ALS [32]. These early developments had provided rich experience and knowledge for the CLS SM Beamline team to develop a much more sophisticated and modernized Cryo-STXM with new hardware and software [33]. The new instrument was indented to measure radiation-sensitive materials, particularly the fluorine-rich proton exchange membrane (PEM) of hydrogen fuel cell membrane electrode assembly (MEA), and the binders, dendrites and solid electrolyte interface (SEI) in LIBs, at 2D and 3D to achieve excellent spectroscopic and tomographic capabilities with minimized radiation damage. Specifically, the CLS Cryo-STXM provides normal incidence and tomographic transmission spectromicroscopic measurements from room temperature to cryogenic temperatures (below −170 °C) for a sample rotation range from −70° to +70° in the Cryo-Tomography mode, as shown in Fig. 2c, d. This is primarily owing to the JEOL 2010 FasTEM Goniometer and the Gatan 630 high-tilt cryo-sample holder in the setup, along with zone plate scanning. The Cryo-STXM can also be converted to the conventional STXM mode after removing the goniometer and cryo-holder for highly efficient spectromicroscopic measurements with the same sample plates interchangeable with the Ambient-STXM, as shown in Fig. 2e, f. Since the instrument was built for close to ultra-high vacuum (UHV) conditions (i.e., 10^{-8} Torr), all samples must be UHV compatible without volatile carbon species or other degassing species.

7.2.1.3 STXM-Ptychography

X-ray ptychography is an established coherent diffractive imaging (CDI) technique that reconstructs the complex transmission function of an object from a series of sample position-dependent diffraction intensities (images) using computational iterative phase retrieval algorithms [34, 35]. Recently, STXM has been demonstrated to be an ideal platform to conduct Ptychography, owing to its nanoscale positioning, flexible and efficient scanning, and ultra-low vibrational noise, and has shown sub-10 nm high spatial resolution for different types of samples with extended physical sizes [36].

Figure 3a shows the schematic of the STXM-Ptychography technique and data reconstruction approach, and Fig. 2b shows the CLS Ambient-STXM integrated a CCD area detector for ptychography experiments. After using the conventional STXM PMT point detector to identify a sample region of interest, the X-ray direct CCD is moved into beam for diffraction images acquisition. Currently, although the quantum efficiency of CCD is superior, other intrinsic limitations, including low frame rate (typically ~1 frame/s), slow data transferring speed (typically 1 to several seconds), and vigorous detector cooling (typically below −40°C), have limited the applications of STXM-Ptychography, especially for the image stack and tomography scans that require significantly longer data acquisition time than that of the conventional STXM. Also, STXM step scanning has to be adopted for stability and synchronization with the CCD image acquisition. To overcome the limitations of CCD, a fast CCD has been developed and used at the ALS COMIC beamline [37, 38], and a new type of soft X-ray direct imaging area detector, i.e., scientific CMOS (sCMOS), has recently been developed and commercialized [39]. In addition to hardware development at elsewhere, we have been using the defocused STXM-Ptychography mode with the beam spot size in micrometers to significantly improve the data acquisition efficiency, meanwhile we can still maintain sub-10 nm spatial resolution [40], and enable ptychography stack and tomography measurements. Figure 3b elucidates the STXM-Ptychography data reconstruction algorithm, i.e., the ptychographical iterative engine (PIE) [41, 42], that we implemented at the CLS SM Beamline. With Python language, libraries, and PyQt framework, a user-friendly STXM-Ptychography data analysis program, named PyPIE, has recently been developed and distributed to users. Figure 3c, d are ptychography

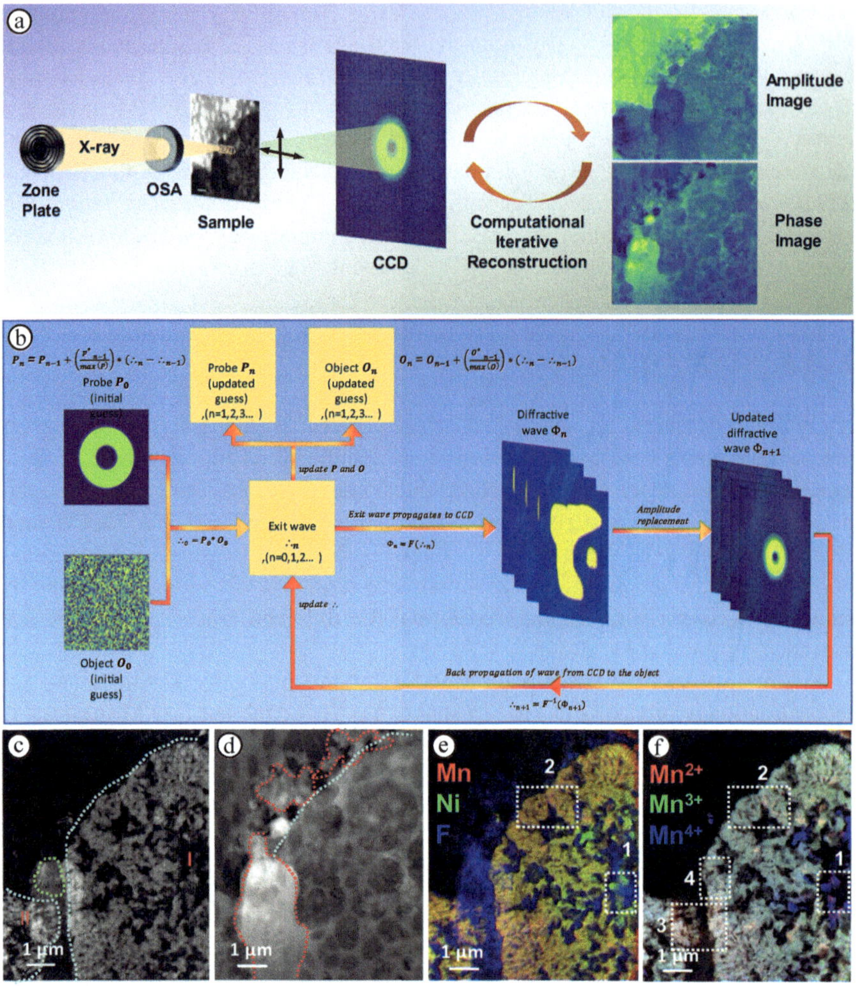

Fig. 7.3 **a** Schematic of the STXM-Ptychography technique; **b** STXM-Ptychography data reconstruction using the ptychographical iterative engine (PIE); **c** ptychography amplitude (optical density) image, and **d** ptychography phase image of a cycled P-treated High-Ni Li-rich cathode thin section obtained at the O K-edge; **d** elemental distribution mapping by ptychography amplitude mode at the Mn L-edge, Ni L-edge and F K-edge, and **e** Mn oxidation state mapping by ptychography amplitude mode at the Mn L-edge. The color dashed lines and boxes are selected regions of interest, for details see the reference [43]. Copyright (2020) American Chemical Society

amplitude (converted to optical density) and phase images, respectively, of a cycled P-treated High-Ni Li-rich cathode thin section obtained at the O K-edge conduced on the CLS SM Beamline Ambient-STXM [43]. Even the imaged size was up to about 10 μm, chemically sensitive and quantitative images with high spatial resolution as fine as 5.6 nm were readily obtained. The ptychography measurement was conducted at 14 photon energies across O K-edge, Mn L-edge, F K-edge, and Ni L-edge to produce elemental distribution mapping, e.g., Fig. 3e, and oxidation state mapping, e.g., Fig. 3f, at the M L-edge. Such high quality elemental/chemical maps allow detailed chemical analysis at local regions of interest to reveal F incorporation into the cathode material lattice and variation in Mn dissolution. With these developments and fantastic results, and the forthcoming next-generation diffraction-limited synchrotron sources, STXM-Ptychography has the full potential to achieve X-ray wavelength-limited high spatial resolution in 2D/3D chemical imaging.

7.2.2 X-PEEM

Photoemission electron microscopy (PEEM) using UV radiation has been developed since 1930s shortly after the invention of electron lenses. The use of X-rays instead of UV, i.e., X-PEEM, was first demonstrated in 1988 by Tonner and Harp [44]. In the following decade, X-PEEM instrumentation had evolved very rapidly, largely owing to contributions from the low-energy electron microscopy (LEEM), including electron energy analyzer and integration with LEEM [45], and aberration correction (AC) [46]. Recently, a novel sophisticated spin-resolved PEEM microscope named NanoESCA (FOCUS GmbH/Omicron) has been installed and operating at the Elettra synchrotron [47]. More detailed history of PEEM development can be found in the literature [48–51]. There are a large number of PEEM microscopes operating in non-synchrotron laboratories, supplied from several commercial vendors. Also, many PEEM instruments with custom design by vendors are operating on synchrotron facilities around the world, either used as a dedicated instrument for X-ray imaging and spectroscopy or integrated with LEEM for extended applications, as listed in Table 7.2. Using synchrotron radiation, X-PEEM has demonstrated to be a very powerful tool in surface science, material science, thin film magnetism, polymer science, geology, biology, etc. [49–51].

The simplest operation mode of X-PEEM is the energy unfiltered mode, in which a single imaging column is used for secondary electrons imaging. Figure 4a, b show such an instrument, i.e., Elmitec PEEM III, at the CLS SM Beamline. Monochromatic X-rays from the synchrotron source and beamline optics are impinged upon a specimen, producing photoelectrons, Auger electrons, and secondary electrons. With proper settings of the X-PEEM microscope, including high electric field between sample and objective lens ($\sim$10 kV/mm), selected-area and contrast apertures, and lens settings, the dominant secondary electrons are focused through a series of electromagnetic lenses and projected onto a microchannel plate (MCP) and phosphor screen detector, which converts the electron image into a photon image, captured by a high-performance visible light camera. Samples are required to be UHV compatible, flat, and conductive surfaces, and can be flash heated to 1800 K or cooled to 120 K by liquid nitrogen. There are some unique advantages for the energy unfiltered X-PEEM: (1) secondary electrons detection is a

Table 7.2 List of the existing operational X-PEEM facilities around the world

Instrument	Facility	Country	X-ray source	X-ray energy range (eV)	Status	Remarks
X-PEEM	ALBA BL24-CIRCE	Spain	EPU	100–2000	Operating since 2012	Elmitec LEEM/ PEEM III
X-PEEM	ALS BL11.0.1.1 PEEM3	USA	EPU	150–2000	Operating since 2007	Custom electrostatic, AC PEEM
X-PEEM	APS BL4IDC	USA	Und-L	500–2500	Operating since 2012	Omicron, Elmitec III
X-PEEM	BESSY II UE49 SMART	Germany	EPU	100–1800	Operating since 2004	Custom design, AC PEEM/ LEEM
X-PEEM	BESSY II UE49	Germany	EPU	100–1800	Operating since 2006	Elmitec III, SPEEM
X-PEEM	BESSY II UE56	Germany	EPU	60–1300	Operating since 2012	Specs
X-PEEM	CLS 10ID-1 SM CaPeRS	Canada	EPU	130–2700	Operating since 2002	Elmitec III
X-PEEM	Diamond I06	UK	EPU	80–2100	Operating since 2012	Elmitec III, LEEM/PEEM
X-PEEM	Elettra BL 1.2L	Italy	EPU	50–1000	Operating since 1999	Elmitec SPELEEM
X-PEEM	Elettra BL 1.2L NanoESCA	Italy	EPU	50–1000	Operating since 2012	Focus Omicron, spin-resolved
X-PEEM	HiSOR BL-5	Japan	BM	40–220	Operating since 2006	Elmitec III
X-PEEM	MAX VI MAXPEEM	Sweden	EPU	30–1500	Operating since 2018	AC-Elmitec III, LEEM/PEEM
X-PEEM	NSLS-II 21-ID-2 ESM-XPEEM	USA	EPU	15–1500	Operating since 2017	AC-Elmitec III, LEEM/PEEM
X-PEEM	NSRRC TLS BL05B2	Taiwan, China	EPU	60–1400	Operating since 2002	Omicron
X-PEEM	PLS 4B1	South Korea	BM	200–1000	Operating since 2000	Omicron IS-PEEM
X-PEEM	SLS SIM	Switzerland	EPU	90–2000	Operating since 2005	Elmitec III, LEEM/PEEM
X-PEEM	SOLARIS PEEM/XAS	Poland	BM	200–2000	Operating since 2018	Elmitec III
X-PEEM	Soleil TEMPO	France	EPU	50–1500	Operating since 2008	Focus Omicron
X-PEEM	Soleil HERMES	France	EPU	70–2500	Operating since 2013	Elmitec III, LEEM/PEEM
X-PEEM	SLRI BL3.2Ub	Thailand	BM	40–1040	Operating since 2015	Elmitec III
X-PEEM	Spring-8 BL17SU	Japan	EPU	250–2000	Operating since 2005	Elmitec SPELEEM
X-PEEM	SSLS SINS	Singapore	BM	50–1200	Operating since 2011	Focus-PEEM
X-PEEM	SSRF BL09	China	EPU	200–2000	Operating since 2012	Elmitec III, LEEM/PEEM

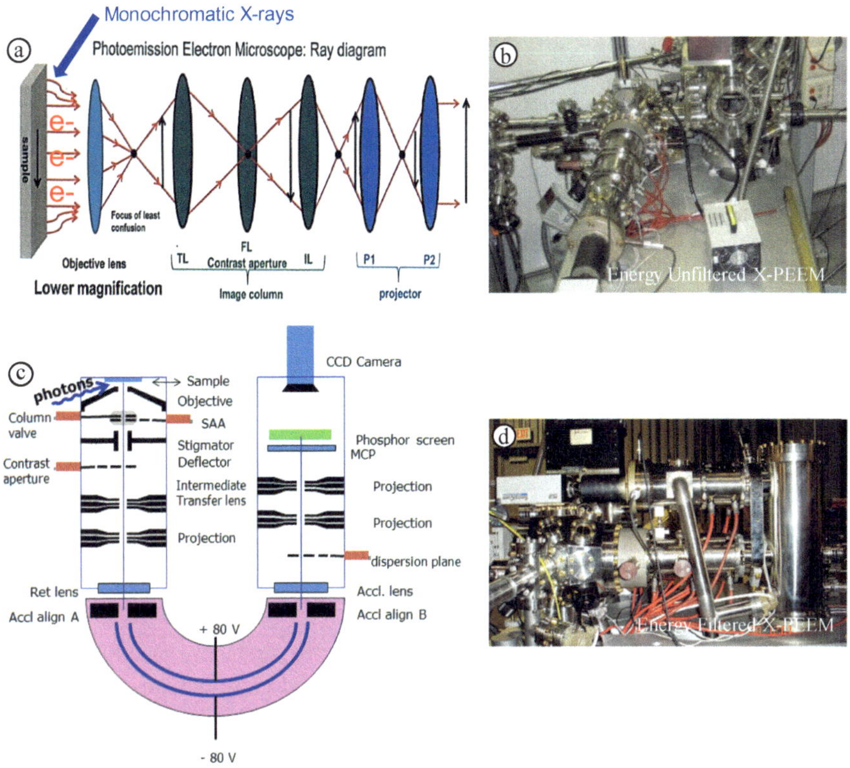

Fig. 7.4 **a** Schematic and **b** photograph of the energy unfiltered Elmitec PEEM III at the CLS SM Beamline; **c** schematic and **d** photograph of the same PEEM in the energy filtered mode. Adapted from Ref. [16] with permission from Cambridge University Press

true measurement of X-ray photoabsorption, which enables XANES spectromicroscopy for samples, also called XAS-PEEM; (2) achieving a balance between spatial resolution (20 ~ 50 nm) and electron signal transmission in the instrument, especially for organic materials (i.e., C, N, O) and other light elements; (3) simplest optics, easy to be tuned and robust in operation and maintenance. Therefore, energy unfiltered X-PEEM is an ideal tool to conduct surface spectromicroscopy for many kinds of materials [49–51], particularly for energy materials. Even for battery electrodes with poor vacuum performance and surface flatness, direct X-PEEM spectromicroscopy of electrode surface has been first achieved at the CLS [52], elucidating electrode surface and interface information, surface conductivity, and electrode state of charge.

The CLS X-PEEM is also equipped with an electron energy analyzer, and when it is installed this can be called energy filtered mode. Figure 4c, d show the schematic and photograph of the energy filtered X-PEEM, respectively. The energy analyzer allows for spatially resolved photoelectron spectroscopy (XPS) and

XPS-based imaging and spectromicroscopy, therefore, the instrument can also be called XPS-PEEM. However, XANES is generally not readily obtained on the energy filtered X-PEEM due to low intensity of photoelectrons and low transmission of secondary electrons through the energy analyzer. A better spatial resolution down to 10 nm can be achieved in this mode [51]. In addition, XPS detection gives much better sensitivity on trace surface elements than X-ray absorption spectroscopy (XAS), although its chemical sensitivity is worse than the latter. Therefore, energy filtered X-PEEM is ideal for probing surface reactions carried out in situ or very top surface sensitivity. As PEEM is limited in spatial resolution by the chromatic and spherical aberrations of the electron lenses, using an aberration corrector can improve the spatial resolution down to 5 nm, eventually approaching to theoretical 1 nm [51]. Meanwhile, electron transmission can be significantly improved by aberration correction, however, it significantly increases instrument complexity, precision level, and tuning efforts.

7.3 Soft X-ray Spectromicroscopy of Energy Nanomaterials

In the past decade varieties of nanomaterials [53–56], including nanotubes and 2D one atomic layer materials such as graphene, BN, and phosphorene, have emerged in the renewable energy applications either as a support or as an active material. Understanding and controlling the chemistry and electronic structure in novel nanomaterials and synergetic hybrids of nanomaterials are the foundation toward their rational design and effective applications in renewable energy. X-ray absorption near edge spectroscopy (XANES) [57] is a powerful method of studying the local chemistry and electronic structure in nanomaterials with unique elemental and orbital specific characteristics. To first approximation, XANES is a replica of bound and quasi-bound unoccupied density of states (UDOS) modified by core-hole interaction, therefore a unique experimental approach of investigating electronic structures in nanomaterials. Furthermore, spatially resolved XANES, i.e., spectromicroscopy, is desirable in electronic structure studies, especially for resolving controversies due to the intrinsic heterogeneities including helicity, size, grain boundaries, defects in nanomaterials. Such heterogeneities are further amplified in functionalized nanomaterials through doping or synergetic bonding in hybrid nanomaterials. Soft X-ray spectromicroscopy [2] produces chemical mapping analogous to that delivered by electron energy loss spectroscopy (EELS) but with advantages in lower radiation damage and the capability to change in the polarization of the incident photons. This section begins with the research on free-standing nitrogen doped CNT, graphene and phosphorene. These studies provide a foundation for evaluating electronic structure changes in novel synergetic bonded nanomaterials including metal oxide decorated CNT and graphene, which are the focus of the second part of this section.

7.3.1 STXM Spectromicroscopy of Electronic Structures in Free-Standing Nanomaterials (N-CNT and Graphene)

Carbon nanotubes (CNTs) have shown great promise in energy conversion and storage, especially when being electronically modified, such as with nitrogen doping (NCNT) [58]. Substituted N and gaseous N_2 in NCNT can be quantitatively mapped out and displayed in Fig. 5a, b, respectively by fitting N K-edge STXM image stack with N reference spectra in Fig. 5d, e. The color composite map of N_2 and substituted N in the individual N-CNTs is displayed in Fig. 5c. A 5 nm layer of adsorbed/intercalated N_2 evenly distributes within and on the inner surface of the tube wall along with highly concentrated trapped N_2 (about 20–45 nm of condensed N_2) in some isolated regions. The pressure of the trapped N_2 is estimated to be ~ 50 atm by scaling to the absorbance of gaseous N_2 with a known pressure. The broadening of the first four vibrational features in N_2 sealed in a N-CNT relative to those of gaseous N_2 supports the high N_2 pressure in N-CNTs (Fig. 5e). This quantitative mapping of N_2 pressure and high quality XNAES are only possible by STXM, since TEM associated with EELS study will damage CNT due to the severe radiation damage. The electronic structure of NCNT can be studied by spatially resolved XANES at N K-edge (Fig. 5d) and C K-edge (Fig. 5f). Though there exists the interference of high concentration N_2, the formation of unsaturated C-N bonds in NCNT is undoubtedly confirmed by the spectroscopic features at 399, 401 and 402.6 eV in the N spectrum [59]. Furthermore, C K-edge spectrum indicates a red shift of the 285 eV feature, the π^* transition in the N-CNT, accompanied by a reduction in intensity, indicating electron doping by the substituted N as being observed in CNT with encapsulated iron particles [60]. Finally, the orientation of condensed nitrogen in N-CNTs is explored by polarized STXM conducted with the linear inclined polarization parallel (+30°) and perpendicular (−60°) to the selected N-CNTs' long axis (Fig. 5g inset). The polarization dependence was evaluated based on the normalized intensity ratio of $I_{\parallel}/I_{\perp}$ for the π^* feature. The polarization maps of N-CNT and N_2 at the C and N π^*, respectively, were generated by subtracting images with the beam **E** vector (−60°) perpendicular to the tube by images with the beam **E** vector (+30°) parallel to the tube. It clearly shows a strong correlation of the polarized N_2 related to the N-CNT tube wall, which hints a certain level electronic structure interaction between intercalated N_2 and carbon matrix in NCNT.

Since the successful isolation of graphene [62] in 2004, two-dimensional (2D) materials, atomically thin nanomaterials with nearly planar morphology, have gained unprecedented attention in the field of energy conversion and storage. This development is largely due to the flexibility in modifying the physical and chemical properties of 2D materials through different strategies such as dimension size and morphology, the crystallographic structure and defects, or by doping with heteroatoms. STXM has demonstrated its unique capabilities in deepening our understanding of 2D materials. The first example we demonstrate here is STXM

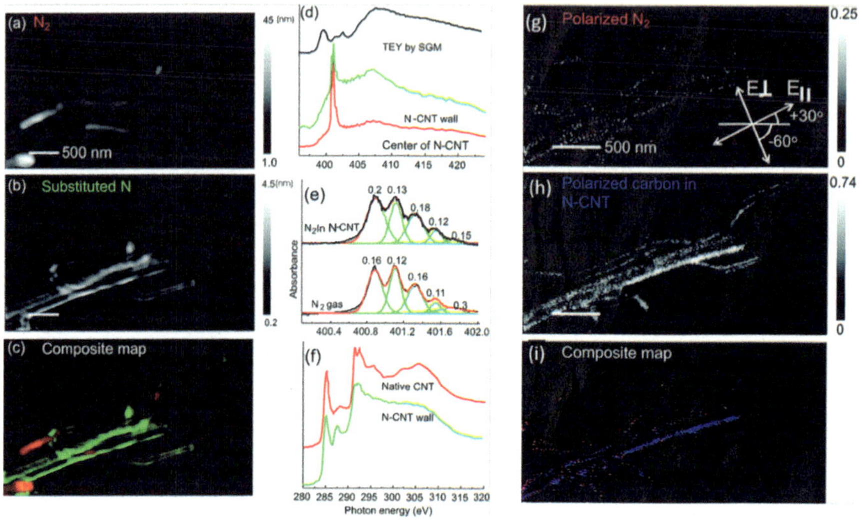

Fig. 7.5 STXM chemical maps of N-CNTs: **a** quantitative chemical map of N_2, and **b** substituted N derived from the N 1 s image stack; the vertical gray scale represents the thickness in nm in (**a**) and (**b**); **c** the colored composite map for relevant components (red: N_2; green: substituted N). The component maps are rescaled individually in each color channel. **d** transmission XANES of the wall and the center regions of a N-CNT (N_2-rich regions) along with N K-edge XANES of N-CNTs recorded in TEY mode; **e** high-resolution N K-edge XANES spectra of N_2 gas and N_2 in N-CNTs obtained by STXM point spectrum (single pixel) scans. The peak width is labeled along with each peak. The fit for the first four peaks is at least semiquantitative. **f** C K-edge XANES of the wall of a N-CNT and a native CNT. Polarization maps of N2 and carbon in N-CNTs derived from the N and C π^* resonance by STXM. **g** the polarization map of N_2 obtained at the N π^* (401.1 eV) by subtracting images with the beam E vector (−60°) perpendicular to the tube by images with the beam E vector (+30°) parallel to the tube; **h** the polarization map of carbon in the N-CNT obtained at the C π^* (285.1 eV); the vertical gray scale represents the absorbance (i.e., optical density); **i** the colored composite map for components (red: polarized N_2; blue: polarized N-CNT). The component maps are rescaled individually in each color channel. Adapted with permission from Ref. [61] Copyright (2010) American Chemical Society

chemical imaging of single and multiple layers of a thermally reduced graphene oxide (rGO) multilayer sheet [63]. Graphene oxide (GO) was made by the modified Hummer's method by exfoliating natural graphite powder. GO was reduced to rGO through high temperature annealing under Ar flow. The regions with 1, 2, 3 and 4 graphene layers in the single sheet can be identified through quantitative analysis of the optical density of the rGO sheet. The result seen in Fig. 6a also clearly reveals the edges and morphology defects in each layer and allows for further locating regions for electronic and chemistry studies. The local electronic and chemical structure of at the thin edge (region 1) and thick center (region 2) has been compared by C K-edge XANES with 30 nm spatial resolution. Higher unoccupied densities of states (UDOS) of carbon σ^* character presents at the edge versus that at the center seen in Fig. 6b. This is the first time of experimental quantifications of UDOS variation as a function of the graphene thickness and is interpreted as the lack of strong layer to layer interaction in the graphene sheet. Actually, the absence of strong interlayer interaction in rGO might be also the driving cause for the rather sharp and intense 291.7 eV (1st σ*). The lack of inter-planar interaction and the consequence of modification in the electronic structure very likely play crucial roles in strengthening the interaction between rGO support and adsorbed or synergistically bonded metal or metal oxide particles for enhanced catalytic capabilities. Very interestingly, the thickness dependent σ* intensity variation is only observed in normal incident XANES where the absorption is at its maximum but not in the 30° XANES (Fig. 6b inset) where the sample is thicker. A noticeable peak centered at 288.5 eV at 30° XANES is absent in the normal incidence XANES. This peak corresponds to defects due to functional groups, doping or interlayer states. The angle dependent nature implies the defects locate within the rGO basal plane, and therefore, it is likely concentrated along the edge. On the contrary, the characteristic peak for single- and double-layer graphene [64] at 283.7 eV is less sensitive to the angle change, which means less texture feature of this state, consistent with randomly distributed site vacancy. The structural difference between the thin edge and thicker center is further examined by the azimuthal linear polarization dependence of C K-edge XANES (Fig. 6c, d) with the normal incident beam. Once again, the edge of the rGO layer presents a more obvious angle dependence which can be due to defects tilting from the basal plane as well as broken symmetry of the edge atoms. This azimuthal effect also may reflect that the sheet terminates at the edge (region 1) with zigzag configuration (1 C atom per six-member ring with infinity symmetry) rather than the armchair (2 C atoms per six-member ring with a symmetry of twofold), since the latter should show a stronger σ* polarization difference with parallel and horizontal ***E*** vector relative to the edge. STXM was also used to map out discrete electronic domains within a single CVD graphene sheet [65]. Through the visualization of the distortions of the π* cloud alongside heterogeneously doped regions, it provides a wealth of detail regarding the extent to which the unoccupied levels of graphene are modified by corrugation, doping and adventitious impurities, as a result of synthesis and processing. STXM on those free-standing graphene samples therefore established a solid foundation for the future work on modifications to the electronic structure in synergistically bonded hybrid nanocomposites. Beyond graphene, other 2D materials such as MXenes, Xenes, transition metal dichalcogenides, and nitrides are also ideal for new types of energy storage applications. XANES or STXM have again shown the power in elucidating the details in the electronic structure in a broad spectrum of 2D materials. Here the nanoscale chemical imaging of phosphene degradation [56] by P K-edge STXM (Fig. 6e) is used as an example to illustrate STXM's application in 2D materials beyond graphene. Phosphorene is single- or few-layered black phosphorus, a promising 2D material for energy storage. However, rapid degradation under ambient condition highly limits its practical applications, which demands a deeper understanding of the oxidization process. Chemical details of the morphological effect and clarified thickness and proximity effects have been confirmed unambiguously for controlling the oxidization process.

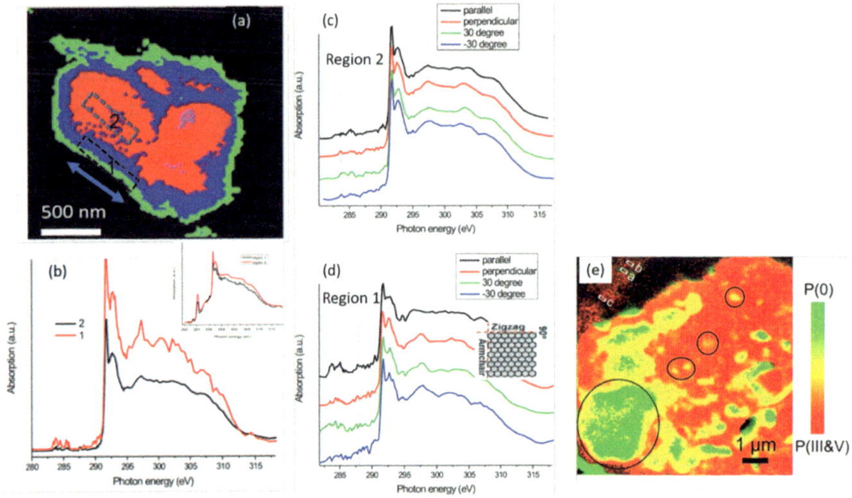

Fig. 7.6 **a** Multiple layer mapping of a multilayer rGO sheet (green: 1 layer, blue: 2 layers, red: 3 layers, purple: 4 layers); **b** C K-edge XANES in region 1 and 2, outlined in (**a**) at normal incidence with circular polarized X-ray beam (inset is XANES at 30 degree angle of incidence relative to normal); linear polarization effects on XANES in region 2 (**c**) and region 1 (**d**) with different linear E vector orientation relative to the selected rGO edge (see the double arrows in (**a**)). Adapted from Ref. [63] with permission from The Royal Society of Chemistry; **e** visualization and quantification of elemental P (false color in green) and oxidized P (red color) in degraded phosphorene by STXM at P K-edge. Adapted from Ref. [56] Copyright (2020) American Chemical Society

7.3.2 STXM Spectromicroscopy of Electronic Structure in Hybrid Nanomaterials (SnO_2-CNT and Co_3O_4/graphene)

Hybrid nanomaterials using CNT or 2D nanomaterials as the support [60, 66] can profoundly improve the performance in energy storage and conversion applications due to the synergetic bonding in the hybrid nanomaterials. STXM in the last decade has been extensively applied to experimentally probe the electronic interaction within varieties of hybrid nanomaterials. Here studies of Tin oxide (SnO_2) nanoparticles (NPs) coated carbon nanotubes (SnO_2-CNTs) and Co_3O_4 coated graphene are used to illustrate how STXM can be used in chemical imaging of the crucial electronic interaction and its resulting impacts.

The electronic interaction between SnO_2 NP and CNT is crucial in stabilizing SnO_2 NP, modifying its electronic structure and leading to desirable functionalities in batteries, fuel cells and supercapacitors. Although X-ray absorption near edge structure (XANES) spectroscopy, Raman spectroscopy and X-ray photoelectron spectroscopy (XPS) [67] can obtain averaged information on the interaction between SnO_2 and CNT in an assembly of SnO_2-CNTs, nanoscale chemical imaging such as STXM is desirable in understanding and controlling the chemistry and electronic structure variation in a single SnO_2-CNT or within the nanocomposite, which is of foremost importance in achieving its promises in single nanostructure sensing devices. In addition, the relative humidity (RH) in the environment is an important parameter as SnO_2 is sensitive to moisture [68], which will impact this nanomaterials performance in many applications. The chemical mapping of individual SnO_2-CNTs was obtained by fitting O K-edge STXM image stacks with the O K-edge XANES of SnO_2 and O-CNTs. The chemical maps thus obtained are shown in Fig. 7a (distribution of oxygen-containing functional groups, i.e., O-CNT) and Fig. 7b (SnO_2 nanoparticles). Higher average O-CNT in the nanocomposite (about 1.3) than that prior to SnO_2 coating (defined as 1) can be found out through detailed intensity analysis of Fig. 7a. This is interpreted as a spectroscopic descriptor of electronic interaction between SnO_2 and O-CNT since there is no any other possibility to oxidize CNT during SnO_2 coating due to the absence of strong oxidant. This increase in absorption "behaves" as if the surface oxidation had increased, being observed also in other CNT-based hybrid nanomaterials [69–71]. The spectroscopic observation here proves that the interaction of SnO_2 NPs with O-CNT goes beyond "physical anchoring" the oxide through the existed carboxylic or other oxygen functional groups on carbon nanostructures as having been proposed in the literature [72]. The increase of the oxygen surface functional group is possibly through interaction of SnO_2 with originally pristine CNT surface through Sn–OO–C bonding as being proved by the XANES at C and O K-edge. Figure 7b displays three main SnO_2-CNT tube features: a sheath structure exists in the upper left corner while the lower right tube appears to be a band or ribbon. The average thickness of SnO_2 coating is about 11 nm for the three SnO_2-CNTs. The green and red colors, corresponding to SnO_2, and oxygen-containing functional groups, respectively, are rescaled individually in each color channel in the color composite map of SnO_2 and oxygen-containing functional groups in the individual SnO_2-CNT in Fig. 7c. The SnO_2 distribution variations among SnO_2-CNT as discussed in Fig. 7b become more noticeable in this figure and three interesting regions with different morphology are selected: region 1 and 2 are core-shell structure while region 3 presents a biaxial side by side morphology. Such morphology variation indicates a surface chemical and electronic structure variation in the O-CNT prior to SnO_2 coating and again highlights the critical role of SnO_2-CNT interaction in dictating the properties of hybrid nanomaterials. The C and O K-edge XANES are used to explore the electronic structure variation among these three regions in Fig. 7e, f, respectively. In addition to the intact graphitic π^* and σ^* transitions in the C K-edge XANES of SnO_2-CNT, which implies that the graphitic framework of the O-CNT remains upon the coating of SnO_2, electronic interaction in the hybrid nanomaterials is justified by the visible spectroscopic changes including that: (1) The π^* transition around 285 eV in SnO_2-CNT is shifted by $\sim$0.2 eV to lower photon energy from that of O-CNT along with a slight broadening; (2) the π^* transition intensity is reduced in SnO_2-CNT compared to O-CNT, which reflects the lower unoccupied DOS of π^* character; and (3) SnO_2-CNT shows significant enhancement in transitions at $\sim$288.3 eV, characteristic of chemical defects (such as carboxylic type COOH groups resulting from the oxidation of CNTs). This agrees with the chemical mapping in Fig. 7a where oxygen functional surface groups were imaged. It is possible that SnO_2 initially grows through the exited carboxylic functional group on O-CNT which "anchors" the SnO_2 and then SnO_2 growth proceeds along the O-CNT surface with increased interaction with O-CNT through possible Sn–OO–C bonding. Interestingly, O-CNT with a smaller diameter (tube 2 and tube 3) has a stronger 288.3 eV resonance intensity variation (Fig. 7e), indicating possible stronger interaction strength in these two regions. A new peak at $\sim$286.5 eV is also observed at the C K-edge XANES of SnO_2-CNT indicating that there is another oxidized C environment such as Sn–O–C (through carbonyl or phenol on O-CNT). It should be noted that although the presence of adventitious carbon from the ambient cannot be totally ruled out, its contribution, which can be a problem in conventional soft X-ray XANES with electron yield, is not an issue in transmission measurement conducted here. The basic O K-edge XANES spectroscopic feature (Fig. 7f) is the same for SnO_2 and SnO_2-CNT, including resonances at $\sim$534 eV (a_g state of O 2p hybridized with Sn 5 s), and the second resonance at $\sim$540 eV (b_u state of O 2p hybridized with Sn 5p) along with a shoulder feature around 537.5 eV possibly due to the surface defects such as oxygen deficiency. Oxygen-containing surface functional groups on the oxidized CNT surface (Fig. 7f inset displays its XANES) also contribute to the O K-edge XANES in SnO_2-CNT but its impact is low due to the relative low concentration. The noticeable difference in the resonance intensity between SnO_2-CNT and SnO_2 at the O K-edge again confirms the interaction between SnO_2 coating and O-CNT. A more intense resonance infers p charge depletion at the O site, resulting from interfacial interaction in the composite. Also agreeing with the chemical imaging and C K-edge XANES, region 3 shows stronger SnO_2-CNT interaction. To understand how electronic structure of SnO_2-CNT impact its performance, water adsorption on single SnO_2-CNT was imaged by STXM at O K-edge. Under the RH = 0.65 in this work, 10–20 nm condensed water was adsorbed on an individual SnO_2/CNT shown in Fig. 7d. Apparently, the water distribution is not homogeneous, reflecting the inhomogeneity of the electronic structure in SnO_2, a consequence of varied SnO_2-CNT interaction as being discussed previously.

Synergetic hybrid nanomaterials such as metal oxides covalently coupled graphene [74], single or dual atom hybrid nanomaterials [75, 76] have shown outstanding performance as energy conversion catalysts, active materials in lithium-ion batteries and supercapacitors. Nanoscaled Co_3O_4 crystals grown on graphene sheets have demonstrated outstanding bifunctional performance in electrocatalysis of oxygen reduction reaction (ORR) and oxygen evolution reaction (OER)—far superior to a commercial Pt/C catalyst and bare Co_3O_4 respectively, with the performance being further enhanced after nitrogen was doped into graphene [53]. As the anode materials of LIBs [77, 78], they have also shown high reversible lithium-ion storage capacity, durable cyclic performance, and good rate capability. The reason behind the enhancement is generally ascribed to the superior properties of graphene (such as high conductivity, large surface area, structural flexibility, mechanical strength, chemical stability). However, the covalent coupling within the

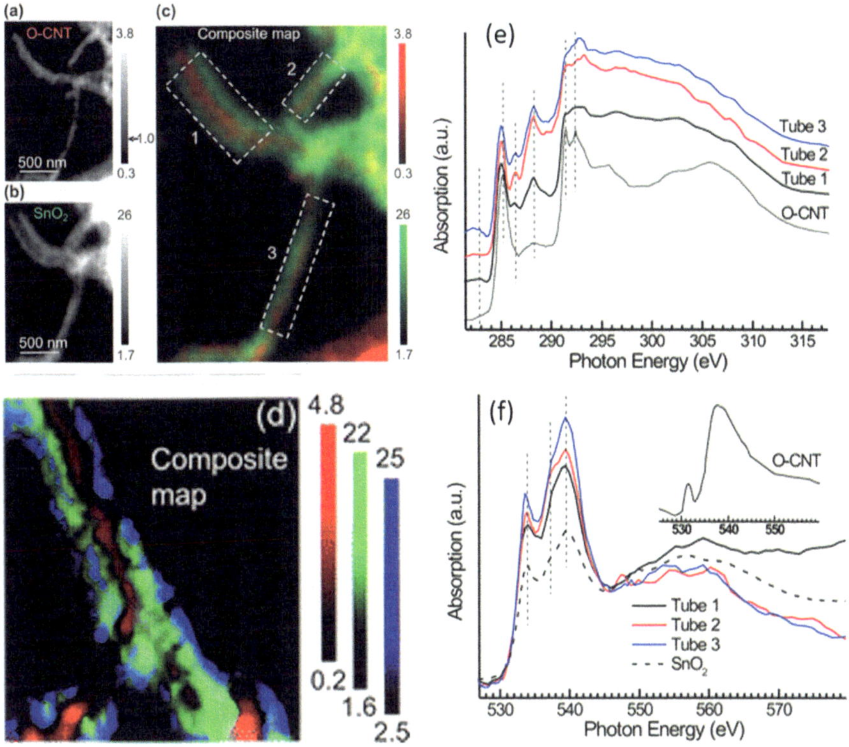

Fig. 7.7 STXM chemical maps from individual SnO_2-CNTs: **a** surface oxygen-containing functional groups on O-CNTs (representative of oxidized CNTs), the gray scale is proportional to the extent of surface oxidation (1 represents the intensity prior to SnO2 coating, intensity greater than 1 shows the effect of oxidation); **b** quantitative chemical map of SnO2, the vertical gray scale represents the thickness in nm; **c** the colored composite map for relevant components (red: O-CNT, green: SnO_2). The component maps are rescaled individually in each color channel. Three white dashed boxes labeled 1–3 are selected regions of interest for XANES comparison; **d** STXM colored composite map for relevant components (red: O-CNT, green: SnO_2, blue: H_2O (l)) of single SnO_2-CNTs under a controlled humidity environment (RH of 0.65 at 26.5 °C); **e** C K-edge XANES and **f** O K-edge XANES of SnO_2-CNTs in different locations as indicated by the white dashed boxes in (**c**) and compared to XANES of oxidized CNT before SnO_2 coating.61 The vertical dashed lines mark the spectral features of interest. Adapted from Ref. [73] with permission from The Royal Society of Chemistry

hybrid nanomaterials, and the following electronic structure modification, have been confirmed by STXM that they play more critical roles in enhancing the performance in the hybrid materials. Chemical imaging and spectroscopic studies of Co_3O_4/N-rmGO hybrid by STXM was performed at the C (Fig. 8a) and N (Fig. 8b) K-edges to image the graphene substrate, as well as at the Co L-edge to image the Co_3O_4 (less bonded Co_3O_4, Fig. 8c) and reduced Co_3O_4 (stronger bonded Co_3O_4, Fig. 8d) using the reference spectra of Co_3O_4 NPs and Co_3O_4/N-rmGO, respectively in Fig. 8j. The quantitative N-rmGO thickness map determined at the C K-edge shows two regions of distinct thickness, i.e. a thin region of $\sim$5 nm thick in the upper part of the sheet, and a thick region of $\sim$ 12 nm in the lower part of the sheet, as outlined by the white dotted lines in Fig. 8a. This suggests the single graphene sheet was folded in the lower part with roughly doubled thickness. Figure 8b clearly shows that N distribution in this hybrid is not proportional to graphene thickness. Quantitative maps of Co_3O_4 and partially reduced Co_3O_4 show that they are separated and the latter is much thinner than the former. Figure 8e, f displays color composite images of Co_3O_4 and reduced Co_3O_4, together with C 1 s and N 1 s derived N-rmGO thickness maps, respectively. The chemical imaging confirms that reduced Co_3O_4 enriches on the thin graphene region with more N doping while Co_3O_4 doesn't show such preference. This is further confirmed by statistic correlation analysis shown in Fig. 8k, l. Since the Co^{3+} reduction in Co_3O_4/N-rmGO is a result of covalent interaction of Co_3O_4 with N-rmGO, the mapped Co^{3+} reduction variation clearly displays the covalent bonding variation on and among individual Co_3O_4/N-rmGO sheets. More details on the covalent bond of anchoring and modifying Co_3O_4 in this hybrid is obtained by comparing spectroscopy at C, N, O and Co edges for the hybrid, free-standing N-rmGO, and free-standing Co_3O_4 in Fig. 8g, h, i and j. For the C K-edge in Fig. 8g, the visible π^* must arise from regions that are not perfectly flat likely due to wrinkling and/or folding of the sheet [65] as being discussed in Fig. 8a. Features between 287 and 291 eV are attributed to various functional groups attached to graphene, as labeled in Fig. 8g. Co_3O_4 hybridization on graphene increased functionalization intensity at these features, suggesting a covalent bond between Co_3O_4 and N-rmGO. The N K-edge XANES of Co_3O_4/N-rmGO and N-rmGO display a more drastic difference, as shown in Fig. 8h. A lower intensity of the π^* features (lower energy shaded region in Fig. 8h) along a stronger $\sigma^*_{C\text{–}N}$ centered at 405.5 eV confirms N doping in the graphene framework [79]. After Co_3O_4 is grown on N-rmGO, the $\sigma^*_{C\text{–}N}$ intensity was significantly reduced. This is likely due to the change of the C–N bond orientation, i.e. from in-plane sp^2 oriented to out of the graphene basal plane sp^3 configuration. This is quite probably a result of the interaction between Co_3O_4 NPs and N-rmGO. This point is further verified by the comparison of O K-edge XANES in Fig. 8i. The significant decrease of the resonance intensity at 531.2 eV (Co 3d/ O 2p) upon Co_3O_4 coated onto N-rmGO indicates an electronic structure modification, probably partial reduction in Co_3O_4 due to the its covalent bonding with N-rmGO. A direct observation of Co_3O_4 electronic structure modification (possible reduction) is clearly shown in the Co L-edge XANES of Co_3O_4/N-rmGO, Co_3O_4 NPs and CoO in Fig. 8j. The change in the peak relative intensity in the hybrid nanomaterials relative to the free-standing Co_3O_4 and CoO suggests possible increase of $(^4T_1)$ $Co^{2+}(O_h)$ species [80] in the hybrid nanomaterials, and again a consequence of the covalent bonding between Co_3O_4 and N-rmGO.

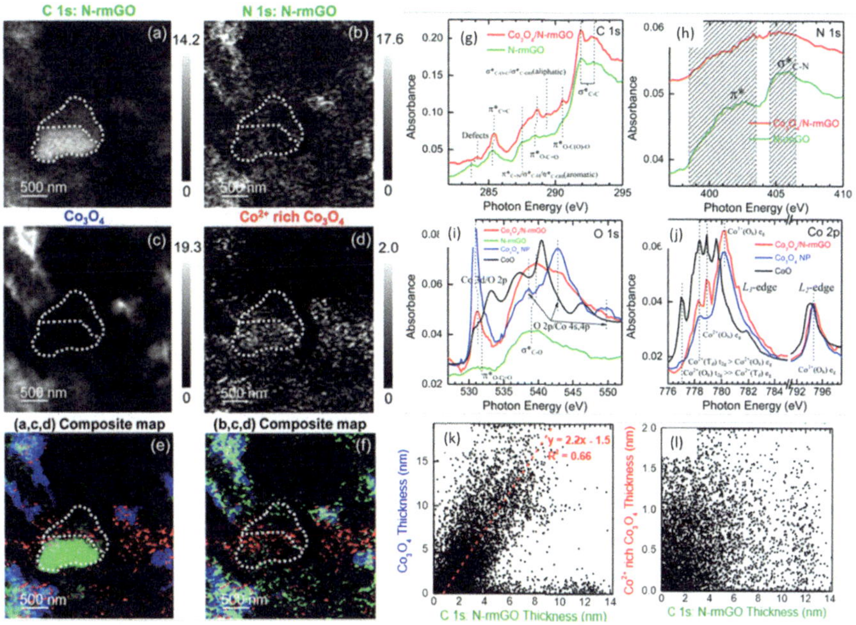

Fig. 7.8 STXM chemical imaging of individual Co_3O_4/N-rmGO sheets on a holey carbon film coated TEM grid, **a** N-rmGO thickness map derived from the C 1 s edge, the enclosed white dotted lines indicate the regions of interest, i.e. a folded graphene sheet, same as below, **b** N-rmGO thickness map derived from the N 1 s edge, **c** Co_3O_4 thickness map, and **d** Co^{2+} rich Co_3O_4 thickness map, all vertical gray scales represent the materials thickness in nm; **e** and **f** color composite maps of (**a**)/(**b**), with (**c**) and (**d**), respectively, green: C 1 s/N 1 s: N-rmGO, blue: Co_3O_4, red: Co^{2+} rich Co_3O_4; STXM-XANES spectra of Co_3O_4/N-rmGO, N-rmGO and Co_3O_4 NPs, compared to the reference XANES spectra of CoO: (g) C K-edge, (h) N K-edge, **i** O K-edge, and **j** Co L-edge XANES spectra; Correlation plot between Co_3O_4 and N-rmGO thickness (**k**) and between Co^{2+} rich Co_3O_4 and N-rmGO thickness (**i**). The red dashed line in (**k**) is a fitted trendline labeled with the linear relationship equation and the correlation coefficient. Adapted from Ref. [81] with permission from The Royal Society of Chemistry

7.4 Soft X-ray Spectromicroscopy Applications in Battery Research

Every year, human being has to spend billions of dollars to deal with the negative impact of climate change. Therefore, it is imperative to reduce greenhouse emissions especially in transportation and power generation to alleviate the problems caused by climate change. Among with other key components in a sustainable, carbon-neutral energy infrastructure such as fuel cell, solar cell, and artificial photosynthesis, long life lithium-ion battery plays a crucial role in enabling the widespread adoption of electric vehicles (EVs) and large-scale electricity generated from intermitting renewable sources such as solar panel and wind turbine. A review by Hitchcock [82] has well outlined the unique advantages of soft X-ray chemical imaging applications in fuel cell and solar cell. Therefore, here we focus on its applications in the battery research. In the past decade, soft X-ray chemical imaging has been applied in studying varieties of battery chemistries in post-mortem, in situ, or in-operando conditions [23, 83, 84] to gain a better understanding of many key elemental processes in a battery, relating to the battery reaction, degradation mechanism, etc. In this section, we summarize soft X-ray chemical imaging application in representative Li-ion cathodes including olivine structured $LiFePO_4$, high voltage spinel $LiNi_{0.5}MnO_4$, and layer-structured $LiCoO_2$.

7.4.1 Soft X-ray Correlative Spectromicroscopy of Chemistry, Transport, and Electronic Structure in $LiFePO_4$ Composite Electrode

Ionic and electronic transport within solid crystals such as in battery electrodes are fundamentally coupled properties, which dictate the insertion electrochemistry, the foundation of modern batteries. Correlative imaging of those transport at surface and bulk of active materials especially in a practical battery electrode is the key to a deeper understanding of the thermodynamic and kinetic properties of electrodes, the core scientific foundations for the rational design of high-performance and long-life batteries, but faces great challenges. Olivine structured $LiFePO_4$ (LFP), a promising "safe" lithium-ion battery cathode because of its structural stability even at a high state of charge (SOC), is also a perfect model for studying ionic and electronic transports, the subsequent phase separation, and their correlations. It has been proposed that non-monotonic Li chemical potential in LFP [85] could drive the ionic transport from Li-poor particles to Li-rich particles during (de)lithiation. However, on one hand, this has not been experimentally validated and its relation with the electronic structures at the surface and bulk is not known; on the other hand, electronic structure especially at the surface, plays a critical role in the lithiation process of LFP. This complexity is further complicated when considering a real-world composite electrode, where complicated interplay exists between all electrode components. Using $LiFe^{2+}PO_4$ (LFP) and $Fe^{3+}PO_4$ (FP) as the reference spectra in Fig. 9c, chemical fitting of both bulk sensitive STXM and surface sensitive X-PEEM of LFP nanorods composite electrode with 50% depth of discharge (DOD, global lithium concentration of $Li_{0.5}FePO_4$) show the distribution of Li-rich (red color in the map) and Li-poor particles (green color) as shown in Fig. 9a, b, respectively [86]. Interestingly, both Li-rich and Li-poor particles, with roughly 1:1 ratio, distribute uniformly without agglomerations, which strongly indicates the possible interparticle lithium-ionic transport [85]. The Fe L_3-edge XANES spectra in the bulk and surface for both Li-rich and Li-poor particles (Fig. 9c) can be extracted from STXM and X-PEEM maps, respectively. Though the local structure inside an LFP particle with mixed phases will be distorted therefore a perfect fitting is extremely difficult to be obtained, the approximation in lithium concentrations are determined by linear combination fitting of LFP and FP reference spectra. The bulk and surface Li concentrations are determined to be $Li_{0.8}FePO_4$ and $Li_{1.0}FePO_4$, respectively in Li-rich particles; the bulk and surface lithium concentrations in the Li-poor particles are similarly determined as $Li_{0.3}FePO_4$. Since the bulk lithium concentrations in Li-rich particles and Li-poor particles are between two thermodynamically stable end-member phases, therefore, both types of particles are taking lithium-ion from electrolyte and electrons from the current collector, as illustrated in Fig. 9d. Such a concurrent lithiation shall correspond to a low stress experienced in individual particles than that under the particle-by-particle lithiation pathway [87] where only very small fraction of particle is under actively (de)lithiation at each given time. Very interestingly, concentration gradient from the surface ($Li_{1.0}FePO_4$) to the bulk ($Li_{0.8}FePO_4$) in the Li-rich particles along with constant concentration profile in Li-poor particles, strongly infer a slower lithiation kinetics in the surface of Li-rich LFP during the ion transport from Li-poor bulk (the source) to Li-rich bulk (the sink) through their surfaces along with the normal concurrent lithiation. It should be emphasized that the proposed interparticle ion transport from Li-poor toward Li-rich particles according to the chemical potential difference will process along with electron transport, therefore only valid for particles with proper electronic connection. This is exactly what being revealed in the non-tact composite electrode in Fig. 9b. To gain insights on the electronic structure dependence of the observed ionic transport, O K-edge XANES between the bulk and surfaces in Li-rich and Li-poor particles are compared (Fig. 9e). Due to the Fe–O octahedral configuration, O K-edge XANES presents pre-edge features between 530 and 535 eV, representing Fe 3d t_{2g} and e_g states, respectively. The inset in Fig. 9e clearly shows the down-shift of the pre-edge energy along with the decrease of the intensity at the surface relative to that in the bulk for both types of particles. This indicates a weaker Fe–O interaction and modified bandgap at the surface, which shall impact the electron transport, therefore change the ionic transport dynamic. In addition, LFP battery degradation has been explored by X-PEEM [84] to correlate LFP agglomeration, binder and conductive additive migration and decomposition, and lithium loss as shown in Fig. 9f. Such a correlative imaging of a degraded LFP composite electrode highlights the interplays among composite electrode components in determining its degradation.

7.4.2 *Unexpected Phase Separation in $Li_{1-x}Ni_{0.5}Mn_{1.5}O_4$ a Thin Porous Composite Electrode*

High voltage spinel $LiNi_{0.5}Mn_{1.5}O_4$ (LNMO) is an attractive high-energy and high-power Li-ion battery cathode chemistry because of its 4.7 V charge–discharge plateau and the three-dimensional lithium diffusion paths [88]. Through the tailoring of the crystalline termination, and the shape and size of LNMO particles, its electrochemical properties such as the lithium transportation coefficient can be

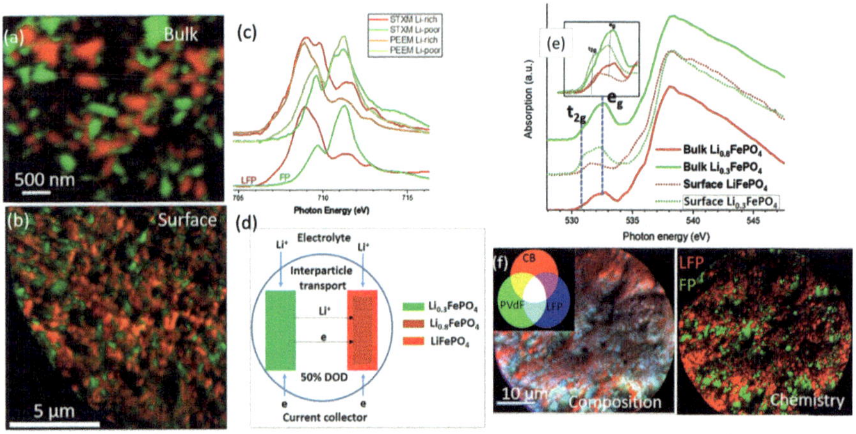

Fig. 7.9 Bulk and surface lithium concentration distribution map obtained by **a** STXM and **b** X-PEEM, respectively. Lithium poor particles are shown as green and lithium rich particles are shown as red in both images. **c** Fe L3-edge XANES of Li-rich and Li-poor particles extracted in the bulk and on the surface extracted from STXM and PEEM stacks, respectively, along with reference spectra of fully lithiated LFP and fully delithiated FP. **d** Detailed Li concentration variation in surface and bulk of Li-rich and Li-poor particles and the proposed lithium transport from Li-poor particles to lithium Li-rich particles are illustrated. **e** O K-edge XANES spectra of surface and bulk phases in Li-rich and Li-poor particles and the enlarged pre-edge region in the inset. Adapted from Ref. [86] with permission from The Royal Society of Chemistry; **f** correlative mapping of composition and lithium loss in a cycled LiFePO4 composite electrode. Adapted from Ref. [84] Copyright (2020) American Chemical Society

effectively tuned. STXM has been successfully applied in revealing the complexity of the interplays of surface morphology, chemistry and local conductivity of LNMO particles using a combination of transmission, fluorescence and electron yield modes [25]. These studies clearly hint that a deep understanding of interplay of these aspects, especially within a real porous composite electrode, shall hold the key toward a rational design of high-performance LNMO battery. As shown in Fig. 7.10 the Ni oxidation state in $Li_{1-x}Ni_{0.5}Mn_{1.5}O_4$ (LNMO) within a composite electrode has been mapped by STXM [89]. The unexpected distinct variations in phase separation among and within individual battery particles have been experimentally correlated to both their morphology and interface structure. In this study, LNMO electrode (LNMO: super P (Timcal): PVDF = 80: 10: 10) was charged in 1 M $LiPF_6$ in the EC/DMC electrolyte to 5.2 V (vs. Li metal) at a 1C rate. This will oxidize the Ni in LNMO to Ni^{4+} state. The electrode from the test coin cell was rinsed with DMC to remove the $LiPF_6$ residue before being fully dried under vacuum. After this, the electrode was stored in an atmospheric environment which experienced the reduction of Ni^{4+} to the lower oxidation state. Then, it was microtomed to thin cross-sections $\sim$200 nm thick, perpendicular to the Al foil current collector. The overall morphology of the electrode thin section is displayed in Fig. 10a. Three regions (red boxes) were selected to obtain Ni oxidation state mapping and XANES by STXM. The top region was close to the separator in the battery, while the bottom region was close to the Al current collector. The morphology maps of the three regions are displayed in Fig. 10b. The brightness is proportional to the Ni concentration (equivalent to the LNMO thickness). Two types of particles are present in these images, i.e., large particles with well-defined shapes, which may correspond to typical spinel morphologies, namely octahedral shapes with (111) facets; and smaller ones that are plate-like shapes with (112) facets. Figure 10c shows the corresponding chemical maps of the phase separation in each region. It was obtained by fitting Ni L-edge STXM by references in Fig. 10d. The references were extracted via principle component analysis (PCA). The PCA analysis is able to identify image pixels with a similar spectral feature, and the average of all similar pixels yields a spectrum that corresponds to a pure or mixed chemical phase including states of Ni^{2+} in $LiNi_{0.5}Mn_{1.5}O_4$ or $HNi_{0.5}MnO_4$, Ni^{4+} in $Ni_{0.5}Mn_{1.5}O_4$, and Ni^{3+} in $H_{0.5}Ni_{0.5}Mn_{1.5}O_4$. These are proton-inserted spinel products, similar to the phases in fully lithiated, fully delithiated and half delithiated LNMO products. They are denoted as Ni^{2+}-like, Ni^{4+}-like and Ni^{3+}-like phases (Fig. 10d). The chemical phase separation in this composite electrode can be visualized by the chemical mapping of the three phases (Ni^{2+}-like, Ni^{3+}-like, and Ni^{4+}-like). The phase separation kinetics are obviously different among the three regions: top two regions, closing to the separator, are mixtures of all the three phases, which are in contrast to that in the bottom region, where a more uniform Ni^{3+}-like phase dominates. Interestingly, in the top regions, large particles with a dominant Ni^{4+}-like phase always have a shell with a Ni^{2+}-like phase. While on the

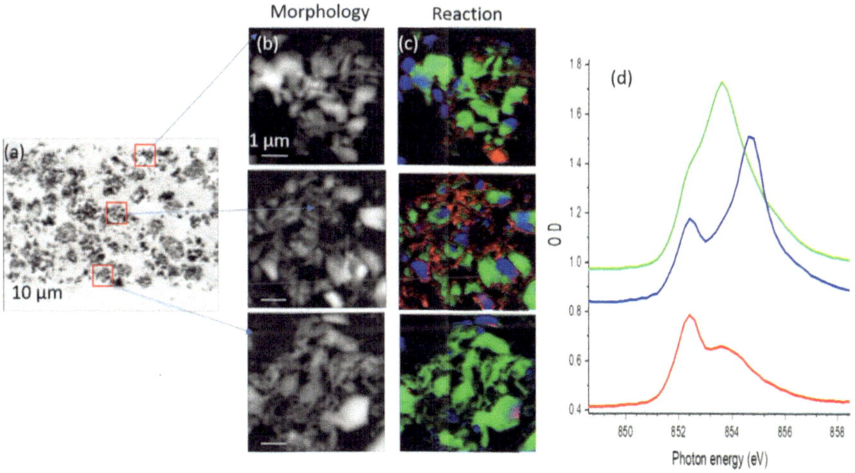

Fig. 7.10 **a** Morphology mapping of the self-discharged fully delithiated LNMO electrode thin section; **b** morphology, **c** chemical map (Ni^{2+}-like shown as red, Ni^{3+}-like shown as green and Ni^{4+}-like shown as blue) of the top, middle and bottom selected regions of the LNMO electrode in (**a**) by SVD fitting of the Ni L3-edge STXM image stack, using the internal fitting references displayed in (**d**). Adapted from the literature with permission from The Royal Society of Chemistry [89]

contrary, the reduction phase, Ni^{3+}-like phase-dominated large particles, don't have such protection shell. This suggests that the Ni^{2+}-like phase is the interface in the form of a shell, a passivation film formed during the electrochemical cycling [88], which can mitigate the reduction of Ni^{4+} in the large octahedral-like particle during storage. The unexpected complex phase separation in a long-term stored LNMO implies that intrinsic differences among LNMO particles in interface, crystal surface termination or strain could impact their performance such as stability, self-discharge, and reaction kinetics much more than commonly accepted.

7.4.3 X-PEEM Spectromicroscopy of Composite $LiCoO_2$ Electrodes Surface Heterogeneities Under Abusive Conditions

$LiCoO_2$ (LCO) was the first commercial Li-ion battery cathode and is still the choice in the portable electronic device applications. It has a layered $NaFeO_2$-type framework with R-3 m space group, therefore it has been used as a model for other emerging layered metal oxide cathodes such as $LiNi_xMn_yCo_zO_2$. Solid-state electrode surface phase heterogeneity in a real-world composite electrode under abusive battery operations is closely related to battery performances including rate capability, degradation and safety and thereby deserves attention from advanced characterization perspective. Thermal runaway [90] of charged or overcharged layered oxide electrodes at high temperatures, a catastrophic failure due to the exothermic reaction between the released active oxygen and the flammable electrolyte, hinders their large-scale applications in electric vehicles. Fundamental understandings of the mechanisms of thermal runaway, especially on the phase heterogeneity and its evolution within a real porous composite electrode, are needed in order to mitigate this problem. Considering the inhomogeneity in cathode particles, it is desirable to track the structural evolution of the same particle during the thermal runaway. This, however, is very challenging since the cathode particles move under heating. X-PEEM with a large field of view (FOV) was found to have the capability of tracking cathode particles during the thermal runaway in a composite electrode [91]. In this study, LCO composite electrode was charged at 0.1 C–4.2 V in a commercial pouch cell and maintained at 4.2 V until the charging current decreased to 0.01 C. This charging procedure resulted in a homogeneous fully charged LCO ($Li_{0.5}CoO_2$) [92] within a composite due to the slow charging rate. A small piece of cathode was rinsed with DMC, and then fully dried under vacuum. In situ heating of the cathode was performed with the resistive heater available inside the X-PEEM sample holder. The morphology, composition, and phase heterogeneity of charged LCO before and after heating are derived from the X-PEEM image stacks at the Co L-edge and the F, and O K-edges. Visualization of the component distribution was achieved by color-coded correlation maps of individual elements as shown in Fig. 11a, b. The morphology and relative locations of multiple particles 1–5 and a large particle and a small particle as shown in Fig. 11a, b confirms the mapping of LCO before and after heating are from the same area. It could not be possible without large FOV PEEM since 10 μm shift of the electrode position after heating was experienced. The surface phase heterogeneity (in Fig. 11c) in charged LCO and its evolution upon heating have been mapped by fitting X-PEEM image stacks with reference spectra in Fig. 11d, which was obtained through principal component analysis (PCA) of the Co L_3-edge X-PEEM image stacks of the charged LCO electrode after heating. Co is clearly reduced upon heating based on the comparison of the high quality Co L_3-edge spectra of the heated LCO electrode to those of Co_3O_4, pristine LCO and charged LCO as shown in Fig. 11d, due to the evolution of O_2. The spectroscopic difference between the two Co species (red and green colored) in the phase mapping suggests more decomposed LCO in the green region than that in the red region. Both well-decomposed (green) and less-decomposed (red) phases co-exist on the flat (001) facet, and on its edges (facets perpendicular to the (001) facet) of the large particles. Both the large and small LCO particles show a similar phase heterogeneity such as the coexistence of well- and less-decomposed phases in a similar domain size. Those observations are reliable since they occurred on identical particles within a real-world composite electrode, which can only be possible by X-PEEM spectro-microscopy. Another example of X-PEEM in imaging LCO under the abusive operation is the chemical imaging of the additive effects (suberonitrile (SUN) and lithium bis(oxalate)borate (LiBOB)) [52] on the interfaces of LCO under high voltage 4.5 V operation in Fig. 11e–g. The elemental mapping identifies two interest regions: (1) a large LCO with very little binder and carbon additives and (2) F enriches on small particles, a mixture of binder and conductive additives (being confirmed by carbon map, not shown here). The dramatic difference in chemistry between these two regions was elucidated by XANES at Co L-edge and F K-edge. The region 1 has both Co^{3+} and Co^{2+} with Co^{3+} being dominant, while the enhanced $\sim$778 eV feature in region 2 suggests dominant Co^{2+} in this region. The coexistence of rich F species in region 2 suggests the existence of CoF_2. This is further confirmed by the F K-edge XANES in Fig. 11f, where the shoulder peak at 685 eV is a spectroscopic feature of CoF_2. Very interestingly, the Co and F chemical mapping of LCO cycled at 4.5 V without additives (not shown here) present very low CoF_2 which is also rich on LCO surface rather than on the surface of mixture of the binder and carbon additives. This difference indicates the root cause of additive effects in stabilizing LCO under a high voltage operation.

7.5 Future Perspective

7.5.1 In Situ STXM

The routine ex situ STXM measurements of energy materials and batteries have shown powerful capabilities in elucidating their chemical, electronic, and morphological structures. However, sample preparation and measurement environment

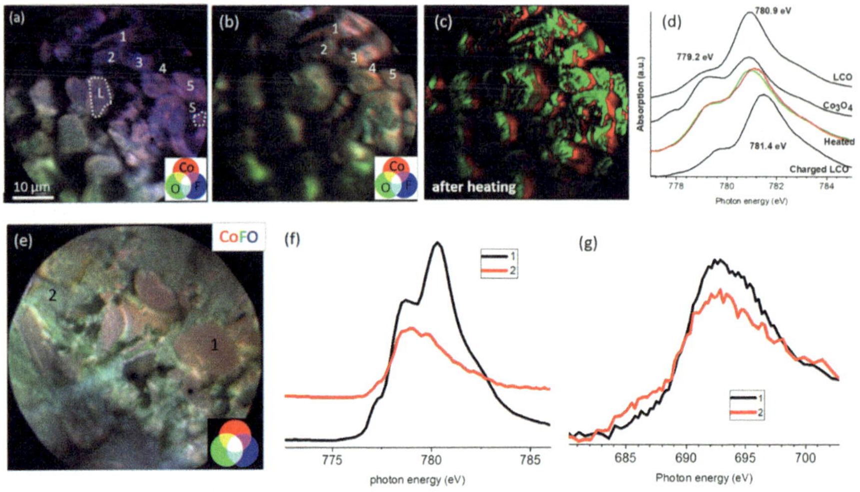

Fig. 7.11 **a** Elemental composition maps of charged LCO, **b** elemental composition maps of charged LCO after heating at 210 °C, **c** phase distribution maps of heated LCO, **d** Co L3-edge XANES of distinct phases in heated LCO. Adapted from Ref. [91] with permission from The Royal Society of Chemistry; **e** elemental composition mapping by X-PEEM, Co, F, O of a LCO electrode cycled between 4.5 and 3 V in LiPF6 with SUN/LiBOB additives, **f** Co L-edge XANES and **g** F K-edge XANES (**d**) of selected regions. Adapted from Ref. [52] with permission from The Royal Society of Chemistry

could introduce artifact or inaccurate characterization of the samples. Therefore, the best approach or the ultimate goal is to conduct in situ/operando measurement of these materials and batteries. There are many in situ techniques available on STXM as briefly introduced in Sect. 2.1.1. Here, we will discuss two most important and representative in situ STXM techniques in more detail, i.e., in situ electrochemistry and heterogenous catalysis, both been performed on the CLS Ambient-STXM.

Figure 12a shows the design and photograph of a 3-electrode in situ electrochemistry device for real-time STXM studies under both static (sealed, electrolyte non-flow) and continuous flow conditions [22]. The device was made using a combination of silicon microfabrication and 3D printing of a custom-designed STXM sample holder. Figure 12b presents a test experiment of copper deposition from an aqueous solution of $CuSO_4$ at the gold working electrode of the device in CLS Ambient-STXM, imaging at the Cu L-edge [22]. Recently, CLS Ambient-STXM has been equipped with a much more sophisticated and professionally designed in situ electromechanical setup from Hummingbird Scientific [23], which allows for continuous electrolyte flow down to sub-micrometer electrolyte thickness, minimized STXM working distance and weight load, multiple electrodes for all electrochemical needs and other functionalities, advanced sealing technology and leak-checking apparatus, etc. We believe this cutting-edge STXM in situ electrochemistry device will significantly boost energy materials and batteries research in the near future.

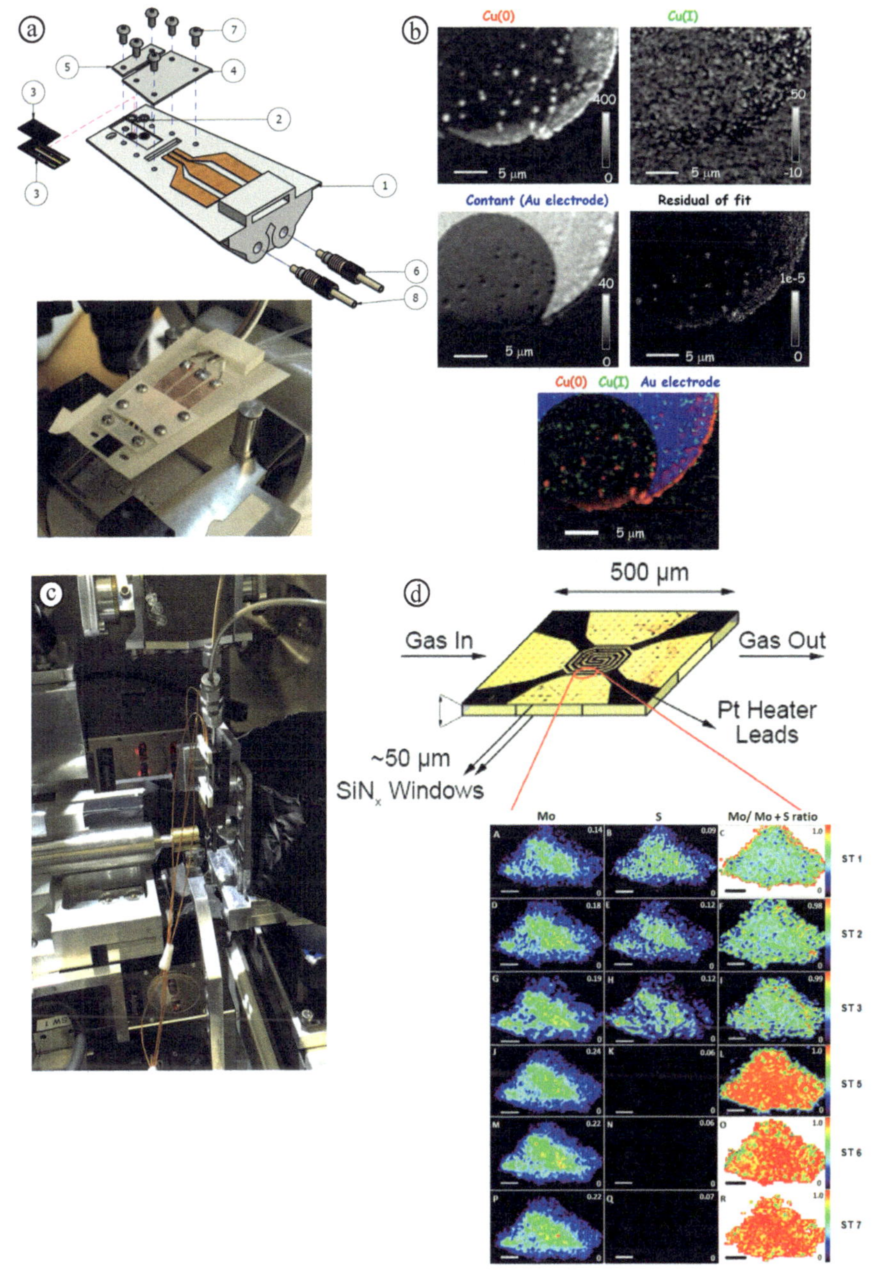
a
b
Cu(0)
Cu(I)
400
0
5 μm
50
-10
Contant (Au electrode)
Residual of fit
40
1e-5
Cu(0) Cu(I) Au electrode
c
d
500 μm
Gas In
Gas Out
Pt Heater Leads
~50 μm
SiNx Windows
Mo
S
Mo/ Mo + S ratio
ST 1
ST 2
ST 3
ST 5
ST 6
ST 7

◂**Fig. 7.12 a** Drawing and photograph of the STXM in situ electrochemistry flow cell, and **b** component maps for Cu(0), Cu(I), constant (Au electrode), residual, and color composite, derived from a singular value decomposition (SVD) fit of a 3-energy Cu 2p stack during in situ flow electrochemical study in CLS Ambient-STXM. Adapted from Ref. [22] with permission from AIP Publishing; **c** Photograph of the in situ heterogeneous catalysis nanoreactor in CLS Ambient-STXM, and **d** schematic of the nanoreactor and STXM in situ catalysis result of an alumina-supported CoMoS catalyst undergoing consecutive temperature (25, 200, or 400 °C) and gas environment (H2, He, or Air) steps. Adapted from Ref. [93], Copyright (2015) American Chemical Society

Figure 12c, d presents the in situ heterogeneous catalysis nanoreactor for gas and solid phase experiments in CLS Ambient-STXM [93]. The device is comprised of a TEM-based microelectromechanical system (MEMS) nanoreactor [94], housed in a custom-designed STXM sample holder. The nanoreactor is functionalized with a micrometer-sized gas-flow channel, electron-transparent windows withstanding a few Bar pressure, and a microheater with resistive heating up to 500 °C and temperature measurement. Figure 12d shows STXM in situ catalysis result of an alumina-supported CoMoS catalyst undergoing a programmed treatment procedure of multiple consecutive steps with temperature varying from 25 to 400 °C and gas environment changing from H_2 to Air to reduce or oxidize the catalyst [93]. Spatial variation of Co, Al, Mo, and S species, particularly the correlation between Co and Mo, was observed in situ and provided insight to the catalyst performance and mechanism. Further development of this setup has been actively pursued, such as enabled to work for low photon energy down to carbon K-edge, coupled with a portable mass spectrometer (MS) for online in-depth product analysis, and improved single/dual chip design [95, 96]. Finally, if the advanced in situ technologies could be combined with the high-resolution STXM-Ptychography in the near future, that will open an unprecedented era of in situ STXM applications.

7.5.2 Next-Generation STXM

With the ground-breaking emergence of the 4th generation diffraction-limited synchrotron sources (DLSR), the X-ray beam properties and quality will be extraordinarily improved. Specifically, compared to the 3rd generation synchrotron, the coherence of DLSR will increase from a few percent to 80–100% in the soft X-ray energy range, and the flux of DLSR will increase by at least 10 times, if considering the solid angle of radiation, the beam brilliance will increase by 1000 times or more. Therefore, the existing STXM will not be competent to operate on DLSR sources and a next-generation STXM is highly demanded to develop in the near future.

In the new STXM to fully utilize beam coherence, ptychography will be the dominant technique that will achieve X-ray wavelength-limited high spatial resolution (i.e., 1–3 nm) in 2D and 3D chemical imaging. To achieve this goal, a dedicated ptychography beamline is required to design and optimize beamline optics performance to reserve or further improve beam properties. Improvement on STXM area detectors is also needed, such as introducing high-performance sCMOS detectors for high frame rate, high data transferring speed, and high quantum efficiency. Finally, complex and sophisticated live-view data analysis software and hardware infrastructure are also highly desired, which include data pipeline, storage, and high-performance computing facility for ptychography data reconstruction. On the other hand, beam flux and brilliance gain will significantly improve XRF and TEY-based measurements and result in much better sensitivity. Also, usable beamline energy range could be largely extended as a result of flux gain. Finally, higher flux and beam quality will push improvement on data acquisition efficiency and implementation of new control hardware like field-programmable gate array (FPGA) and dynamic digital linearization (DDL), and development of new control and data acquisition software using powerful open source libraries in Python.

The above huge performance enhancement requires a completely new design of the STXM microscope to accommodate multiple functions, including ptychography, transmission, XRF, TEY, and possibly X-ray diffraction (XRD) and X-ray photon correlation spectroscopy (XPCS). In addition, controlled sample environment needs to be substantially improved to be compatible with the base UHV environment for all in situ devices and techniques. Finally, the new STXM layout will allow convenient modular connection to large analytical instruments, such as X-ray emission spectrometer (XES), XPS, and modular scanning electron microscope (SEM). Table 7.3 lists a detailed comparison of STXM on the 3rd and 4th generation synchrotron, and Fig. 13a illustrates the preliminary design of the next-generation STXM based on the CLS Cryo-STXM. Currently, the CLS Cryo-STXM is undergoing major upgrade to include ptychography, nanotomography, and XRF, as partially shown in Fig. 13b. These efforts toward the next-generation STXM will assure the STXM technology to stay at cutting-edge and innovative for growing demands on materials and devices characterization in many fields.

7.5.3 *Upgraded X-PEEM*

To achieve better performance on the existing 3rd generation synchrotron, the CLS X-PEEM is currently under a major upgrade. The X-PEEM will be rotated vertically by about 90 degree to improve beam spot uniformity and flux density 5 times. The upgrade preliminary design, as shown in Fig. 13c, integrates both the energy unfiltered and energy filtered X-PEEM mode designed by Elmitec, together with new KB refocusing optics from another vendor, but the actual upgrade will depend on available funding. Switching between the energy unfiltered and filtered X-PEEM mode will be enabled by an electromagnetic prism. The new KB optics will produce a finer and more uniform beam spot. Finally, improved working environment around X-PEEM will be considered for better stability and spatial resolution.

Table 7.3 Comparison of STXM on the 3rd and 4th generation synchrotron

	3rd Generation Synchrotron STXM	4th Generation Synchrotron STXM
Energy Range	**Narrow energy range**: soft X-ray (typically 200–2500 eV)	**Wide energy range**: cover from soft X-ray to tender X-ray (50–10,000 eV)
Spatial Resolution	**Low spatial resolution**: conventional: 30–50 nm, ptychography: 5–10 nm	**High spatial resolution**: conventional: 10–20 nm, ptychography: 1–3 nm
Sensitivity	**Low sensitivity**: conventional: 0.1–1%; soft X-ray XRF/TEY-STXM: 100–1000 ppm	**High sensitivity:** conventional: 0.05–0.1%; soft X-ray XRF/TEY-STXM: 10–100 ppm; tender X-ray XRF-STXM: 1–10 ppm
Functions	**Limited functions**: dominated by conventional transmission 2D and 3D STXM, optional with XRF, TEY, and ptychography	**Multi-functional**: dominated by 2D and 3D ptychography, highly integrated with transmission, XRF, TEY, and possibly XRD and XPCS
Sample environment	**Limited sample environment**: low vacuum up to 10^{-6} Torr, ambient or low vacuum in situ devices	**Controlled sample environment**: UHV: 10^{-8} Torr, high vacuum in situ devices and goniometer plus cryo-holder all based on TEM technologies
Control software and hardware	**Aged control software and hardware**: developed around 15 years ago	**Advanced control software and hardware**: advanced data acquisition and fast scanning, i.e., FPGA and DDL; interferometer position sampling for vibration/distortion-free scanning; overall data acquisition and scanning efficiency increased by 10–100 times
Data Analysis	**Simple data analysis software**: real-space image, spectromicroscopy, and spectral data analysis	**Complex and sophisticated data analysis software and hardware**: data pipeline, storage, and high-performance computing facility for reciprocal-space ptychography data reconstruction and live-view
Compatibility	**Low compatibility**: no compatibility with other analytical instruments; can add a laser system for initial STXM alignment	**High compatibility**: modular connection to large analytical instruments, such as XES, XPS, SEM; having an advanced laser system for STXM alignment, software development, sample preview, and user training

The before-mentioned upgrades will also benefit significantly from a DLSR synchrotron source in the future. Specifically, the substantially increased beam flux and brilliance will significantly increase the yield of photoelectrons and secondary electrons, making both X-PEEM modes have a much stronger signal to noise ratio for detection, particularly for the energy filtered mode. Therefore, a wide range of

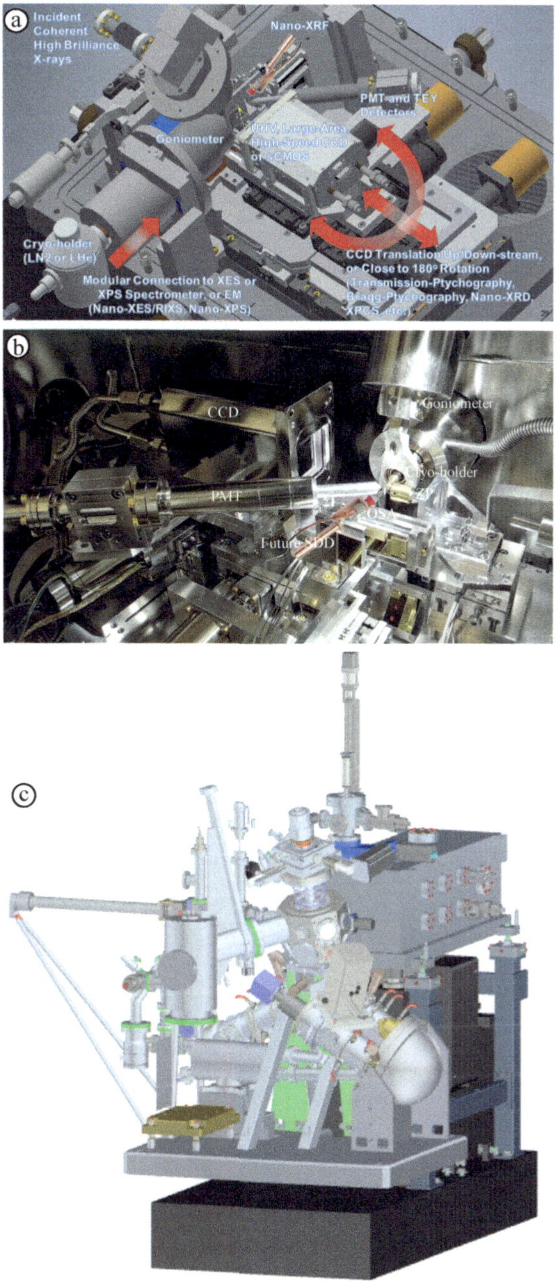

Fig. 7.13 **a** Preliminary design of the next-generation STXM based on the CLS Cryo-STXM, and **b** photograph of the Cryo-STXM being under upgrade to include ptychography and XRF; **c** preliminary design of the upgraded CLS X-PEEM by Elmitec integrating the energy unfiltered and energy filtered mode, together with new KB refocusing optics

samples including insulating inorganic and organic materials could be measured by the XPS-PEEM, which was not possible on a 3rd generation synchrotron. In addition, better quality and full coherence of the DLSR beam will further improve energy resolution of the monochromatic beam, in general, producing better controlled photoemission and resulting in much improved spatial resolution. Finally, the coherent beam can also produce a Fresnel diffraction pattern from sample surface topographic features, which can be used to deduce the height and shape of the features through diffraction fringes analysis and simulation [97]. Therefore, the overall performance, efficiency, and functions of X-PEEM will be largely improved on DLSR.

In summary, STXM and X-PEEM provide X-ray imaging and spectroscopy over a wide range of probing depths, transverse length scales and spatial resolutions with outstanding chemical sensitivity and site-specificity. Continuing developments of STXM and X-PEEM, especially related to the next-generation DLSR sources, will make them the ideal and powerful tools to characterize energy materials and batteries.

References

1. Hitchcock, A.P.: Handbook of Nanoscopy. Wiley-VCH Verlag & Co, KGaA (2012)
2. Hitchcock, A.P.: Soft X-ray spectromicroscopy and ptychography. J. Electron Spectrosc. Relat. Phenom. **200**, 49–63 (2015)
3. Stuckey, J.W., Yang, J., Wang, J., Sparks, D.L.: Advances in scanning transmission X-ray microscopy for elucidating soil biogeochemical processes at the submicron scale. J. Environ. Qual. **46**, 1166–1174 (2017)
4. Jacobsen, C.: X-ray Microscopy. Cambridge University Press (2019)
5. Rarback, H., et al.: X-ray Microscopy Series in Optical Sciences. Springer (1984)
6. Jacobsen, C., et al.: Quantitative imaging and microanalysis with a scanning soft X-ray microscope. Phys. Med. Biol. **32**, 431–437 (1987)
7. Rarback, H., et al.: Scanning x-ray microscope with 75-nm resolution. Rev. Sci. Instrum. **59**, 52–59 (1988)
8. Ade, H., et al.: Chemical contrast in X-ray microscopy and spatially resolved XANES spectroscopy of organic specimens. Science **258**, 972–975 (1992)
9. Jacobsen, C., Wirick, S., Flynn, G., Zimba, C.: Soft X-ray spectroscopy from image sequences with sub-100 nm spatial resolution. J. Microsc. **197**, 173–184 (2000)
10. Shu, D., Siddons, D.P., Rarback, H., Kirz, J.: Two-dimensional laser interferometric encoder for the soft X-ray scanning microscope at the NSLS. Nucl. Instrum. Methods Phys. Res., Sect. A **266**, 313–317 (1988)
11. Kilcoyne, A.L.D., et al.: Interferometer-controlled scanning transmission X-ray microscopes at the advanced light source. J. Synchrotron Radiat. **10**, 125–136 (2003)
12. Howells, M., Jacobsen, C., Warwick, T.: Science of Microscopy. In: Hawkes, P.W., Spence, J.C.H. (eds) 835–926 (2007)
13. Bluhm, H., et al.: Soft X-ray microscopy and spectroscopy at the molecular environmental science beamline at the Advanced Light Source. J. Electron Spectrosc. Relat. Phenom. **150**, 86–104 (2006)
14. Ade, H., Hitchcock, A.P.: NEXAFS microscopy and resonant scattering: composition and orientation probed in real and reciprocal space. Polymer **49**, 643–675 (2008)

15. Kaznatcheev, K.V., et al.: Soft X-ray spectromicroscopy beamline at the CLS: commissioning results. Nucl. Instrum. Methods Phys. Res. Sect. A **582**, 96–99 (2007)
16. Wang, J.: Applications of soft X-ray spectromicroscopies to lithium ion battery analysis. Microsc. Microanal. **24**, 416–419 (2018)
17. Hernández-Cruz, D., Hitchcock, A.P., Tyliszczak, T., Rousseau, M.E., Pézolet, M.: In situ azimuthal rotation device for linear dichroism measurements in scanning transmission x-ray microscopy. Rev. Sci. Instr. **78**, 033703 (2007)
18. Obst, M., Wang, J., Hitchcock, A.P.: Soft X-ray spectro-tomography study of cyanobacterial biomineral nucleation. Geobiology **7**, 577–591 (2009)
19. Hegde, M., Farvid, S.S., Hosein, I.D., Radovanovic, P.V.: Tuning manganese dopant spin interactions in single GaN nanowires at room temperature. ACS Nano. **5**, 6365–6373 (2011)
20. Leontowich, A.F.G., Hitchcock, A.P.: Experimental investigation of beam heating in a soft X-ray scanning transmission X-ray microscope. Analyst **137**, 370–375 (2012)
21. Wang, J., et al.: Effect of humidity on individual SnO_2 coated carbon nanotubes studied by in situ STXM. J. Electron Spectrosc. Relat. Phenom. **184**, 296–300 (2011)
22. Prabu, V., et al.: Instrumentation for in situ flow electrochemical scanning transmission X-ray microscopy (STXM). Rev. Sci. Instr. **89**, 063702 (2018)
23. Lim, J., et al.: Origin and hysteresis of lithium compositional spatiodynamics within battery primary particles. Science **353**, 566–571 (2016)
24. Förster, J.D., et al.: MIMiX: a multipurpose in situ microreactor system for X-ray microspectroscopy to mimic atmospheric aerosol processing. Atmos. Meas. Tech. **13**, 3717–3729 (2020)
25. Zhou, J., Wang, J., Cutler, J., Hu, E., Yang, X.-Q.: Imaging the surface morphology, chemistry and conductivity of $LiNi_{1/3}Fe_{1/3}Mn_{4/3}O_4$ crystalline facets using scanning transmission X-ray microscopy. Phys. Chem. Chem. Phys. **18**, 22789–22793 (2016)
26. Howells, M.R., et al.: An assessment of the resolution limitation due to radiation-damage in X-ray diffraction microscopy. J. Electron Spectrosc. Relat. Phenom. **170**, 4–12 (2009)
27. Huang, X., et al.: Soft X-ray diffraction microscopy of a frozen hydrated yeast cell. Phys. Rev. Lett. **103**, 198101 (2009)
28. Frano, A., et al.: Orbital control of noncollinear magnetic order in nickel oxide heterostructures. Phys. Rev. Lett. **111**, 106804 (2013)
29. Stahl, C., et al.: Magnetic X-ray microscopy at low temperatures—visualization of flux distributions in superconductors. AIP Conf. Proc. **1696**, 020031 (2016)
30. Maser, J., et al.: Soft X-ray microscopy with a cryo scanning transmission X-ray microscope: I. Instrumentation, imaging and spectroscopy. J. Microsc. **197**, 68–79 (2000)
31. Wang, Y., Jacobsen, C., Maser, J., Osanna, A.: Soft X-ray microscopy with a cryo scanning transmission X-ray microscope: II. Tomography. J. Microsc. **197**, 80–93 (2000)
32. Beetz, T., et al.: Apparatus for X-ray diffraction microscopy and tomography of cryo specimens. Nucl. Instr. Methods Phys. Res. Sect. A **545**, 459–468 (2005)
33. Leontowich, A.F.G., et al.: Cryo scanning transmission x-ray microscope optimized for spectrotomography. Rev. Sci. Instr. **89**, 093704 (2018)
34. Rodenburg, J.M.: Advances in Imaging and Electron Physics, vol. 150, pp. 87–184 (2008)
35. Thibault, P., et al.: High-resolution scanning X-ray diffraction microscopy. Science **321**, 379–382 (2008)
36. Shapiro, D.A., et al.: Chemical composition mapping with nanometre resolution by soft X-ray microscopy. Nat. Photonics **8**, 765–769 (2014)
37. Denes, P., et al.: A fast, direct x-ray detection charge-coupled device. Rev. Sci. Instr. **80**, 083302 (2009)
38. Shapiro, D.A., et al.: The COSMIC imaging beamline at the advanced light source: a new facility for spectro-microscopy of nano-materials. Microsc. Microanal. **24**, 8–11 (2018)
39. Desjardins, K., et al.: Backside-illuminated scientific CMOS detector for soft X-ray resonant scattering and ptychography. J. Synchrotron Radiat. **27**, 1577–1589 (2020)
40. Wang, C., et al.: Background noise removal in x-ray ptychography. Appl. Opt. **56**, 2099–2111 (2017)

41. Rodenburg, J.M., Faulkner, H.M.: A phase retrieval algorithm for shifting illumination. Appl. Phys. Lett. **85**, 4795–4797 (2004)
42. Maiden, A.M., Rodenburg, J.M.: An improved ptychographical phase retrieval algorithm for diffractive imaging. Ultramicroscopy **109**, 1256–1262 (2009)
43. Sun, T., et al.: Soft X-ray ptychography chemical imaging of degradation in a composite surface-reconstructed Li-rich cathode. ACS Nano. (2020)
44. Tonner, B.P., Harp, G.R.: Photoelectron microscopy with synchrotron radiation. Rev. Sci. Instr. **59**, 853–858 (1988)
45. Schmidt, T., et al.: SPELEEM: combining LEEM and spectroscopic imaging. Surf. Rev. Lett. **5**, 1287–1296 (1998)
46. Fink, R., et al.: SMART: a planned ultrahigh-resolution spectromicroscope for BESSY II. J. Electron Spectrosc. Relat. Phenom. **84**, 231–250 (1997)
47. Tusche, C., et al.: Spin resolved photoelectron microscopy using a two-dimensional spin-polarizing electron mirror. Appl. Phys. Lett. **99**, 032505 (2011)
48. Bauer, E.: Photoelectron spectromicroscopy: present and future. J. Electron Spectrosc. Relat. Phenom. **114–116**, 975–987 (2001)
49. Feng, J., Scholl, A.: Science of Microscopy. In: Hawkes, P.W., Spence, J.C. (eds), pp. 657–695 (2007)
50. Bauer, E.: A brief history of PEEM. J. Electron Spectrosc. Relat. Phenom. **185**, 314–322 (2012)
51. Bauer, E.: Surface Microscopy with Low Energy Electrons (2014)
52. Wang, J., Ji, Y., Appathurai, N., Zhou, J., Yang, Y.: Nanoscale chemical imaging of the additive effects on the interfaces of high-voltage $LiCoO_2$ composite electrodes. Chem. Commun. **53**, 8581–8584 (2017)
53. Liang, Y.Y., et al.: Co_3O_4 nanocrystals on graphene as a synergistic catalyst for oxygen reduction reaction. Nat. Mater. **10**, 780–786 (2011)
54. Georgakilas, V., et al.: Functionalization of graphene: covalent and non-covalent approaches derivatives and applications. Chem. Rev. **112**, 6156–6214 (2012)
55. Sun, S.H., et al.: Single-atom catalysis using Pt/graphene achieved through atomic layer deposition. Sci. Rep. **3** (2013)
56. Li, W.H., et al.: Phosphorene degradation: visualization and quantification of nanoscale phase evolution by scanning transmission X-ray microscopy. Chem. Mater. **32**, 1272–1280 (2020)
57. Chen, J.G.: NEXAFS investigations of transition metal oxides, nitrides, carbides, sulfides and other interstitial compounds. Surf. Sci. Rep. **30**, 1–152 (1997)
58. Gong, K., Du, F., Xia, Z., Durstock, M., Dai, L.: Nitrogen-doped carbon nanotube arrays with high electrocatalytic activity for oxygen reduction. Science **323**, 760–764 (2009)
59. Zhou, J.G., et al.: Electronic structure and luminescence center of blue luminescent carbon nanocrystals. Chem. Phys. Lett. **474**, 320–324 (2009)
60. Chen, X.Q., et al.: Visualizing electronic interactions between iron and carbon by X-ray chemical imaging and spectroscopy. Chem. Sci. **6**, 3262–3267 (2015)
61. Zhou, J.G., et al.: Imaging nitrogen in individual carbon nanotubes. J. Phys. Chem. Lett. **1** (2010)
62. Novoselov, K.S., et al.: Electric field effect in atomically thin carbon films. Science. **306**, 666–669 (2004)
63. Zhou, J.G., et al.: Nano-scale chemical imaging of a single sheet of reduced graphene oxide. J. Mater. Chem. **21**, 14622–14630 (2011)
64. Pacile, D., Papagno, M., Rodriguez, A.F., Grioni, M., Papagno, L.: Near-edge x-ray absorption fine-structure investigation of graphene. Phys. Rev. Lett. **101**, (2008)
65. Schultz, B.J., et al.: Imaging local electronic corrugations and doped regions in graphene. Nat. Commun. **2**, 372 (2011)
66. Gong, M. et al.: An advanced Ni-Fe layered double hydroxide electrocatalyst for water oxidation. J. Am. Chem. Soc. **135**, 8452–8455 (2013)
67. Zhou, J.G., et al.: An X-ray absorption, photoemission, and Raman study of the interaction between SnO_2 nanoparticle and carbon nanotube. J. Phys. Chem. C. **113**, 6114–6117 (2009)

68. Kuang, Q., Lao, C.S., Wang, Z.L., Xie, Z.X., Zheng, L.S.: High-sensitivity humidity sensor based on a single SnO_2 nanowire. J. Am. Chem. Soc. **129**, 6070 (2007)
69. Zhou, J.G., et al.: Nanoscale chemical imaging and spectroscopy of individual RuO_2 coated carbon nanotubes. Chem. Commun. **46**, 2778–2780 (2010)
70. Liang, Y.Y., et al.: Oxygen reduction electrocatalyst based on strongly coupled cobalt oxide nanocrystals and carbon nanotubes. J. Am. Chem. Soc. **134**, 15849–15857 (2012)
71. Gong, M., et al.: Nanoscale nickel oxide/nickel heterostructures for active hydrogen evolution electrocatalysis. Nat. Commun. **5** (2014)
72. Wepasnick, K.A., et al.: Surface and structural characterization of multi-walled carbon nanotubes following different oxidative treatments. Carbon **49**, 24–36 (2011)
73. Zhou, J.G., Wang, J.A., Fang, H.T., Sham, T.K.: Structural variation and water adsorption of a SnO_2 coated carbon nanotube: a nanoscale chemical imaging study. J. Mater. Chem. **21**, 5944–5949 (2011)
74. Liang, Y.Y., et al.: Covalent hybrid of spinel manganese-cobalt oxide and graphene as advanced oxygen reduction electrocatalysts. J. Am. Chem. Soc. **134**, 3517–3523 (2012)
75. Sun, W., et al.: Engineering of nitrogen coordinated single cobalt atom moieties for oxygen electroreduction. ACS Appl. Mater. Interfaces **11**, 41258–41266 (2019)
76. Yang, Y., et al.: O-coordinated W-Mo dual-atom catalyst for pH-universal electrocatalytic hydrogen evolution. Sci. Adv. **6** (2020)
77. Wu, Z.S., et al.: Graphene anchored with Co_3O_4 nanoparticles as anode of lithium ion batteries with enhanced reversible capacity and cyclic performance. ACS Nano. **4**, 3187–3194 (2010)
78. Qi, Y., Zhang, H., Du, N., Yang, D.R.: Highly loaded CoO/graphene nanocomposites as lithium-ion anodes with superior reversible capacity. J. Mater. Chem. A. **1**, 2337–2342 (2013)
79. Zhou, J.G., et al.: Fe-N bonding in a carbon nanotube-graphene complex for oxygen reduction: an XAS study. Phys. Chem. Chem. Phys. **16**, 15787–15791 (2014)
80. Csiszar, S.I., et al.: Controlling orbital moment and spin orientation in CoO layers by strain. Phys. Rev. Lett. **95** (2005)
81. Wang, J., Zhou, J., Hu, Y.F., Regier, T.: Chemical interaction and imaging of single Co_3O_4/graphene sheets studied by scanning transmission X-ray microscopy and X-ray absorption spectroscopy. Energy Environ. Sci. **6**, 926–934 (2013)
82. Hitchcock, A.P., Toney, M.F.: Spectromicroscopy and coherent diffraction imaging: focus on energy materials applications. J. Synchrotron Radiat. **21**, 1019–1030 (2014)
83. Li, Y.Y., et al.: Dichotomy in the lithiation pathway of ellipsoidal and platelet $LiFePO_4$ particles revealed through nanoscale operando state-of-charge imaging. Adv. Func. Mater. **25**, 3677–3687 (2015)
84. Zhou, J.G., Wang, J., Hu, Y.F., Lu, M.: Chemical imaging of nanoscale interfacial inhomogeneity in $LiFePO_4$ composite electrodes from a cycled large-format battery. ACS Appl. Mater. Interfaces. **9**, 39336–39341 (2017)
85. Dreyer, W., et al.: The thermodynamic origin of hysteresis in insertion batteries. Nat. Mater. **9**, 448–453 (2010)
86. Lu, M., et al.: Correlative imaging of ionic transport and electronic structure in nano Li0.5FePO4 electrodes. Chem. Commun. **56**, 984–987 (2020)
87. Li, Y., et al.: Current-induced transition from particle-by-particle to concurrent intercalation in phase-separating battery electrodes. Nat. Mater. **13**, 1149 (2014)
88. Zhou, J., et al.: Electronic structure variation of the surface and bulk of a $LiNi_{0.5}Mn_{1.5}O_4$ cathode as a function of state of charge: X-ray absorption spectroscopic study. Phys. Chem. Chem. Phys. **16**, 13838–13842 (2014)
89. Lu, M., Wang, J., Fang, H., Hu, Y., Zhou, J.: Unexpected phase separation in $Li_{1-x}Ni_{0.5}Mn_{1.5}O_4$ within a porous composite electrode. Chem. Commun. **54**, 4152–4155 (2018)
90. Sharifi-Asl, S., et al.: Facet-dependent thermal instability in $LiCoO_2$. Nano Lett. **17**, 2165–2171 (2017)
91. Lu, M., Mao, Y.Z., Wang, J., Hu, Y.F., Zhou, J.G.: Surface heterogeneity in $Li_{0.5}CoO_2$ within a porous composite electrode. Chem. Commun. **54** (2018)

92. MacNeil, D.D., Dahn, J.R.: The reactions of $Li_{0.5}CoO_2$ with nonaqueous solvents at elevated temperatures. J. Electrochem. Soc. **149**, A912–A919 (2002)
93. Al Samarai, M., et al.: Unraveling the redox behavior of a CoMoS hydrodesulfurization catalyst: a scanning transmission X-ray microscopy study in the tender X-ray range. J. Phys. Chem. C. **119**, 2530–2536 (2015)
94. Creemer, J.F., et al.: Atomic-scale electron microscopy at ambient pressure. Ultramicroscopy **108**, 993–998 (2008)
95. van Ravenhorst, I.K., et al.: Capturing the genesis of an active fischer-tropsch synthesis catalyst with operando X-ray nanospectroscopy. Angew. Chem. Int. Ed. **57**, 11957 (2018)
96. van Ravenhorst, I.K., et al.: In situ local temperature mapping in microscopy nano-reactors with luminescence thermometry. ChemCatChem **11**, 5505–5512 (2019)
97. Schmidt, T., et al.: Optical layout of a beamline for photoemission microscopy. J. Synchrotron Radiat. **6**, 957–963 (1999)

Chapter 8
Principles and Applications of Industrial X-ray Computed Tomography

Fanpeng Kong, Qingsong Liu, Wei Zhao, and Jiajun Wang

8.1 Brief Introduction of Industrial X-ray Computed Tomography

X-ray as a high energy electromagnetic wave was firstly found by Wilhelm Rontgen in 1895, exhibiting high feasibility in imaging due to its powerful penetration. Hounsfied designed the first X-ray-computed tomography (X-ray-CT) machine, which produced the detailed two-dimensional (2D) image through a two-dimensional tomographic reconstruction algorithm such as Filtered Back Projection (FBP). Clinical CT has been extensively applied to medicine, but insufficient spatial resolution makes it hard to reveal microstructure. High-resolution peripheral quantitative CT improved the spatial resolution as high as 150 μm. Different from Fan Beam X-ray source used in clinical CT, X-ray Micro-CT using cone-beam X-ray source provided much higher spatial resolution. More importantly, it enables there construction of the 3D internal structure of objects without destructivity and prior preparation. A series of 2D projections at different angles could be obtained when the X-ray source is circularly rotated around the sample. The reconstructed 3D image is produced by using the Feldkamp-David-Kress (FDK) algorithm. Micro-CT systems have been successfully created as a simple desktop instrument by assembling modern technology including X-ray sources, detectors, computer systems and image processing systems. However, it is worthy of noting that it is impossible for a single micro-CT machine covering all ranges of spatial resolution, which is determined by the X-ray voltage and current. With the rapid development of nanotechnology, there is a

F. Kong (✉) · Q. Liu · W. Zhao · J. Wang (✉)
School of Chemistry and Chemical Engineering, Harbin Institute of Technology, Harbin 150001, China
e-mail: fpkong@hit.edu.cn

J. Wang
e-mail: jiajunhit@hit.edu.cn

J. Wang (ed.), *Advanced X-ray Imaging of Electrochemical Energy Materials and Devices*, https://doi.org/10.1007/978-981-16-5328-5_8

strong need for non-destructive testing instruments that providing inner structure at nano-scale. X-ray nano-CT as a new generation of CT could achieve high spatial resolution as high as 200 nm by using a nano-focal spot source.

Compared with Magnetic resonance imaging (MRI), positron emission tomography (PET), fluorescence imaging (FI), and photoacoustic imaging (PI), X-ray CT exhibits its particular advantage and non-destructively provides spatially structural and tomographic information, which are of high importance to understand physical, chemical and biological properties of objects. Up to now, CT technology has been of the essence in many fields, such as drug, energy, biology, and so on [1–3].

8.2 The Principle and Structure of Industrial X-ray Microscopy

The X-ray CT device usually consists of X-ray source, radiation detector and collimator, data acquisition system, sample scanning mechanical system, computer system, and auxiliary system so on [4–7]. The principle of X-ray CT technique is based on the interaction between X-ray and object, such as photoelectric effect, compton effect, electron pair effect, and so on. If the object is a uniform material, the interaction can be described by the Lambert-Beer law [8]:

$$I = I_0 e^{-fL}$$

where I_0 and I are the intensity of incident and transmitted X-ray, L is the thickness of the uniform material, and f is the absorption coefficient of the uniform material for X-ray. However, samples are usually non-uniform materials with different absorption coefficients in the actual research, so the Lambert-Beer law can be generalized as:

$$I = I_0 e^{-\int_L f(x)dx}$$

L is the path where the X-ray passes through the material, and $f(x)$ represents the absorption coefficient at the point x of the path L.

$$g = \ln\left(\frac{I_0}{I}\right) = \int_L f(x)dx$$

g is called the projection after the X-ray through the material. The value is equal to the line integral of the line attenuation coefficient on the X-ray path. If projections at multiple angles can be obtained, the corresponding function-density distribution can be obtained by the inverse projection of the integrand function $f(x)$. The process is to reconstruct the image from the projection. In short, CT is the process that the detector turns the received attenuated signal of the X-ray through the material to the digital signal, and the digital signal processed by the computer is displayed as the reconstructed image.

8.3 The Application of Industrial X-ray Microscopy in Medicine

In the past decades, a significant number of efforts have been devoted to the development of medical science, mainly including the treatment and prevention of physiological diseases as well as improvement of physical health. However, some diseases still cannot be efficiently cured up to now, such as many kinds of cancer, grimly killing millions of people every year. The lack of cognition of tumor cells including morphology, structure, and microenvironment, directly determining proliferation and apoptotic progress of cancer, strongly limits the emerge of feasible therapies and drugs. Recently, clinic non-invasive X-ray CT technique with enough spatial resolution has been used to successfully image the bone, blood vessel, and liver at the cellular scale, unambiguously revealing their distribution, tomography and operation condition [9]. Therefore, high-resolution X-ray CT is of great importance for understanding and therapy of cancer mainly through imaging tumor cells at different stages during pharmacotherapy and physicotherapeutics.

Recently, photothermal therapy (PTT) as a novel tumor treatment, has received considerable attention from researchers due to its minimal invasiveness and better outcomes. During PTT, different normal cells, the tumor cells with a lower tolerance to temperature can be efficaciously killed through increasing temperature to a suitable therapy value achieved by the near-infrared (NIR) light [10]. In addition, photothermal agents adsorbing on the surface of cancer cells effectively convert captured NIR light into heat, playing a pivotal role in killing cancer cells. It has been revealed that Prussian blue with PH-dependent stability and high safety to the human body is an ideal candidate for the photothermal agent. Prussian blue based on PTT is found to reduce the neuroblastoma burden due to the increased infiltration of lymphocytes and T cells to the tumor. Also, the tumor regression can be completely realized and long-term survival can be greatly prolonged for the tumor-bearing mice through X-ray CT, after the treatment of the combination of PB-based PTT and anti-cytotoxic T-lymphocyte-associated protein. Zheng et al. found that FePt nanoparticles could remarkably limit the HeLa cells progression using tumor-bearing mice [11]. Besides, gold nanorods (GNRs) with a high absorption coefficient in the NIR region and a better heat generation rate has been confirmed to be an efficient photothermal agent and CT contrast agent. Liu et al. fabricated a novel multi-functional probe composed of gold nano-star core and doxorubicin-loaded mesoporous silica shells, showing great potential in the early diagnosis and treatment of cancer [12]. As shown in in-vivo CT imaging (Fig. 8.1), the probe after 2 h intratumoral injection could enrich the tumor site and realize the

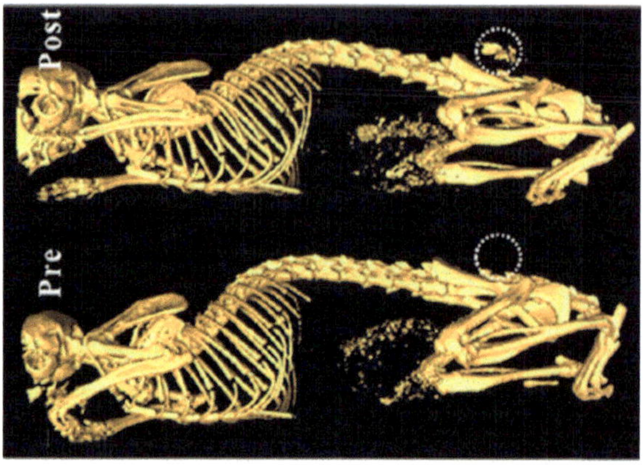

Fig. 8.1 CT image of mice before and after intratumoral injection with probe. Adapted from Ref. [13] Copyright (2019) Wiley-VCH

clear imaging of tumors compared with the same site before injection. Chen et al. developed a nano-bismuth sphere cluster as a CT contrast agent and photothermal agent considering that its high X-ray absorption coefficient and excellent photothermal properties [13]. During the in-vivo experiment, the probe was injected into a tumor-bearing mouse when the tumor grew to a certain volume. As shown in Fig. 8.2, the intensity of the probe increased and reached the maximum at 12 h, while almost vanished at 24 h, indicating PTT was able to kill the tumor cell within a short period of time.

Bone fragility such as osteoporosis has been mainly resulted from the failed material or structural adaptions to mechanical stress, which is a current research focus to clarify the mechanism related to bone loss and bone failure [14]. Osteocyte plays an important role in bone modeling and remodeling process through orchestrating the equilibrium between osteoclast and osteoblast activity. The osteocyte called "the unrecognized side of bone tissue" mainly resides in lacunae and is interconnected through canaliculi, forming a lacunae-canaliculi network, denoted as LCN. Although some microscopic techniques such as scanning electron microscopy (SEM) and transmission electron microscopy (TEM) have been utilized to image the LCN, they only provide the 2D image, making it challenging to deeply understand the complex network. X-ray imaging is highly suited for the analysis of bone at different scales due to its high penetration ability, which has successfully revealed the tomography of trabecular bone. Varga et al. found that two dominant oscillating and twisted plywood patterns coexisted in osteon cells through an auto-correlation-based orientation measure [15]. Langer et al. adopted X-ray phase nano-tomography to detail study the 3D organization of LCN over cells in the osteonal and interstitial tissue, as shown in Fig. 8.3a, b [16]. The anatomical features of LCN, such as the morphology of lacunae and canaliculi and the relationship between LCN and the cement line are clearly revealed. As shown in Fig. 8.3c, d, a fully connected network on the side of the cement line and the complete absence of

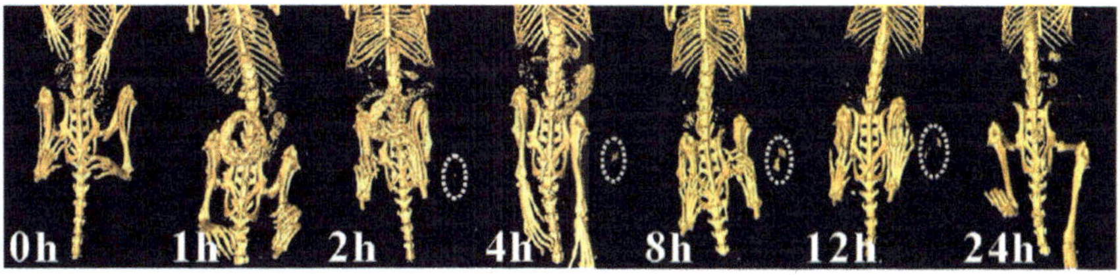

Fig. 8.2 CT image of tumor-bearing BALB/c mouse before and after injection. Adapted from Ref. [13] Copyright (2019) Wiley-VCH

canaliculi on the other side is observed. In addition, the density of tissues relative to the background can be directly measured and the cement line exhibits a higher mass density compared with the osteonal and interstitial tissue. Hesse et al. further used phase-contrast nano-CT evidenced that the spatial distribution of mass density in bone tissue was mainly determined by the canalicular network morphology [15].

Ni et al. used X-ray CT to evaluate the therapy of developed drugs for bone loss and found that nano-encapsulated enantiomer of (+) promethazine was a highly effective anti-bone loss drug [17]. It is also confirmed that long-term application of thalidomide could reduce the plaque growth and VV neovascularization revealed by Nano-CT because thalidomide inhibited proliferation and migration of human coronary artery endothelial cells (HCAEC) dose-dependently. Izutsu et al. employed X-ray CT to study the tomography of lyophilized protein solids, demonstrating that the effect of post-freeze annealing and controlled nucleation was of importance for the quality of pharmaceutical [18]. X-ray CT is also of the essence for the dentology, scaffold, biomaterials, and so on. The strategy on the improvement of sensitivity also received extensive attention mainly by incorporating novel accessories and developing a new sample treatment method. For instance, Dudark et al utilized ethanol fixation and large-area photon-counting detector to increase the sensitivity of CT for soft tissue [19]. Unprecedented high precision of CT detection is successfully realized by using liver-targeted iodinated nano-emulsions. Jpshua et al. proposed an automated segmentation to address the traditional limitation of human-centric segmentation [2].

In summary, the widespread application of X-ray CT in medicine has been confirmed to be a great help for the understanding of infection and immune. It is also highly of importance for characterizing the therapy effect of physical and chemical treatment, obviously promoting the development of therapy method with high safety and efficiency. Nevertheless, the X-ray CT with higher spatial resolution and in-situ detect function is still highly desirable for studying medicine and life-science at a more micro-scale and revealing their intrinsic mechanism.

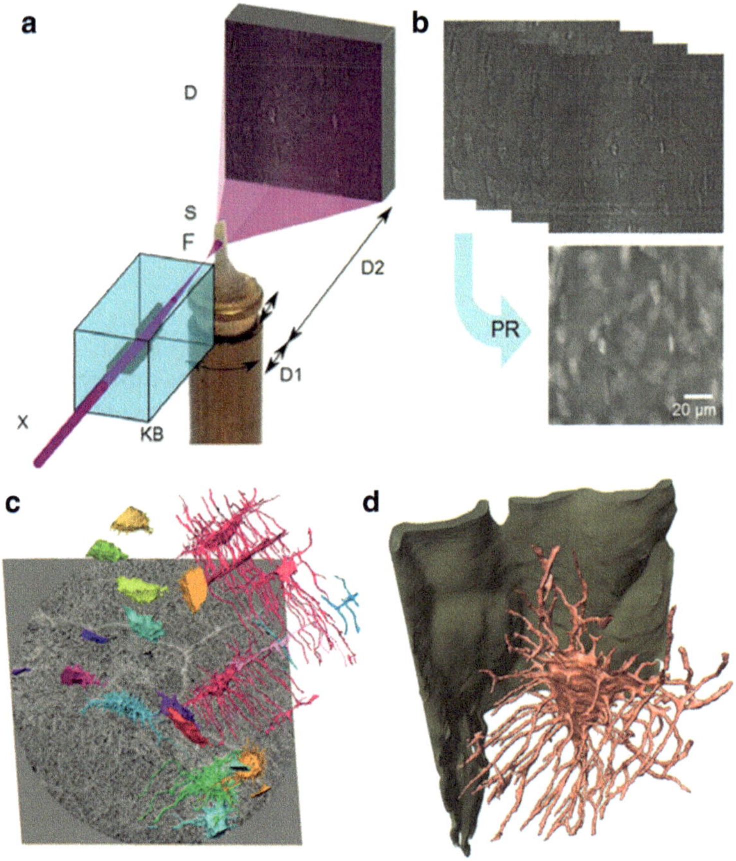

Fig. 8.3 **a** Schematic of the experimental setup. The X-ray beam is monochromatized and focused into a focal spot by X-ray reflective optics. The sample is positioned on a translation-rotation stage downstream of the focus and imaged onto a stationary detector. Due to the resulting divergent beam, different spot-sample distances and different free-space propagation distances imply different magnification factors on the detector. **b** Images were recorded at four focus-to-sample distances over a complete turn of the sample at 2999 projection angles. The images were used to reconstruct the phase shift at each angle, which was used as input to a tomographic reconstruction algorithm to reconstruct the 3D local mass density. **c** Rendering of osteocyte lacunae and canaliculi in the whole imaged volume overlayed over the bottom slice shown in grayscale. Colors correspond to connected components and grayscale to mass density. Note the difference in structure in the interstitia and osteon: the connected cells are all in the osteonal tissue, the others in the interstitial. The canaliculi are considerably reduced in the interstitia. **d** Zoom on the highlighted lacuna in A showing the interaction between the canaliculi [pink] and the cement line [green], and branching of the canaliculi. Adapted from Ref. [16] Copyright (2014) Springer Link

8.4 The Application of Industrial X-ray Microscopy in Renewable Energy

X-ray computed tomography has shown great promise in the research of renewable energy [20–24]. The spatial structural information of different components in the sample can be distinguished as intensity contrast mode is directly related to the density and atomic coefficients of components. The intensity contrast mode in X-ray CT mainly includes the absorption contrast mode and Zernike phase-contrast mode [25–27]. Under the absorption mode, different components of the sample have different absorption of X-rays, leading to the contrast of light and dark on the image [28–30]. The intensity and thickness of samples directly determine the absorption coefficient. In addition, the element with high atomic numbers shows a higher absorption coefficient. It is worth noting that X-ray CT using absorption mode cannot distinguish the specimens with similar densities, especially with the lower density. In this case, the phase-contrast mode technology can greatly increase the contrast of the collected image by collecting the difference in the refractive index of the X-ray, not just the difference in the absorption coefficient. Also, different from other three-dimensional characterization techniques, X-ray CT images the internal and interface structure without dead angles, showing the widespread application in energy storage and conversion devices [8, 31].

The cycling performance and multiplier performance of lithium-ion batteries are dependent on the 3D microstructure of lithium-ion batteries in a large extent [32]. Therefore, it is necessary to detect the 3D microstructure of the electrode and evaluate the heterogeneity of the microstructure and its impact on the battery performance through CT technology. By using nano-CT technology, Huang et al. [33]. achieved the interior observation of $LiFePO_4$/carbon cathode materials by imaging the three-dimensional morphology (Fig. 8.4). The research process mainly includes fine sample preparation, data collection, and data preprocessing. Finally, through the corresponding processing software (Image J, Avizo), the quantitative calculation of the pore content is realized. With the aid of the three-dimensional rendering distribution, it is proved that there are a small number of independent pores inside the electrode, which do not contribute to the battery capacity. Similarly, the researchers also used CT technology to conduct a large number of studies on the internal porosity, active material distribution, and active surface area of the electrode under different compaction densities.

Understanding of the decay mechanism of lithium-ion batteries is another research hotspot in the field of energy. CT technology has been widely used in lithium-ion battery performance degradation due to its non-destructive testing characteristics [34, 35]. Figgemeier et al. [36] used CT technology to study the three-dimensional structural information of the batteries before and after aging and further analyze the structural changes of the batteries during cycling. CT results revealed that organic residues and deposits are the main reason for the decrease of the porous anode of the aging battery. On the cathode, crushing of cathode particles and current collector corrosion are also the reasons for the decline of battery

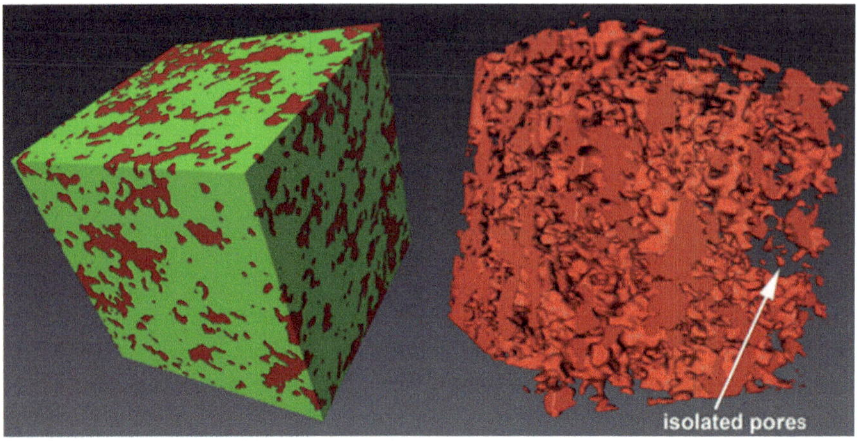

Fig. 8.4 Pore distribution of the cropped cube of sample after the segmentation (left) in the compact LFP/C cube, and (right) the distribution of pore network. Adapted from Ref. [33] Copyright (2015) Elsevier

capacity and the increase of impedance (Fig. 8.5). In the detection of soft-pack batteries, Bearckmans et al. [37] used X-ray CT technology to study the behavior for rapid decay of the pouch cells containing silicon alloy anodes and found that there was obvious mechanical deformation between the electrode layers. It is speculated that rapid decay mainly originated from the decomposition of FEC to CO_2 during the electrochemical cycle. Additionally, the decomposition of FEC will also produce LiF, which will form a more stable SEI film, significantly increasing the cycle stability and generating CO_2 at the same time. However, the generated CO_2 will hinder the transmission of lithium ions, causing loss of active surface area and decay of battery capacity. Based on the aforementioned phenomenon, the disadvantages arising from the decomposition of FEC exceed the benefits. With the help of CT technology, the structure changes before and after the electrode aging, particle fragmentation, pore blockage, and current collector corrosion can be effectively investigated.

Drawing the electrode microstructure in 3D is necessary to evaluate the unevenness of the microstructure and its impact on battery performance. However, using CT technology to identify carbon and binder (CBD) in the electrode is still challenging. In the usual research, pores and CBD are regarded as a single phase (pore-CBD phase), assuming that it is filled with electrolyte, which may lead to a decrease in the representative microstructure and mass transfer performance. In order to accurately correlate the microstructure with performance, more practical solutions are high of necessity. The Paul R. Shearing group [38] developed a complete 3D model with microstructure resolution, using a novel X-ray nano-computed tomography (CT) dual-scan superimposition technology, which captures the characteristics of the CBD (Fig. 8.6). This technology combines an imaged low-attenuation CBD and a high-attenuation $LiNi_{1/3}Mn_{1/3}Co_{1/3}O_2$ positive

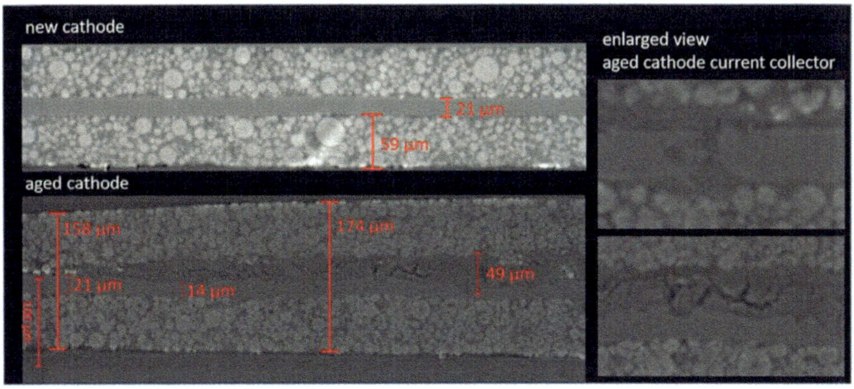

Fig. 8.5 X-ray CT image of the aged cathode. The cracked particles are visible as well as the oxidized current collector. Adapted from Ref. [36] Copyright (2019) Elsevier

electrode (NMC111). Through a 3D microstructure-resolved physics-based battery model with high fidelity, the complete 3D microstructure reconstruction, metrological characterization, and modeling to clarify the interaction between microstructure and electrochemical performance are fully revealed, thereby emphasizing the effect of global and local inhomogeneities on electrochemical state variables (such as SoL, charge transfer, electrolyte concentration, and electrochemical potential).

Different from second rechargeable batteries, such as lithium-ion battery and sodium-ion battery, proton exchange membrane fuel cells (PEMFCs) could directly convert the carried hydrogen at the anode and oxygen at the cathode into electricity [39, 40]. As a promising renewable energy conversion device, PEMFCs are attracting significant amounts of attention because of their sustainability and environmental friendliness [41]. High loading of Pt at the cathode is necessary to accelerate the sluggish kinetic of oxygen reduction reaction (ORR), severely impeding the commercialization of PEMFCs due to costliness and scarcity of Pt [42]. In the past decade, numerous efforts have been dedicated to the design and fabrication of electrocatalysts with high ORR activity evaluated by rotating disk electrode (RDE) technique, such as alloy, core-shell, and nano-frames. Unfortunately, the above electrocatalysts delivered the almost unchanged even worse performance in comparison with commercial Pt/C, in a membrane electrode assembly (MEA) involved transportation, water, and heat management [43]. X-ray CT has been commonly accepted as a powerful technique to reveal critical factors limiting the performance of PEMFCs due to its 3D tomography.

Hot pressure is a conventional technique to assembly the MEA consisted of gas diffusion layers (GDLs), catalysts layers (CLs), and membrane [44]. MEA has been 3D imaged to investigate the effect of hot-pressure temperature on the quality of MEA and its degradation behavior. It was confirmed that 130 °C is the optimized hot-pressure temperature, resulting in the improved bond with CLs [45].

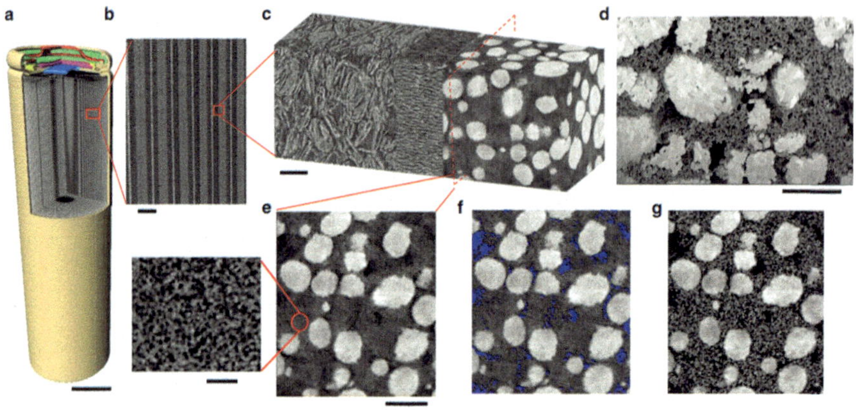

Fig. 8.6 Hierarchical structure of the LiB. **a** Volume rendering of the reconstructed cylindrical battery scanned by X-ray micro-CT (accelerating voltage180 kV, exposure time 1 s and voxel size 12.9 μm). The metal shell (brown), top button (red), venting disk (green), crimp plate (pink), seal insulator (yellow), and current collector (blue) are shown, with a corner cut displaying the internal structure of the battery; **b** magnified virtual slice to show the periodic layered structure of the cell; **c** X-ray nano-CT image showing (from left to right) the graphite anode, polyolefin separator, and NMC cathode; **d** SEM image showing the CBD morphology alongside the secondary NMC particles, wherein the crystallography of the primary particles is seen; **e** virtual slice of the reconstructed 3D electrode, showing a blurry phase comprising the CBD and micro-pores (white: active material particles; dark gray: macro-pores; light gray: blurry mixed phase of micro-pores and CBD). Inset image shows the higher resolution scan of the CBD phase; **f** macro-pores (blue) are highlighted and retained with the active particle in the dual-scan superimposition process; **g** the resultant full microstructure of the NMC cathode by X-ray CT. The scale bars in (**a**, **b**) represent 10 mm and 240 μm respectively, 10 μm in (**c**, **d**, **e**) and 2 μm in the inset of (**e**). Adapted from Ref. [38] Copyright (2020) Springer Nature

Higher-temperature such as 170 °C above the Nafion glass transition temperature, causes severe mechanism deformation, while lower temperature for hot pressure leads to incomplete contact between CLs, GDLs, and membrane [46].

The gas diffusion layer (GDL), consisted of carbon paper and microporous layer (MPL), provides a place for the transportation of electron, reactant, and production. The carbon paper is composed of fiber substrate and poly(tetrafluoroethylene) (PTFE). And the MPL is the mixture of carbon powder and PTFE and deposited on the carbon paper through the roll-press and slurry-coat method [47]. 3D rendering of the GDL indicated that the MPL could penetrate into the carbon paper through the slurry coat in comparison with the separated layers via roll-pression. The structural data such as average fiber diameter and pore distribution has been successfully revealed by 3D reconstruction. Also, the permeability, effective diffusivity and capillary radii also could be obtained from the reconstructed 3D model. It is further revealed that the pores ($\sim$15 μm) in the carbon paper is larger than the pores ($\sim$500 nm) in MPL, suggesting the different transportation mechanism for liquid and gas in carbon and MPL [48]. The gases such as oxygen may travel through the carbon paper to the reaction sites and the generated water mainly travel through the MPL. The water management of GDL can be reasonably controlled by the incorporation of PTFE with high hydrophobicity. Moosavi et al. confirmed that GDL with 10% PTFE greatly enhanced the water transportation without the effect on the air diffusivity through high-resolution computed tomography [49]. Besides, the addition of MPL and increased pores size of GDL both is helpful for the water permeability by diminishing the water accumulation.

The catalyst layer (CL) is composed of the per-fluorosulfonate acid ionomer as hydrogen ion conductivity, catalysts, and PTFE providing enough pores for the permeability of reactants. The visualization of CL has been successfully realized through X-ray CT, where metal-containing material is detected as a bright phase, in contrast well with the carbon-based materials. X-ray CT has demonstrated that the deposition method of CL on the MPL directly controls the micro- and macro-strain and contact resistance, which determine the transportation and charge transfer process. Pokhrel et al. used the three-dimensional multi-length scale X-ray CT to study the microstructure of CL at the agglomerate scale and revealed the main failure model of CL [50]. As shown in Fig. 8.7a, the mud cracks are observed in the BOL image possibly due to the existence of the stress and strain resulting from solvent evaporation. However, prevalent mud cracks are observed in the EOL image, where the CL exhibited an isolated island structure. The micro-length scale analysis shows a five-fold enhancement in crack size and about 60% decrease in thickness in the EOL CL in comparison with the BOL, suggesting the severe carbon corrosion during operation. The complementary nano-length scale also indicated a great reduction in porosity and effective diffusion. The phase-segregated 3D structure of CL at BOL (Fig. 8.7a) displayed a highly homogeneous distribution of Pt/C and ionomer across the whole CL, in contrast to the inhomogeneity in EOL at the EOL due to the carbon corrosion (Fig. 8.7b). In addition, the ionomer dominated structure has increased from 20 to 50% in volume fraction, highly inhibiting the gas transportation and leading to the striking degradation of PEMFCs. The large voids were also found in the uneven catalysts layer in the cross-sectional view of the reconstructed image of EOL MEAs, compared with that of the BOL image as shown in Fig. 8.7c, d. The formed cracks led to the formation of gas crossover in a membrane, which grow gradually from a circular pinhole to elongated shapes in multi-directions.

Shawn et al. used the nano-CT with a resolution of 50 nm to investigated the size and form of the catalyst particle agglomerates and pore spaces [51]. It was confirmed that the reconstructed 3D X-ray CT image showed great agreement with MIP. More importantly, the pore distribution determined by the nano-CT contributed to the evaluation of water transportation and Knudsen diffusion effects during PEMFCs operation. And the agglomerate size distribution supports to evaluate the dispersion of catalyst ink and mass transportation. Siddharth et al. found that the kinetics and mass transportation were highly dependent on the Nafion loading through polarization curves. The 50 and 60 wt% Nafion cathodes showed high kinetics in comparison with 35 wt% electrode [52]. However, 60% Nafion resulted in a severe electrode flooding and further strikingly reduced performance at high current region due to limited mass transportation. Nano-CT was

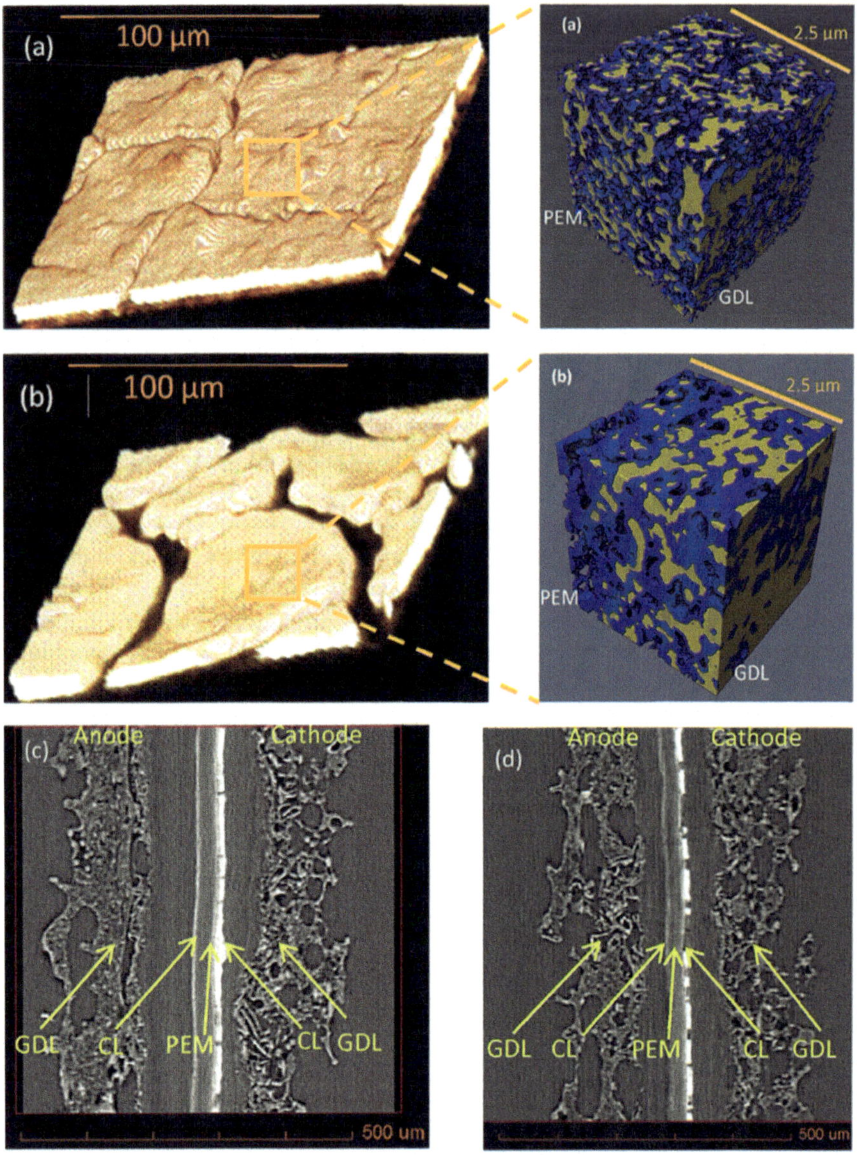

Fig. 8.7 **a**3D rendering of the virtually extracted CLs of MEA (BOL) and the phase-segregated 3D structure. **b** The 3D rendering of the virtually extracted CLs of MEA (EOL) and the phase-segregated 3D structure. 2D virtual cross-section of the BOL-MEA (**c**) and EOL-MEA (**d**). Adapted from Ref. [50] Copyright (2016) Elsevier

then conducted to study the effect of ionomer content on the size and distribution of hierarchical pore and resolve the corresponding electrode tomography's impact on PGM-free cathode performance [53]. As shown in Fig. 8.8, the 3D reconstructed image of 50 wt% Nafion cathode electrode displayed a high uniformity in the distribution of Nafion within nanoparticles. Large dense agglomerates of Nafion were formed without the infiltration into nanoparticles for the 35 wt% Nafion cathode electrodes, resulting in inferior kinetics due to the poor conductivity of porton. In contrast, a great thick layer of Nafion was found to be covered on the catalysts for the 60 wt% Nafion electrodes, yielding the highly hydrophilic pores and limited oxygen diffusion. Recently, the effect of the interplay between PGM-free catalyst's primary size and ionomer on fuel cell performances was unraveled by the nano-CT. It was found that the size of primary nanoparticles should be large enough to permit uniform ionomer thin films throughout the surrounding pores, while oversize particles may impact the intraparticle diffusion of reactants.

In summary, CT technology has been widely used in the field of structural characterization and aging analysis for energy storage and conversion devices. CT technology can perform material characterization in three-dimensional space, ensuring the true internal structure of the energy storage and conversion devices in a non-destructive manner. The devices do not need to be disassembled, and the three-dimensional perspective of X-ray CT technology can effectively analyze the cause for the inactivation of devices. Based on the results of 3D CT testing, it serves as an important basis for process optimization and control. This kind of CT analysis mode is attracting widespread attention in the energy field.

8.5 The Application of Industrial X-ray Microscopy in Geology

The X-ray CT technology can clearly reflect the evolution of the microstructure of the detected particles according to the change in the gray level of the image. X-ray CT technology has been widely used in the field of geoscience [54, 55]. In the earlier works, X-ray CT technology mainly focused on the imaging of flooding, but lacked the quantifying information about the CO_2-enhanced oil recovery imaging for mobility control and three-phase flow. Recently, X-ray CT technology is used to reconstruct and characterize quantitatively the static nano- and micro-structure of the rock, sandstone, concrete, and so on, including pore structure, crack structure, and aggregate structure [56–58]. In addition, the evolution of macroscopic cracks and pore structure during dynamic loading is also studied in detail [59–62].

Skarzynski et al. [63] conducted an experimental study on the three-dimensional fracture evolution of concrete, and obtained full three-dimensional high-resolution damage maps of the microstructure of concrete under different loading cycles through X-ray CT technology. The irregular thin cracks with the width of 0.04 mm, mainly occurred in the interface transition zones between the cement matrix and aggregate particles and propagated to the cement matrix after 10,000 cycles. After 30,000 cycles, the internal cracks further propagated, and the number of detected

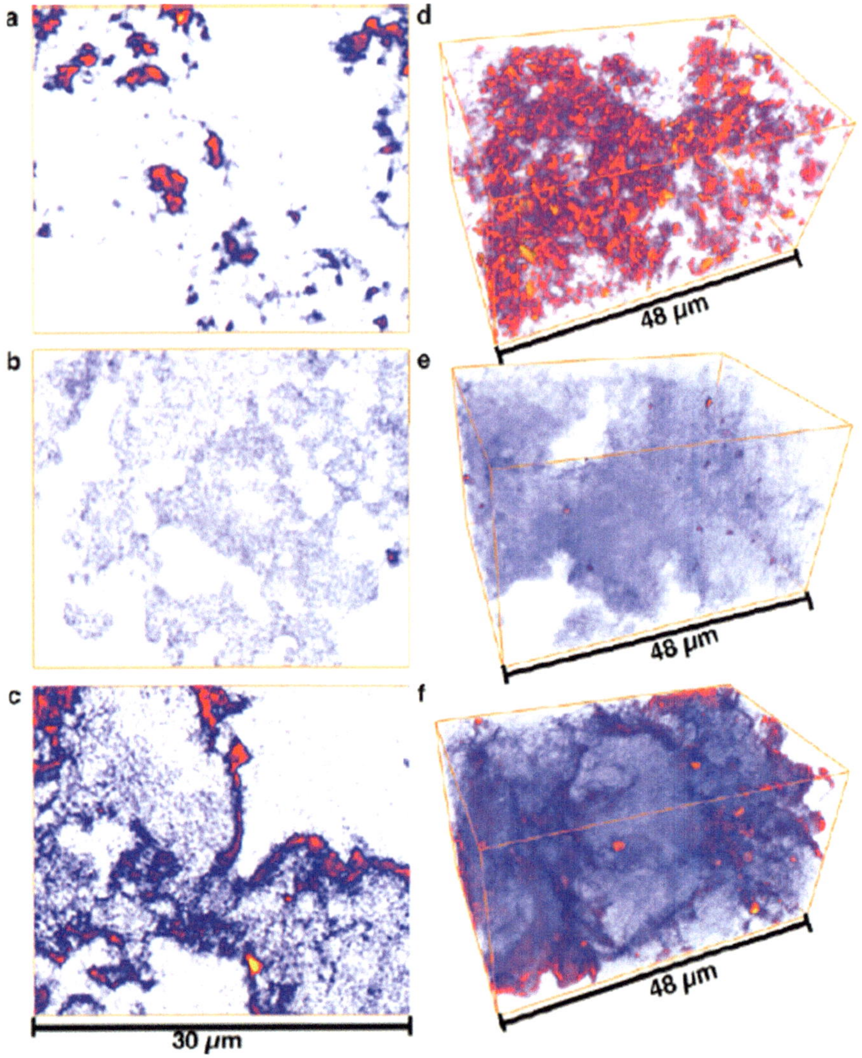

Fig. 8.8 Segmented slices and 3D rendering Nafion domain in absorption contrast imaging mode. In-plane slice of the PGM-free catalyst with **a** 35, **b** 50, and **c** 60 wt% Nafion loading and 3D volume renderings of Nafion distribution for an electrode with Nafion loading of **d** 35, **e** 50, and **f** 60 wt% Nafion loadings. Intensity of color shows the local density of Nafion plus some minor Fe agglomerations. Adapted from Ref. [52] Copyright (2016) American Chemical Society

broken aggregate particles is up to 7. The greatest crack width in the entire volume achieved as wide as 0.16 mm. After 60,000 cycles, the vertical macro-cracks along the lateral edge further developed and the maximum width of cracks and the number of broken aggregate particles is 0.42 mm and 13, respectively. After 70,000 cycles, macro-crack intersected the concrete specimen. And on the second lateral surface, the upper part of the area along the horizontal edge of a macro-cracks, a tilt macro-crack appeared and the maximum crack width is 0.72 mm. The number of broken aggregate particles increased up to 28. The quantitative analysis of the largest crack volume of pores and cracks shows that in the final loading stage between 60,000 and 70,000 cycles as shown in Fig. 8.9, the volume of crack increases by 2.26% significantly, accounting for 40% of the total volume of the crack growth. In contrast, closed pores volume decreased from 1.34 to 1.24%, while opening pores volume increased from 4.39 to 6.75%. The width of cracks is not uniformly distributed but increased by 0.1–0.2 mm, even increased by 0.3 mm in some special specimen areas.

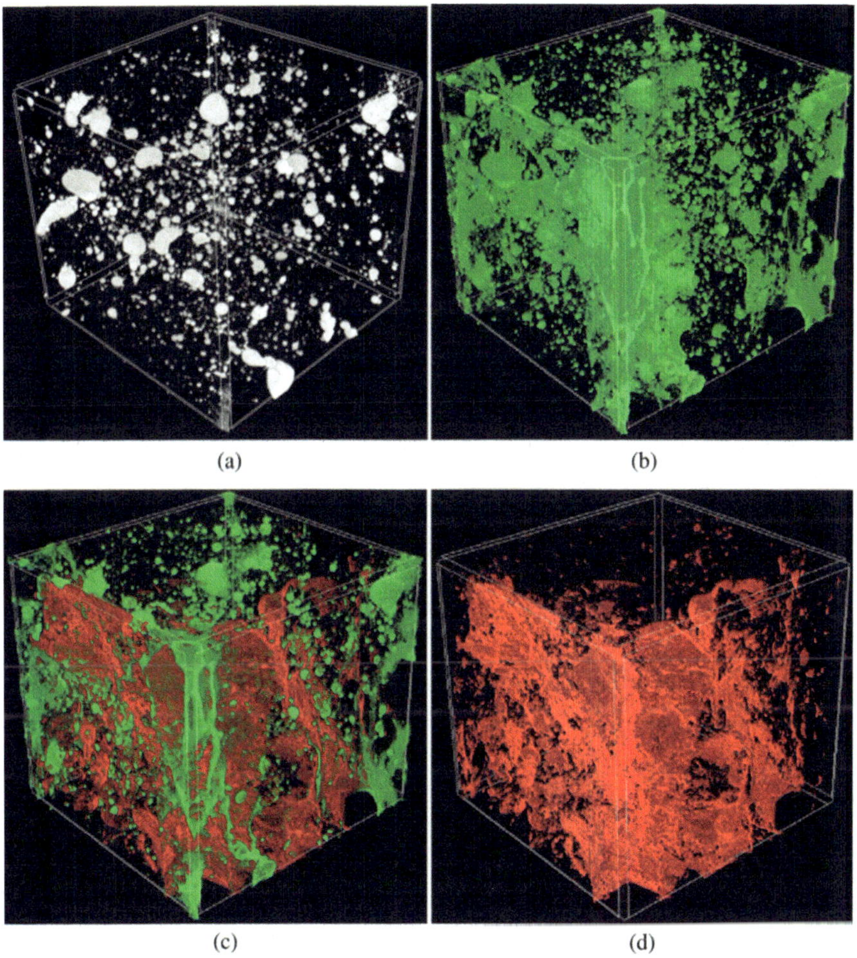

Fig. 8.9 Distribution of pores and cracks in specimen: **a** non-cracked concrete specimen (white color), **b** after N4 = 60,000 cycles (green color), **c** after N5 = 70,000 loading cycles (green and red color) and **d** between N4 = 60,000 and N5 = 70,000 cycles (red color). Adapted from Ref. [63] Copyright (2019) Elsevier

Yang et al. [61] used X-ray CT technology and rock triaxial test system to study the strength and deformation characteristics of composite rock materials composed of hard and weak rock materials under different confining pressures. The three-dimensional volume rendering is in good agreement with the actual surface crack images of the transversely isotropic composite material specimens. The distribution and damage of internal cracks can be clearly observed from the vertical and horizontal cross-sectional images drawn from the three-dimensional volume. Xu et al. [64] used miniature X-ray CT and laboratory experimental devices to study the bearing mechanism of model piles embedded in synthetic soft sandstone rock under compression loads. For each pile load test, five sets of high-resolution image data under different pile head displacements were obtained, and qualitatively explained the sliding and shearing mechanism of the rough body between the pile-rock interface and the end bearing mechanism of the pile. Fan et al. [65] studied the influence of thermal effects on the volume porosity of granite microstructures through X-ray CT technology and three-dimensional (3D) image reconstruction. It was found that obvious new cracks formed at 500 °C, while the formed cracks tended to expand and further aggregate when the temperature was higher than 600 °C. The anisotropy coefficient of granite was demonstrated to be stable when it reached 1000 °C, therefore it can be regarded as an isotropic material. Tim et al. [66] overcome the limitation of long scan time of X-ray microcomputer tomography in the dynamic measurement process and obtained the computer tomography within minutes. The developed novel X-ray CT technique enabled researchers to study the dynamics of porous fracture related to ice crystallization under in-situ freeze-thaw cycles. It is of great significance to understand the freeze-thaw cycle brought about by climate change for the exploitation of the soil, architecture, and archaeological sites.

In addition, in the field of geoscience, the micropore structure of reservoir, such as oil, gas, and CO_2, is related to geometry, size, distribution, and interrelation of pore and throat. Due to the existence of ultra-fine pore throats in tight oil reservoirs, higher requirements for advanced characterization methods are highly urgent. X-ray CT technology can quickly provide comprehensive and non-destructive imaging at a larger-scale, and can reveal the microscopic pore characteristics of digital cores. Song et al. [67] used CT scanning and 3D reconstruction technology to characterize the pore throat structure of tight oil layers, and combined with SEM technology to show multiple pore types in tight oil layers. The tight rock samples were taken from the Chang 7 formation in Xunyi county of Ordos Basin, China. The distribution of different types of connected pores is shown in 3D reconstructed image. Different colors represent different connected types of pores, and the same color represents the same connected pores as shown in Fig. 8.10 where skeleton structure is not presented. The pore connectivity analysis showed heterogeneity in scale, type, and distribution of these pores where the isolated pores originating from the core sample is large, while the proportion of visible pores is small. In addition, the types of pores are dominated by intergranular residual and feldspar pores show tubular shape and

Fig. 8.10 3D reconstruction image of different types of connected pores in tight core at micron scale. Adapted from Ref. [67] Copyright (2018) Springer Nature

tubers. Large pore throat structure and good connectivity pores provide a good channel for tight oil migration. In the quantitative analysis of micron porosity, the main storage space of tight oil, the porosity is more than 93%.

Ben Callow et al. [68] studied basalt rock samples with CO_2 storage capacity utilizing X-ray CT technology. The 3D distribution maps of various pores of samples of basalt reservoirs were obtained by performing 3D reconstruction imaging characterization of core samples. Combined with the fluid mechanic's experiment of the same sample, it was observed that basalt had greater absolute vertical permeability and showed that the CO_2 storage potential of basalt reservoirs was 0.33 Gt CO_2, while the CO_2 storage potential along the active rift zone in Iceland is estimated to be 6100 Gt CO_2.

In summary, CT technology has been widely used in the field of geoscience because of the advantages of three-dimensional, non-destructive, and high-resolution. It can obtain clear two-dimensional and three-dimensional images in multiple resolutions and different views. CT technology can quantitatively describe the size, distribution, and connectivity of pores and cracks, Therefore, it is of great significance to the microstructure dynamic change of rock, sandstone, concrete, and the prediction of reservoir content.

8.6 The Application of Industrial X-ray Microscopy in Biology

Advances in biological research methods often promote the development of corresponding fields [69–71]. Traditional biological research methods are mainly based on the methods of tissue dissection and sectioning under the optical microscope or electron microscope. These methods are accompanied by the destruction of biological tissues and cannot truly restore the internal structure of organisms. In recent years, with the development of X-ray computed tomography, it has become possible to obtain 3D information of the inside and outside of biological samples without damage [72–74]. According to different imaging mechanisms, biological samples (such as muscle fat) have a small absorption of X-rays, and the contrast obtained on the detector is very low, resulting in a decrease in spatial resolution. The X-ray phase contrast presents great promise for the image of weakly absorbing substances with smaller absorption coefficients such as muscles [75]. Comparing with the resolution of traditional X-ray absorption imaging technology w, phase-contrast imaging technology showed higher spatial resolution up to 0.35 μm, making it possible to observe the finer structure of living things [76].

Phase-contrast micro-CT technology has made great progress to study the three-dimensional morphology of different structures of living organisms [77]. For example, the use of phase-contrast micro-CT technology can non-destructively obtain 3D information such as the brain muscles and bones of insects [78], which is of great significance to the study of the systematic evolution of organisms. Li et al. [79] used X-ray phase-contrast micron CT imaging technology to compare the relationship between the upper jaw, upper jaw muscle, and headshell of different representative groups of paederus, pointing that the upper jaw was the insect feeding process (Fig. 8.11). The structure used to hold or pierce food was mainly controlled by the adductor maxillary muscles, which was connected to the headshell. The use of X-ray micro-CT imaging technology to collect three-dimensional information of biological endoskeleton and muscle increases the reliability of the study of biological system evolution based on morphological information and is gradually applied to the study of various group system evolution.

Many internal structures of small organisms are usually only tens of micrometers, leading to the internal three-dimensional morphological structures cannot be distinguished by the traditional methods [80]. With the development of nano-resolved X-ray Zernike phase-contrast CT technology, the three-dimensional information collection of samples has advanced from the order of micrometer resolution to the order of nanometer resolution, and the three-dimensional structure reconstruction of biological samples has also reached the nanometer level. Zhang et al. [81] used the principle of X-ray Zernike phase-contrast imaging, by adding a phase ring to the focal plane of the zone plate to convert the phase change of X-rays after passing through the sample into light intensity changes, thereby obtaining a nano-resolved 3D high-contrast structure of the female spermatheca cochleate and subapex of the male flagellum in the species Aleochara verna (Fig. 8.12). And

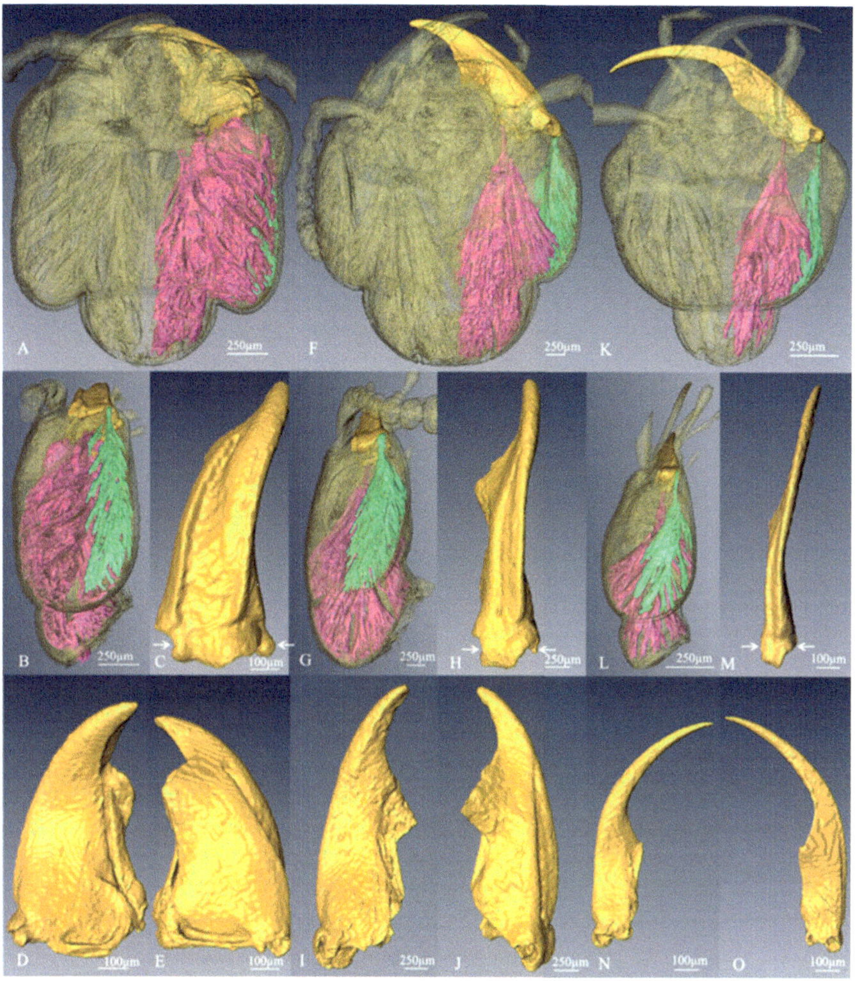

Fig. 8.11 Three-dimensional reconstructions of Noddia sp. (**a–e**), Creophilus maxillosus (**f–j**), and Hesperosoma sp. (**k–o**). a, b, c, d, k, l Head, transparent view; mandible and mandible muscles, shade view; a, f, k dorsal view; b, g, l lateral view. c–e, h–j, m–o. Adapted from Ref. [79] Copyright (2011) Springer Nature

according to the 3D results, the sperm transmission mechanism of P. praecox is speculated, which provides new clues for the evolution of insect genitalia.

X-ray CT imaging technology has high-throughput screening capabilities, which can realize the statistical study of structural differences between normal and mutant genes. Larabcll et al. used a full-field X-ray microscope to image the yeast cells, which are widely used to study the genes [82]. In order to achieve imaging, the yeast was quickly frozen after hydration treatment and then placed on the hollow capillary sample holder. The image shooting conditions are 517 eV, and the

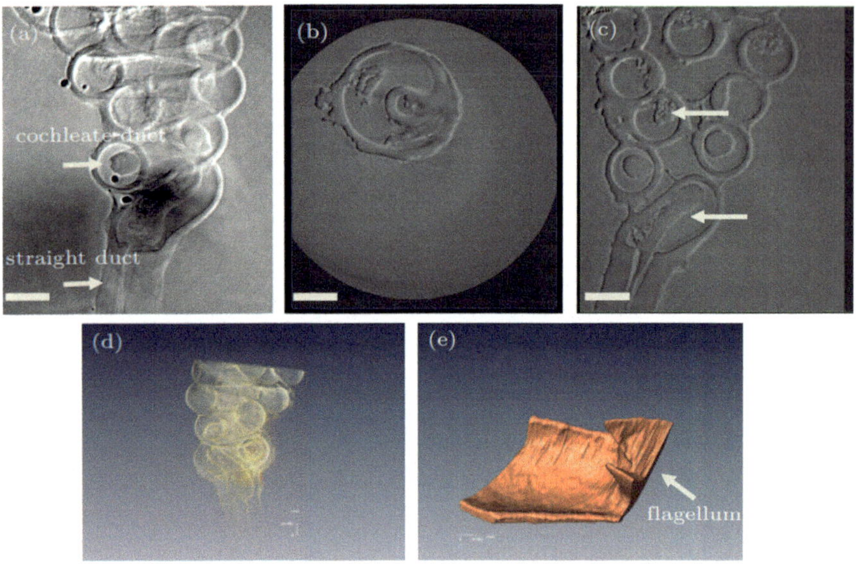

Fig. 8.12 (color online) The 3D structure of the female spermatheca cochleate and subapex of the male flagellum in the species Aleochara verna. **a** Projective image of the cochleate duct. The transverse slice (**b**), and coronal slice (**c**) of the cochleate duct, **d** 3D reconstruction of the cochleate duct, **e** 3D reconstruction of the subapex of the flagellum. The scale bar is 10 μm. Adapted from Ref. [81] Copyright (2013) IOP Publishing Ltd

rotation angle range and step size are $\pm$ 90° and 2°, respectively. Figure 8.13a is a two-dimensional image of a pair of yeast cells that have not yet been divided. Figure 8.13b is a 3D image of all the two-dimensional images reconstructed by yeast cells. According to the size of the attenuation coefficient, the cell structure is distinguished by different colors. The same color corresponds to the same organelle type.

In summary, X-ray microscopy has its unique advantages over optical fluorescence and electron microscopy when characterizing biological cell tissues: (1) The sample preparation is simple (no fluorescent labeling, no sectioning processing required), and it can be frozen in water to image cells or biological structures in their native state. (2) It can achieve non-destructive two-dimensional and three-dimensional imaging of the entire cell or biological structure, with good image contrast. (3) The imaging time is short. (4) The imaging resolution is high where the two-dimensional and three-dimensional imaging resolution is about 12–15 nm and 50 nm, respectively [83]. With the development of biomorphology and advanced chemistry, new requirements have been put forward for the development of X-ray imaging technology. Time-resolved, high-resolution, high-contrast dynamic X-ray imaging has become a hot spot for scientists. The realization of related technologies will have a huge impact on the development of disciplines such as biomorphology and reproductive chemistry.

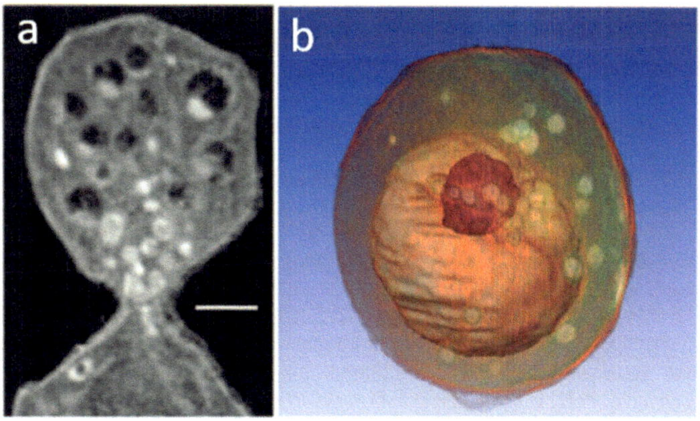

Fig. 8.13 Tomography of whole yeast cells using TXM. **a** Computer-generated section through a tomographic reconstruction of the raw data shown in (**a**). The scale bar is 0.5 mm. **b** An edge enhancement gradient algorithm was used to volume segment the three-dimensional data used to produce this volume-rendered image, showing the nucleus (purple), vacuole (pink) and lipid droplets (white). Adapted from Ref. [82] Copyright (2005) Elsevier

References

1. Mu, L., Yuan, Q., Tian, C., Wei, C., Zhang, K., Liu, J., Pianetta, P., Doeff, M.M., Liu, Y., Lin, F.: Propagation topography of redox phase transformations in heterogeneous layered oxide cathode materials. Nat. Commun. **9**(1), 2810 (2018)
2. Swee, J.K.Y., Sheridan, C., de Bruin, E., Downward, J., Lassailly, F., Pizarro, L.: Medical Imaging 2014: Computer-Aided Diagnosis, vol. 9035 (Eds: Aylward, S., Hadjiiski, L.M.) (2014)
3. Kozaki, T., Suzuki, S., Kozai, N., Sato, S., Ohashi, H.: Observation of microstructures of compacted bentonite by microfocus X-ray computerized tomography (micro-CT). J. Nucl. Sci. Technol. **38**(8), 697–699 (2001)
4. Sasov, A.: Medical Imaging 2001: Physics of Medical Imaging, vol. 4320, pp. 711–718. (Eds: Antonuk, M.J. Yaffe M.J.) (2001)
5. Sasov, A.: Developments in X-Ray Tomography Iii, vol. 4503, pp. 282–290. (Ed: Bonse, U.) (2002)
6. Sasov, A., Pauwels, B., Bruyndonckx, P.: Developments in X-Ray Tomography Vii, vol. 7804 (Ed: Stock, S.R.) (2010)
7. Nachtrab, F., Firsching, M., Speier, C., Uhlmann, N., Takman, P., Tuohimaa, T., Heinzl, C., Kastner, J., Larsson, D.H., Holmberg, A., Berti, G., Krumm, M., Sauerwein, C.: Developments in X-Ray Tomography Ix, Vol. 9212 (Ed: Stock, S.R.) (2014)
8. Bailey, J.J., Heenan, T.M.M., Finegan, D.P., Lu, X., Daemi, S.R., Iacoviello, F., Backeberg, N.R., Taiwo, O.O., Brett, D.J.L., Atkinson, A., Shearing, P.R.: Laser-preparation of geometrically optimised samples for X-ray nano-CT. J. Microsc. **267**(3), 384–396 (2017)
9. Nesterets, Y., Gureyev, T., Stevenson, A., Pogany, A., Wilkins, S., Kincaid, R., Ye, H., Vogelsang, L., Lipson, E., Coman, L., Fourmaux, S., Kieffer, J.-C., Krol, A.: Medical Imaging 2008: Physics Of Medical Imaging, Pts 1–3, vol. 6913 (Eds: J. Hsieh, E. Samei) (2008)

10. Yue, L., Dai, Z., Chen, X., Liu, C., Hu, Z., Song, B., Zheng, X.: Development of a novel FePt-based multifunctional ferroptosis agent for high-efficiency anticancer therapy. Nanoscale **10**(37), 17858–17864 (2018)
11. Hu, Z., Wang, S., Dai, Z., Zhang, H., Zheng, X.: A novel theranostic nano-platform (PB@FePt-HA-g-PEG) for tumor chemodynamic-photothermal co-therapy and triple-modal imaging (MR/CT/PI) diagnosis. J Mater Chem B. **8**(24), 5351–5360 (2020)
12. An, J., Yang, X.-Q., Cheng, K., Song, X.-L., Zhang, L., Li, C., Zhang, X.-S., Xuan, Y., Song, Y.-Y., Fang, B.-Y., Hou, X.-L., Zhao, Y.-D., Liu, B.: In vivo computed tomography/photoacoustic imaging and NIR-triggered chemo–photothermal combined therapy based on a gold nanostar-, mesoporous silica-, and thermosensitive liposome-composited nanoprobe. ACS Appl. Mater. Interf. **9**(48), 41748–41759 (2017)
13. Xuan, Y., Yang, X.Q., Song, Z.Y., Zhang, R.Y., Zhao, D.H., Hou, X.L., Song, X.L., Liu, B., Zhao, Y.D., Chen, W.: High-security multifunctional nano-bismuth-sphere-cluster prepared from oral gastric drug for CT/PA dual-mode imaging and chemo-photothermal combined therapy in vivo. Adv. Funct. Mater. **29**(18) (2019)
14. Hagen, C.K., Endrizzi, M., Towns, R., Meganck, J.A., Olivo, A.: A preliminary investigation into the use of edge illumination X-ray phase contrast micro-CT for preclinical imaging. Mol. Imaging Biol. **22**(3), 539–548 (2020)
15. Hesse, B., Varga, P., Langer, M., Pacureanu, A., Schrof, S., Maennicke, N., Suhonen, H., Maurer, P., Cloetens, P., Peyrin, F., Raum, K.: Canalicular network morphology is the major determinant of the spatial distribution of mass density in human bone tissue: evidence by means of synchrotron radiation phase-contrast nano-CT. J. Bone Miner. Res. **30**(2), 346–356 (2015)
16. Peyrin, F., Dong, P., Pacureanu, A., Langer, M.: Micro- and nano-CT for the study of bone ultrastructure. Curr. Osteoporos. Rep. **12**(4), 465–474 (2014)
17. Ni, Q., Dixon, H., Gutierrez, G., Bi, L., Qin, Y.-X.: Characterization of synthesized NANO-encapsulated drug for bone loss on hind limb suspension rat model by NMR and micro-CT. Adv. Biosci. Bioeng. (New York, NY) **1**(1), 1–7 (2013)
18. Izutsu, K.-I., Yonemochi, E., Yomota, C., Goda, Y., Okuda, H.: Studying the morphology of lyophilized protein solids using X-ray micro-CT: effect of post-freeze annealing and controlled nucleation. AAPS PharmSciTech **15**(5), 1181–1188 (2014)
19. Dudak, J., Zemlicka, J., Karch, J., Patzelt, M., Mrzilkova, J., Zach, P., Hermanova, Z., Kvacek, J., Krejci, F.: High-contrast X-ray micro-radiography and micro-CT of ex-vivo soft tissue murine organs utilizing ethanol fixation and large area photon-counting detector. Sci. Rep. **6**, 30385 (2016)
20. Gonzalez, J.F., Antartis, D.A., Chasiotis, I., Dillon, S.J., Lambros, J.: In situ X-ray micro-CT characterization of chemo-mechanical relaxations during Sn lithiation. J. Power Sources **381**, 181–189 (2018)
21. Sun, Y., Zhao, Y., Zhang, H., Zhang, C.: Visualization of gas diffusion-sorption in coal: a study based on synchrotron radiation nano-CT. Geofluids (2020)
22. Tichter, T., Andrae, D., Schneider, J., Gebhard, M., Hilger, A., Manke, I., Roth, C.: Real-space simulation of cyclic voltammetry in carbon felt electrodes by combining micro X-ray CT data, digital simulation and convolutive modeling. Electrochimica Acta 353 (2020)
23. Cordes, N.L., Havrilla, G.J., Usov, I.O., Obrey, K.A., Patterson, B.M.: Non-destructive elemental quantification of polymer-embedded thin films using laboratory based X-ray techniques. Spectrochim. Acta Part B-Atomic Spectro. **101**, 320–329 (2014)
24. Zhao, C., Wada, T., De Andrade, V., Gursoy, D., Kato, H., Chen-Wiegart, Y.-C.K.: Imaging of 3D morphological evolution of nanoporous silicon anode in lithium ion battery by X-ray nano-tomography. Nano Energy **52**, 381–390 (2018)
25. Jiang, Z., Li, J., Yang, Y., Mu, L., Wei, C., Yu, X., Pianetta, P., Zhao, K., Cloetens, P., Lin, F., Liu, Y.: Machine-learning-revealed statistics of the particle-carbon/binder detachment in lithium-ion battery cathodes. Nat. Commun. **11**(1) (2020)

26. Liu, Q., Gao, J., Cao, C., Yin, G., Jiang, Z., Ge, M., Xiao, X., Lee, W.-K., Wang, J.: Insights into enhanced sodium ion storage mechanism in Fe_3S_4: the coupling of surface chemistry, microstructural regulation and 3D electronic transport. Nano Energy **62**, 384–392 (2019)
27. Wang, J., Yang, J., Tang, Y., Liu, J., Zhang, Y., Liang, G., Gauthier, M., Chen-Wiegart, Y.-c. K., Banis, M.N., Li, X., Li, R., Wang, J., Sham, T.K., Sun, X.: Size-dependent surface phase change of lithium iron phosphate during carbon coating. Nat. Commun. **5** (2014)
28. Wang, J., Chen-Wiegart, Y.-c.K., Wang, J.: In situ three-dimensional synchrotron X-Ray nanotomography of the (De) lithiation processes in tin anodes**. Angewandte Chemie-International Edition **53**(17), 4460–4464 (2014)
29. Babu, S.K., Chung, H.T., Zelenay, P., Litster, S.: Resolving electrode morphology's impact on platinum group metal-free cathode performance using nano-CT of 3D hierarchical pore and ionomer distribution. ACS Appl. Mater. Interfaces **8**(48), 32764–32777 (2016)
30. Wang, J., Chen-Wiegart, Y.-c.K., Eng, C., Shen, Q., Wang, J.: Visualization of anisotropic-isotropic phase transformation dynamics in battery electrode particles. Nat. Commun. **7** (2016)
31. Higashi, K., Takao, S., Samjeske, G., Matsui, H., Tada, M., Uruga, T., Iwasawa, Y.: Visualization and understanding of the degradation behaviors of a PEFC Pt/C cathode electrocatalyst using a multi-analysis system combining time-resolved quick XAFS, three-dimensional XAFS-CT, and same-view nano-XAFS/STEM-EDS techniques. Phys. Chem. Chem. Phys. **22**(34), 18919–18931 (2020)
32. Mueller, V., Bernhard, R., Wegener, J., Pfeiffer, J., Roessler, S., Scurtu, R.-G., Memm, M., Danzer, M.A., Wohlfahrt-Mehrens, M.: Evaluation of scalable porous si-rich si/c composites with low volume expansion in coin cells to prismatic cell formats. Energy Technol. **8**(7) (2020)
33. Jiao, L.A., Li, X., Ren, L.L., Kong, L.Y., Hong, Y.L., Li, Z.W., Huang, X.B., Tao, X.F.: 3D structural properties study on compact LiFePO(4)s based on X-ray computed tomography technique. Powder Technol. **281**, 1–6 (2015)
34. Sun, N., Liu, Q., Cao, Y., Lou, S., Ge, M., Xiao, X., Lee, W.-K., Gao, Y., Yin, G., Wang, J., Sun, X.: Anisotropically electrochemical-mechanical evolution in solid-state batteries and interfacial tailored strategy. Angewandte Chemie-International Edition **58**(51), 18647–18653 (2019)
35. Zhang, F., Lou, S., Li, S., Yu, Z., Liu, Q., Dai, A., Cao, C., Toney, M.F., Ge, M., Xiao, X., Lee, W.-K., Yao, Y., Deng, J., Liu, T., Tang, Y., Yin, G., Lu, J., Su, D., Wang, J.: Surface regulation enables high stability of single-crystal lithium-ion cathodes at high voltage. Nat. Commun. **11**(1), (2020)
36. Rahe, C., Kelly, S.T., Rad, M.N., Sauer, D.U., Mayer, J., Figgemeier, E.: Nanoscale X-ray imaging of ageing in automotive lithium ion battery cells. J. Power Sources 433 (2019)
37. Berckmans, G., Lysander, D.S., Algirdas, K., Akos, K., Mario, M., Michael, K., Peter, A., Jelle, S., Margret, W.-M., Andreas, P., Joris, J., Joeri, V.M., Noshin, O., 1 a.N.O.: Electrical characterization and micro x-ray computed tomography analysis of next-generation silicon alloy cells. World Electric Vehicle J. **9**(43) (2018)
38. Lu, X., Bertei, A., Finegan, D.P., Tan, C., Daemi, S.R., Weaving, J.S., O'Regan, K.B., Heenan, T.M.M., Hinds, G., Kendrick, E., Brett, D.J.L., Shearing, P.R.: 3D microstructure design of lithium-ion battery electrodes assisted by X-ray nano-computed tomography and modelling. Nature Commun. **11**(1) (2020)
39. Kong, F., Liu, S., Li, J., Du, L., Banis, M.N., Zhang, L., Chen, G., Doyle-Davis, K., Liang, J., Wang, S., Zhao, F., Li, R., Du, C., Yin, G., Zhao, Z., Sun, X.: Trimetallic Pt-Pd-Ni octahedral nanocages with subnanometer thick-wall towards high oxygen reduction reaction. Nano Energy 64 (2019)
40. Strasser, P., Gliech, M., Kuehl, S., Moeller, T.: Electrochemical processes on solid shaped nanoparticles with defined facets. Chem. Soc. Rev. **47**(3), 715–735 (2018)

41. Kong, F., Ren, Z., Banis, M.N., Du, L., Zhou, X., Chen, G., Zhang, L., Li, J., Wang, S., Li, M., Doyle-Davis, K., Ma, Y., Li, R., Young, A.P., Yang, L., Markiewicz, M., Tong, Y., Yin, G., Du, C., Luo, J., Sun, X.: Active and stable Pt-Ni alloy octahedra catalyst for oxygen reduction via near-surface atomical engineering. ACS Catal. **10**(7), 4205–4214 (2020)
42. He, Y., Liu, S., Priest, C., Shi, Q., Wu, G.: Atomically dispersed metal–nitrogen–carbon catalysts for fuel cells: advances in catalyst design, electrode performance, and durability improvement. Chem. Soc. Rev. (2020). https://doi.org/10.1039/c9cs00903e
43. Kong, F., Banis, M.N., Du, L., Zhang, L., Zhang, L., Li, J., Doyle-Davis, K., Liang, J., Liu, Q., Yang, X., Li, R., Du, C., Yin, G., Sun, X.: Highly stable one-dimensional Pt nanowires with modulated structural disorder towards the oxygen reduction reaction. J. Mater. Chem. A. **7**(43), 24830–24836 (2019)
44. Rabis, A., Rodriguez, P., Schmidt, T.J.: Electrocatalysis for polymer electrolyte fuel cells: recent achievements and future challenges. ACS Catal. **2**(5), 864–890 (2012)
45. Meyer, Q., Mansor, N., Iacoviello, F., Cullen, P.L., Jervis, R., Finegan, D., Tan, C., Bailey, J., Shearing, P.R., Brett, D.J.L.: Investigation of hot pressed polymer electrolyte fuel cell assemblies via X-ray computed tomography. Electrochim. Acta **242**, 125–136 (2017)
46. Meyer, Q., Hack, J., Mansor, N., Iacoviello, F., Bailey, J.J., Shearing, P.R., Brett, D.J.L.: Multi-scale imaging of polymer electrolyte fuel cells using X-ray micro- and nano-computed tomography, transmission electron microscopy and helium-ion microscopy. Fuel Cells. **19**(1), 35–42 (2019)
47. Rama, P., Liu, Y., Chen, R., Ostadi, H., Jiang, K., Zhang, X., Fisher, R., Jeschke, M.: An X-ray tomography based lattice boltzmann simulation study on gas diffusion layers of polymer electrolyte fuel cells. J. Fuel Cell Sci. Technol. **7**(3) (2010)
48. Terao, T., Inoue, G., Kawase, M., Kubo, N., Yamaguchi, M., Yokoyama, K., Tokunaga, T., Shinohara, K., Hara, Y., Hara, T.: Development of novel three-dimensional reconstruction method for porous media for polymer electrolyte fuel cells using focused ion beam-scanning electron microscope tomography. J. Power Sources **347**, 108–113 (2017)
49. Moosavi, S.M., Niffeler, M., Gostick, J., Haussener, S.: Transport characteristics of saturated gas diffusion layers treated with hydrophobic coatings. Chem. Eng. Sci. **176**, 503–514 (2018)
50. Pokhrel, A., El Hannach, M., Orfino, F.P., Dutta, M., Kjeang, E.: Failure analysis of fuel cell electrodes using three-dimensional multi-length scale X-ray computed tomography. J. Power Sources **329**, 330–338 (2016)
51. Litster, S., Epting, W.K., Wargo, E.A., Kalidindi, S.R., Kumbur, E.C.: Morphological analyses of polymer electrolyte fuel cell electrodes with nano-scale computed tomography imaging. Fuel Cells. (2013). https://doi.org/10.1002/fuce.201300008
52. Komini Babu, S., Chung, H.T., Zelenay, P., Litster, S.: Resolving electrode morphology's impact on platinum group metal-free cathode performance using nano-CT of 3D hierarchical pore and ionomer distribution. ACS Appl. Mater. Interfaces **8**(48), 32764–32777 (2016)
53. Uddin, A., Dunsmore, L., Zhang, H., Hu, L., Wu, G., Litster, S.: High power density platinum group metal-free cathodes for polymer electrolyte fuel cells. ACS Appl. Mater. Interfaces **12** (2), 2216–2224 (2020)
54. Brisard, S., Chae, R.S., Bihannic, I., Michot, L., Guttmann, P., Thieme, J., Schneider, G., Monteiro, P.J.M., Levitz, P.: Morphological quantification of hierarchical geomaterials by X-ray nano-CT bridges the gap from nano to micro length scales. Am. Miner. **97**(2–3), 480–483 (2012)
55. Elkhoury, J.E., Shankar, R., Ramakrishnan, T.S.: Resolution and limitations of X-ray micro-CT with applications to sandstones and limestones. Transp. Porous Media **129**(1), 413–425 (2019)
56. Wang, J.P., Andò, E., Charrier, P., Salager, S., Lambert, P., François, B.: Micro-scale investigation of unsaturated sand in mini-triaxial shearing using X-ray CT. Géotechnique Letters **9**(4), 269–277 (2019)
57. Pratesi, G., Caporali, S., Loglio, F., Giuli, G., Dzikova, L., Skala, R.: Quantitative study of porosity and pore features in moldavites by means of X-ray micro-CT. Materials **7**(4), 3319–3336 (2014)

58. Moriya, J., Aizawa, M., Yoshimura, H.: 10th International Conference on X-Ray Microscopy, vol. 1365, pp. 407–410 (Eds: McNulty, I., Eyberger, C., Lai, B.) (2011)
59. Bugani, S., Camaiti, M., Morselli, L., Van de Casteele, E., Janssens, K.: Investigating morphological changes in treated vs. untreated stone building materials by x-ray micro-CT. Anal. Bioanal. Chem. **391**(4), 1343–1350 (2008)
60. Fukuda, D., Maruyama, M., Nara, Y., Hayashi, D., Ogawa, H., Kaneko, K.: Observation of fracture sealing in high-strength and ultra-low-permeability concrete by micro-focus X-ray CT and SEM/EDX. Int. J. Fract. **188**(2), 159–171 (2014)
61. Yang, S.-Q., Yin, P.-F., Huang, Y.-H., Cheng, J.-L.: Strength, deformability and X-ray micro-CT observations of transversely isotropic composite rock under different confining pressures. Eng. Fract. Mech. **214**, 1–20 (2019)
62. Cho, S.-H., Cho, S.-K., Kim, S.-K., Choen, D.-S., Synn, J.-H., Katsuhiko, K.: 3-D Visualization of dynamic loading induced damage of rock-like materials using Micro-focus X-ray CT scanner. Geosyst. Eng. **11**(3), 51–56 (2008)
63. Skarżyński, Ł, Marzec, I., Tejchman, J.: Fracture evolution in concrete compressive fatigue experiments based on X-ray micro-CT images. Int. J. Fatigue **122**, 256–272 (2019)
64. Xu, J., Haque, A., Gong, W., Gamage, R.P., Dai, G., Zhang, Q., Xu, F.: Experimental study on the bearing mechanisms of rock-socketed piles in soft rock based on micro X-ray CT analysis. Rock Mech. Rock Eng. **53**(8), 3395–3416 (2020)
65. Fan, L.F., Gao, J.W., Wu, Z.J., Yang, S.Q., Ma, G.W.: An investigation of thermal effects on micro-properties of granite by X-ray CT technique. Appl. Therm. Eng. **140**, 505–519 (2018)
66. De Kock, T., Boone, M.A., De Schryver, T., Van Stappen, J., Derluyn, H., Masschaele, B., De Schutter, G., Cnudde, V.: A pore-scale study of fracture dynamics in rock using X-ray micro-CT under ambient freeze-thaw cycling. Environ. Sci. Technol. **49**(5), 2867–2874 (2015)
67. Song, L., Ning, Z., Duan, L.: Research on reservoir characteristics of Chang7 tight oil based on nano-CT. Arabian J. Geosci. **11**(16) (2018)
68. Callow, B., Falcon-Suarez, I., Ahmed, S., Matter, J.: Assessing the carbon sequestration potential of basalt using X-ray micro-CT and rock mechanics. Int. J. Greenhouse Gas Control **70**, 146–156 (2018)
69. Devarrewaere, W., Foque, D., Heimbach, U., Cantre, D., Nicolai, B., Nuyttens, D., Verboven, P.: Quantitative 3D shape description of dust particles from treated seeds by means of X-ray micro-CT. Environ. Sci. Technol. **49**(12), 7310–7318 (2015)
70. Donato, S., Pacile, S., Colombo, F., Garrovo, C., Dal Monego, S., Macor, P., Tromba, G., Biffi, S.: Meniscal ossicles as micro-ct imaging biomarker in a rodent model of antigen-induced arthritis: a synchrotron-based X-ray pilot study. Sci. Rep. **7** (2017)
71. Parlanti, P., Cappello, V., Brun, F., Tromba, G., Rigolio, R., Tonazzini, I., Cecchini, M., Piazza, V., Gemmi, M.: Size and specimen-dependent strategy for x-ray micro-ct and tem correlative analysis of nervous system samples. Sci. Rep. **7** (2017)
72. Dudak, J., Zemlicka, J., Karch, J., Patzelt, M., Mrzilkova, J., Zach, P., Hermanova, Z., Kvacek, J., Krejci, F.: High-contrast X-ray micro-radiography and micro-CT of ex-vivo soft tissue murine organs utilizing ethanol fixation and large area photon-counting detector. Sci. Rep. **6** (2016)
73. Garcia, F.H., Fischer, G., Liu, C., Audisio, T.L., Economo, E.P.: Next-generation morphological character discovery and evaluation: an X-ray micro-CT enhanced revision of the ant genus Zasphinctus Wheeler (Hymenoptera, Formicidae, Dorylinae) in the Afrotropics. Zookeys 33–93 (2017). https://doi.org/10.3897/zookeys.693.13012693
74. Vavrik, D., Kytyr, D., Muehleder, S., Vopalensky, M., Benes, P., Pichotka, M., Kumpova, I., Koudelka, P.: High resolution X-ray micro-CT imaging of fibrin scaffold using large area single photon counting detector. J. Instrum. **13** (2018)
75. Friis, E.M., Crane, P.R., Pedersen, K.R., Bengtson, S., Donoghue, P.C.J., Grimm, G.W., Stampanoni, M.: Phase-contrast x-ray microtomography links cretaceous seeds with Gnetales and Bennettitales. Nature **450**(7169), 549-U511 (2007)

76. Hwu, Y., Tsai, W.L., Lai, B., Je, J.H., Fecher, G.H., Bertolo, M., Margaritondo, G.: Using photoelectron emission microscopy with hard-X-rays. Surf. Sci. **480**(3), 188–195 (2001)
77. Hasslinger, P., Vass, V., Dejaco, A., Blanchard, R., Orlygsson, G., Gargiulo, P., Hellmich, C.: Coupling multiscale X-ray physics and micromechanics for bone tissue composition and elasticity determination from micro-CT data, by example of femora from OVX and sham rats. Int. J. Comput. Methods Eng. Sci. Mech. **17**(3), 222–244 (2016)
78. Zhang, K., Li, D.-E., Zhu, P., Yuan, Q., Huang, W., Liu, X., Hong, Y., Gao, G., Ge, X., Zhou, H., Wu, Z.: 3D visualization of the microstructure of Quedius beesoni Cameron using micro-CT. Anal. Bioanal. Chem. **397**(6), 2143–2148 (2010)
79. Li, D., Zhang, K., Zhu, P., Wu, Z., Zhou, H.: 3D configuration of mandibles and controlling muscles in rove beetles based on micro-CT technique. Anal. Bioanal. Chem. **401**(3), 817–825 (2011)
80. O'Sullivan, J.D.B., Behnsen, J., Starborg, T., MacDonald, A.S., Phythian-Adams, A.T., Else, K.J., Cruickshank, S.M., Withers, P.J.: X-ray micro-computed tomography (CT): an emerging opportunity in parasite imaging. Parasitology **145**(7), 848–854 (2018)
81. Zhang, K., Li, D.E., Hong, Y.L., Zhu, P.P., Yuan, Q.X., Huang, W.X., Gao, K., Zhou, H.Z., Wu, Z.Y.: Penetrating view of nano-structures in Aleochara verna spermatheca and flagellum by hard X-ray microscopy. Chinese Phys. B. **22**(7), 076801 (2013)
82. Le Gros, M.A., McDermott, G., Larabell, C.A.: X-ray tomography of whole cells. Curr. Opin. Struct. Biol. **15**(5), 593–600 (2005)
83. Chao, W.L., Harteneck, B.D., Liddle, J.A., Anderson, E.H., Attwood, D.T.: Soft X-ray microscopy at a spatial resolution better than 15nm. Nature **435**(7046), 1210–1213 (2005)

Chapter 9
Machine Learning in X-ray Imaging and Microscopy Applications

Guo-Xu Zhang

9.1 Introduction

In materials science, microscopic imaging provides abundant information for understanding the relationships between the structure of materials and their properties. Among various spectroscopic techniques, X-ray spectroscopy is the method of choice to determine the structure of a compound, along with the determination of properties such as bond lengths and bond angles. An X-ray belongs to the group of high-energy electromagnetic radiation, with wavelengths shorter than visible light. Generally speaking, X-ray spectroscopy is a term for several spectroscopic techniques for characterization of materials by using X-ray light sources. Besides, there are soft X-rays and hard X-rays, differing in photon energies, and the latter is widely used to image the inside of objects due to their penetrating ability. Note that different applications, such as X-ray crystallography, medical computed tomography (CT), airport security, use different parts of the X-ray spectrum. For extended discussion, we refer the reader to previous chapters of this book.

X-ray imaging has become an essential detection technique in a wide range of applications since its discovery in 1895. Today, X-ray tube sources based on imaging systems have been broadly used in medical diagnosis, public security, and safety screening, materials characterization and manufacturing, earth science, to mention just a few. The most relevant concept to understand how X-ray imaging works is the behavior of X-rays when they interact with matter [1], as outlined in detail in previous chapters. As a matter of fact, X-rays lose a certain amount of energy when they pass through matter. The energy loss is relevant to the absorption behavior of the material. This is the main principle of conventional X-ray imaging,

G.-X. Zhang (✉)
School of Chemistry and Chemical Engineering, Harbin Institute of Technology, West Dazhi 92, 150001 Harbin, P. R. China
e-mail: zhanggx@hit.edu.cn

J. Wang (ed.), *Advanced X-ray Imaging of Electrochemical Energy Materials and Devices*, https://doi.org/10.1007/978-981-16-5328-5_9

i.e., measuring the amount of energy loss. Due to the fact that energy loss differs for different materials (material-dependent), a certain contrast can be detected from an X-ray image.

Images generated with X-rays are often known as radiographs (or skiagraphs). Radiography has been applied in medical, academic, and industrial areas, where it is crucial for diagnosis and nondestructive testing of substances for defects. The first medical use was reported by Röntgen shortly after his discovery of X-rays. Figure 9.1 shows the hand of Röntgen's wife and one can clearly recognize the ring she was wearing on her annular finger. In addition, X-rays have the ability to cause fluorescence in most materials. Thus, one can analyze these emissions to determine the chemical elements of an image object. This type of radiography is referred as to fluoroscopy. It is based upon the same imaging technique, with the photographic plate replaced by a fluorescent screen (The interested reader is referred to Chap. 6). X-rays are also widely used in X-ray crystallography, in which diffraction patterns are generated. The three-dimensional structure of a crystal can be determined by analyzing the internal reflections of a diffraction pattern.

Another use of radiography is X-ray CT, which is an advanced form of X-rays and provides cross-sectional images inside a sample. It is well suited to characterizing a material's integrity and three-dimensional (3D) image reconstructions [2]. The uses of synchrotron light sources have offered new opportunities for X-ray microscopy since 1980s. Synchrotron X-ray tomography offers a way of visualizing 3D interior structure of real objects nondestructively and with a high spatial

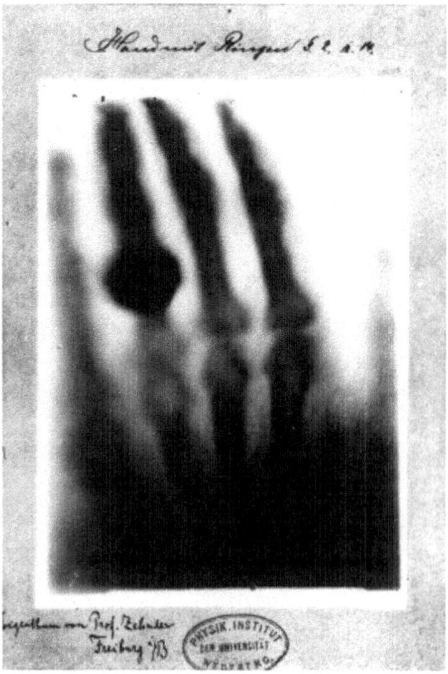

Fig. 9.1 A print of one of the first X-rays by Wilhelm Röntgen

resolution [3, 4]. This allows for the detailed microstructural analysis of many different types of materials, such as rechargeable batteries. The reader is referred to Chap. 6 for extended discussion.

Today, X-ray imaging systems have been widely used. While many challenges still remain, particularly in recognizing and extracting all relevant information from the images. It is known that raw experimental microscopy images are typically associated with high-level noise and distortions. Furthermore, there are unavoidable drawbacks, such as insufficient spatial and temporal resolution, as well as, low contrast of these systems [5]. It is clear that efforts are required to combine with other techniques for X-ray imaging analysis. Machine learning (ML), a subset of artificial intelligence (AI), is the study of computer algorithms that improve automatically with experience [6]. Figure 9.2 depicts the relationship of deep learning, machine learning, and artificial intelligence. ML has gained great popularity and is widespread in both the natural sciences and engineering. Indeed, the importance of data-driven science, the fourth paradigm of science [7], can hardly be overestimated. Here we will focus on how ML, basically deep learning (DL), can be utilized in the X-ray imaging domain, together with some latest progress in the following sections.

9.2 Introduction to Machine Learning

Machine learning studies computer algorithms whose performance improves with data [6]. More specifically, ML algorithms build statistical models based on sample data (termed as "training data") and optimize the algorithms to make predictions or decisions without an explicit concept of underlying principles being encoded [8]. Generally speaking, the success of ML and AI methods lies in the fact that

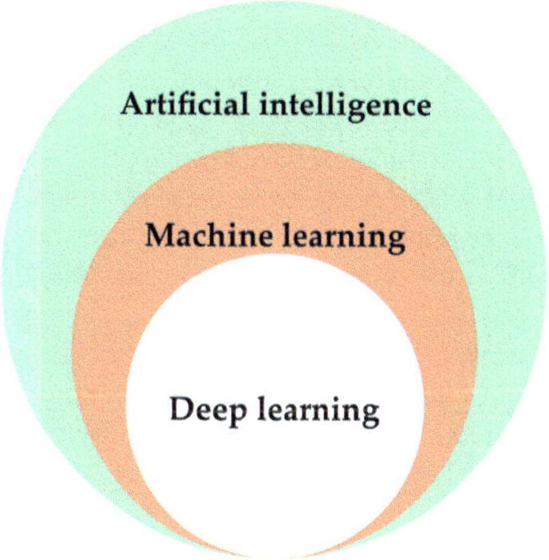

Fig. 9.2 The relationship of deep learning, machine learning, and artificial intelligence (Reproduced from Ref. [59])

computers can handle the problems with much larger and higher dimensional data evolved. This allows for multi-scale modelings of materials science, as ML applies to all scales. ML has enormously become the method of choice in a multidisciplinary research field. As a matter of fact, most of us (directly or indirectly) live with ML algorithms in our daily life, such as email filtering, computer vision, and smartphone.

The discipline of machine learning is closely related to computational statistics, which focuses on the design of algorithm for implementing statistical methods on computers. Many statistical methods for ML have been developed for decades. The pioneering ML research can be traced back to 1950s. Marvin Minsky and Dean Edmonds built the first neural network machine in 1951, namely the SNARC [9] (Stochastic neural analog reinforcement calculator). Since then, the term "artificial intelligence" was proposed. Today's AI boom has been enormously facilitated by the rapid development of hardware for information storage and processing, together with the growth of machine learning methods [10].

Generally speaking, machine learning can be divided into three broad categories, depending on the nature of available data [10]. Supervised learning is the most mature and powerful one, and it accounts for the majority of ML studies in the physical sciences [11]. Supervised learning is suited to situations where a ML model is trained on example-input and desired-output pairs for unseen inputs, that is, the goal is to derive a function that maps inputs to outputs. Unsupervised learning, on the other hand, utilizes only input data for training a learning algorithm, but no output data are given. This type of learning can be a goal in itself, for example, discovering hidden trends, patterns, or clustering in the data. The third category is called reinforcement learning, representing a rapidly emerging field with promising applications in tasks that require machine creativity [12]. In reinforcement learning, a model is trained by mimicking a dynamic environment in which a long-term goal is targeted and the computer algorithm improves its performance by directly maximizing cumulative rewards. Reinforcement learning differs from supervised learning as no "correct" input-output pairs are needed. Instead, the focus of the training is to find a balance between exploration and exploitation by a long-term reward [13]. This category of learning has been shown to be promising in the exploration of compound and material spaces and the search for new molecules with objective-reinforced generative adversarial networks [12, 14].

ML techniques have been applied to the discovery and design of various energy materials, ranging from energy storage to energy conversion. ML models have a major impact on the following research objectives in materials science: materials discovery, advancing materials modeling, and understanding materials phenomena [12, 15, 16]. For example, there have been a number of studies on stability (e.g., formation energies and energy above hull) [17–20] of energy materials by using ML algorithms, along with other properties such as adsorption energies, reaction barriers, bandgaps, ionic conductivity, to mention just a few. Another point to highlight is that ML methods are often used to predict properties that are computationally very demanding by using first-principles approaches, among which density-functional theory (DFT) [21] has become the method of choice for obtaining ground-state properties of molecules and materials. ML has shown its potential to reveal novel chemical insights, enabling the improvement of structure prediction algorithms that generate better "guesses" for novel structures/compositions [22–24]. Last but not least, recent ML methods have received great attention for their potential to enhance experimental characterization and interpretation. As a matter of fact, this research area benefits from the rapid development of graphics processing units (GPUs) [25, 26] and GPU codes, which work well on parallel deep neural network computations. We refer the reader to some recent literature on the interpretation and labeling of experimental images, e.g., scanning transmission electron microscopy (STEM), [27, 28], X-ray diffraction (XRD) [29], X-ray absorption near-edge structure (XANES) [30].

In summary, the discipline of ML is rapidly developing, and its applications have become ubiquitous in all sciences. As of 2020, deep learning has become the dominant approach for much ongoing work [31], which will be the topic of next section.

9.3 Deep Learning for Image Segmentation

As introduced in previous chapters, in recent years there has been significant progress in the field of X-ray microscopy, including transmission X-ray microscopy (TXM), scanning transmission X-ray microscopy (STXM), scanning photoelectron microscopy, micro-X-ray fluorescence (μ-XRF) spectroscopy, and synchrotron radiation X-ray tomography microscopy (SRXTM). However, many challenges still remain, particularly in recognizing and extracting all relevant information from the images. This is because every pixel of the image is a degree of freedom, thus leading to extremely high dimensionality of the information space [32]. Furthermore, there are unavoidable drawbacks, such as insufficient spatial and temporal resolution, low contrast of these systems, and time-consuming issues. With the rapid growth of microstructure image databases, the above issues can be tackled by combining data-driven approaches, such as ML, as illustrated in Fig. 9.3 (Adapted from Ref. [12]).

In materials science, ML techniques are becoming an increasingly active field of research to describe relationships between the structure of materials and their properties. Here the structure can be at nano- and microscale, i.e., atomic structure, crystal, surface facet, interfacial structure, morphology, and so on. The knowledge of structure-property correlations allows to tune the parameters of synthesis, thus controlling the structure of materials for new applications [33]. This is basically a fundamental question for many types of materials including energy materials, such as batteries, capacitors, fuel cells, and solar cells. Eventually, all ML methods aim to achieve a balance between accuracy and efficiency for predictions that replace more expensive computational, experimental, or human-driven techniques.

Deep learning is a class of ML algorithms based on artificial neural networks (ANN) with representation learning [34, 35]. It is defined in terms of multiple

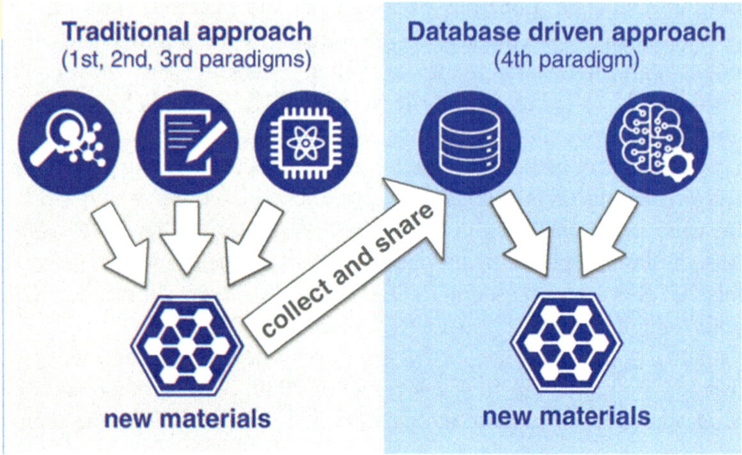

Fig. 9.3 Illustration of the traditional approach (referred to as 1st, 2nd, and 3rd paradigms and symbolized by the three icons at the top of the left panel) and data-driven approach (referred to as 4th paradigm) for materials discovery. Adapted with permission of Ref. [12]. Copyright 2019, Wiley

layers of neurons that progressively extract higher-level features from the raw input, as shown in Fig. 9.4 (cf Ref. [36]). Deep learning has gained tremendous attention in various fields in sciences, due to its revolutionary performance in complex tasks including speech recognition, object detection, drug discovery, to mention just a few, as listed in Fig. 9.5. The focus of this chapter is on one of the most critical areas of computer vision: image analysis, particularly deep learning-based approaches for image segmentation of energy materials.

Segmentation is a method of choice for partitioning two-dimensional (2D) or three-dimensional (3D) space into visually distinct segments with certain characteristics. This term is often used in the context of digital images, where the spatial information is featured by using picture elements (pixels) in 2D and volume elements (voxels) in 3D spaces [14, 37]. In practice, image segmentation aims to recognize objects and boundaries (lines, curves, etc.) in images. More specifically, image segmentation is the process of assigning a label to every pixel in an image, and those pixels with the same label share certain nature [37]. As a result, this image process simplifies rich images into a few meaningful intensities or components, leading to a set of segments that collectively cover the entire image or a set of contours extracted from the image [38]. With the increasing precision of instrumentation, image processing is becoming fast-growing research topic in materials science. Some typical tasks of segmentation include the determination of the material phases that are present in image data, and the detection and extraction of single particles, grains, or fibers [39]. The quality of the segmentation has no doubt become crucial for the subsequent analysis of microstructure and macroscopic properties of a given material.

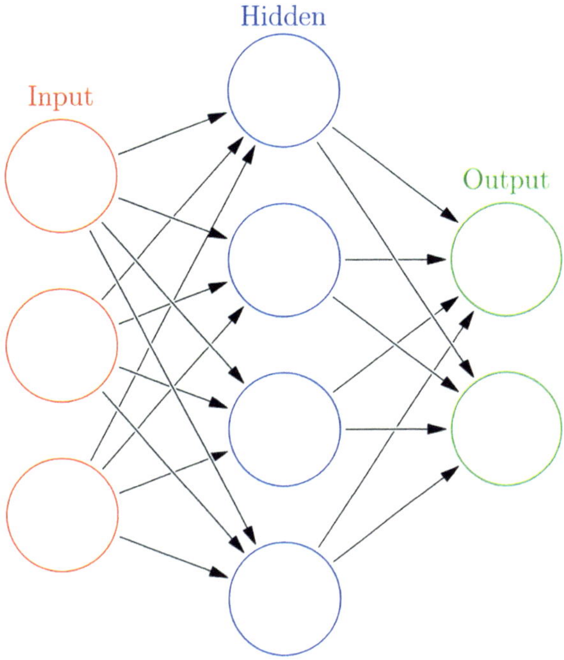

Fig. 9.4 Artificial neural network with layer coloring. Licensed under the Creative Commons Attribution-Share Alike 3.0 Unported license [https://en.wikipedia.org/wiki/File:Colored_neural_network.svg]

Numerous promising insights have been proposed in deep learning architectures for image segmentation since 2012, where specifically convolutional neural networks (CNNs) [40] gain an increasing popularity. Guo and co-authors [41] provided a review of deep learning-based semantic segmentation of images and divided the literature into three categories: region-based, fully convolutional network (FCN)-based, and weakly supervised segmentation methods. Image segmentation differs from image classification or object recognition. There are two types of image segmentation, i.e., semantic segmentation and instance segmentation. More specifically, semantic segmentation is an approach detecting, for every pixel, a belonging class of the object. On the other hand, instance segmentation is an approach that identifies, for every pixel, a belonging instance of the object, that is, it detects each distinct object of interest in the image [14]. Here we give a short overview of deep learning models for image analysis. Region-CNN (R-CNN) [42] has been successfully used for determining bounding boxes around objects of interest for the task of object detection in 2D images. Semantic segmentation benefits from FCN proposed by Long et al. in 2014 [43]. The key idea in FCN-based methods is that they learn a mapping from pixels to pixels without extracting the region proposals. R-CNN family was strongly influenced by FCN, resulting in Fast R-CNN [44] and Faster R-CNN [45]. In 2017 He et al. [46] extended Faster R-CNN for instance segmentation, namely Mask R-CNN, by adding a branch for predicting segmentation masks on each Region of Interest (RoI), along with a class label and a bounding box. Figure 9.6 shows the

Fig. 9.5 Applications of deep learning in various fields. Adapted with permission of Ref. [35]. Copyright 2018, ACM Journals

framework of Mask R-CNN for instance segmentation. Mask R-CNN has been extensively employed for multi-task segmentation models for a wide range of application areas [47–49]. In 2020, it has been applied to study the microstructure of lithium-ion battery cathodes [50]. Extended applications of deep learning models for X-ray imaging will be introduced in the last section.

9.4 Workflow of Using Deep Learning on Image Analysis

A typical workflow for deep learning analysis on images can be described in Fig. 9.7 (Adapted from Ref. [16]). Generally speaking, there are five steps. The first step in any ML study is task specification and analysis. It refers to identifications of the task category and prediction targets of the ML model. This stage is often considered to be the most essential part, as the determination of the goal should be

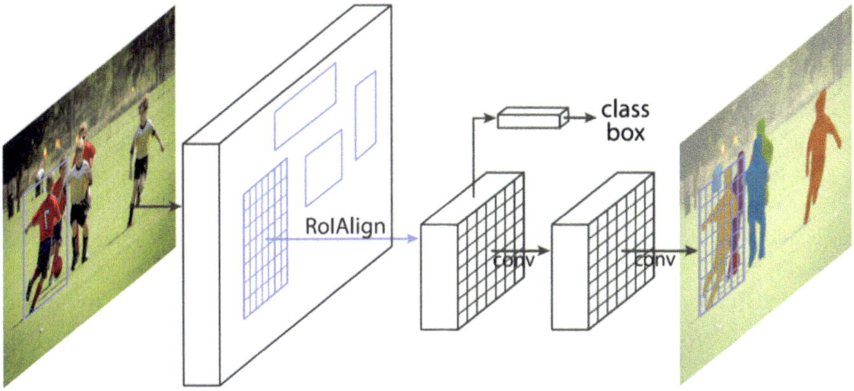

Fig. 9.6 The Mask R-CNN framework for instance segmentation. Adapted from Ref. [46]. Copyright 2017, IEEE

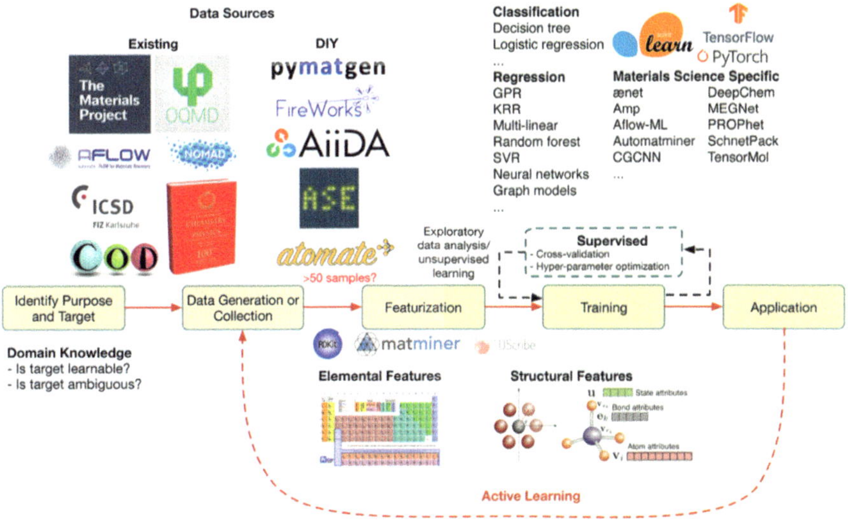

Fig. 9.7 A schematic representation of five steps of machine learning models, starting from identification of purpose to data collection, featurization, model building, and eventually application. Adapted from Ref. [16]. Copyright 2020, Wiley

potentially learnable from available information, including structure and composition of materials, quantities measured from experiments, images, just to mention a few. It is noteworthy to mention that the wrong choice of a prediction target can give rise to ML models with large errors or uncertainties. The second step is data generation (data collection), i.e., training data either from experiments or computations. Note that training data can be self-generated, or taken from open-source platforms. Some of the databases are given as follows: the Materials Project's software suite comprising the Python Materials Genomics (pymatgen) materials analysis library [51], novel materials discovery (NOMAD) repository [52], AFLOW [53], and AiiDA [54]. The third stage is to design the model based on the task and the available dataset. The next step is feature extraction (featurization), which helps to reduce the number of resources required by extracting characteristic patterns from the outputs of the trained model. For the task of imaging, features (or descriptors) include properties like corners, edges, regions of interest points, ridges, and so on. Feature extraction allows to facilitate the subsequent learning and generalization steps. The final step is model selection and training for making predictions on properties or performance, verifying assumptions, or observing specific phenomena. The overall objective of the utilization of deep learning is to model the connection between experimental data and the target properties, and then using the connection to reconstruct or predict the performance. Deep learning is known to be flexible in the choice of the number of parameters, thus being suitable for big-data problems. During the training procedure, hyperparameters of the model are optimized by iteratively minimizing the loss function. For regression models, the mean absolute error (MAE), mean square error (MSE), and root mean square error (RMSE) are used as common model loss or prediction error metrics. For classification models, a surrogate loss function is typically employed. For a more extended discussion, we refer the interested reader to standard textbooks and recent reviews on the topic.

9.5 Applications of Deep Learning for X-ray Imaging in Materials Science

X-rays are ideal for the microscopic investigation of materials due to their short wavelength and penetrating character. In materials science, X-ray imaging at the nano- and microscale thus can provide abundant information on complex materials, such as crystal structure, composition, bonding, and so on. However, the interpretation and labeling of experimental X-ray images (e.g., STEM, XRD, XANES, SRXTM, etc.) are today still mostly painstakingly carried out by human experts. The discipline of machine learning is rapidly developing, and its applications have become ubiquitous in all sciences. Here, we present the latest progress in applying ML techniques to X-ray imaging and relevant applications for energy-related fields.

ML approaches have shown their potential to enhance experimental characterization, though they received far less attention than other applications. In 2019, Liu et al. [30] studied the XANES spectra of copper oxide clusters (measured in operando conditions of the methanation reaction) in combination with a CNN model. They have demonstrated that their ML model can be trained on theoretical spectra and utilized to "invert" experimental XANES data to obtain structural descriptors, i.e., coordination numbers of Cu–Cu pairs. This novel study allows us to better understand the structure, composition, and function relationship in catalysis. Different structural motifs of copper oxide clusters can be distinguished, along with average cluster size reliably evaluated.

Deep learning has shown its capability of identifying crystal structures. Ziletti et al. [55] have proposed a deep learning neural network model to classify the crystal structures into eight space groups. They calculated 100,000 diffraction patterns from the 3D atomic arrangement and then used them to train the deep learning model. They demonstrated their ML model outperforms other packages significantly regardless of defects in the crystal structures (though the current model is limited to only eight crystal systems). A reliable identification of lattice symmetry plays an essential role in materials characterization and analytics. From a theoretical point of view, knowing the symmetry group of the crystal structure is a critical first step for efficient sampling in reciprocal space and reducing the computational cost for matrix operations [55]. On the experimental side, the identification of symmetry group information from the spectroscopy data can enhance characterization and interpretation. The visualization of the CNN model attentive response maps are shown in Fig. 9.8 (Adapted from Ref. [55]). Park and co-authors [56] trained a deep learning model based on the CNN to classify XRD patterns in terms of crystal system, extinction group, and space group. They collected 150,000 powder XRD patterns and used them as input for the CNN architecture. They have reported a reliable accuracy of 81.14%, 83.83%, and 94.99% for the space group, extinction group, and crystal system, respectively [56].

X-ray tomography or X-ray computed tomography, is a characterization technique that can offer different types of 3D information, such as the microstructure, the defects, and the crystallography of most types of materials with sub-micron resolution. X-ray tomography microscopy is a powerful technique to visualize and quantify the morphology of electrodes. In 2020, Jiang and co-authors [50] have carried out a ML-assisted study on statistics of the particle-carbon/binder

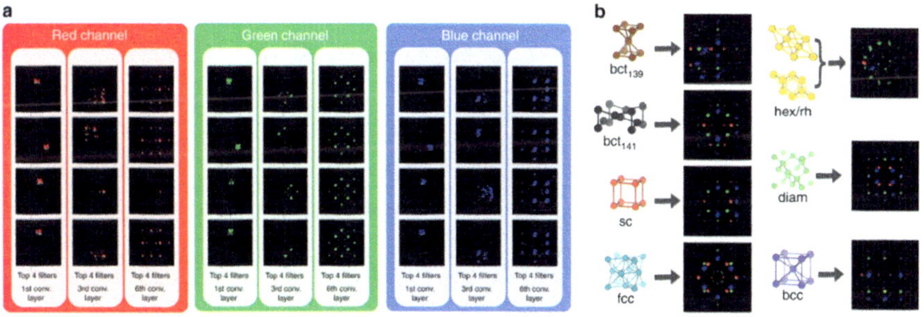

Fig. 9.8 Visualization of the convolutional neural network (ConvNet) attentive response maps. **a** Attentive response maps from the top four most activated filters of the first, third and last convolutional layers for the simple cubic class. **b** Sum of the last convolutional layer filters for all seven crystal classes: the ConvNet learned crystal templates automatically from the data. Such attentive response maps allow to identify the parts of the image which are the most important in the classification decision. Adapted from Ref. [55]. Copyright 2018, Nature

detachment in lithium-ion battery cathodes, by combining with high-resolution hard X-ray nano-tomography characterization and numerical modeling. The Mask R-CNN was utilized in their work for image segmentation. They demonstrated the importance of precisely quantifying the evolving nature of the battery electrode's microstructure with ML models. They have reported that over 650 unique particles (with different size, shape, position, and degree of cracking) were successfully identified and isolated automatically from the imaging data, as shown in Fig. 9.9. In 2019, Furat et al. [39] have employed ML approaches for the segmentation of tomographic image data of functional materials, including Ibuprofen tablets, mineral particle systems, lithium-ion batteries, etc. Various kinds of applications were discussed, in which tomographic image data depicting microstructures of materials are semantically segmented by combining ML methods and conventional image processing steps.

Deep learning methods can also be used in X-ray scattering imaging. Wang et al. [57] have applied the CNN architecture for automatically analyzing X-ray scattering images. In their study, synthetic X-ray scattering images were simulated.

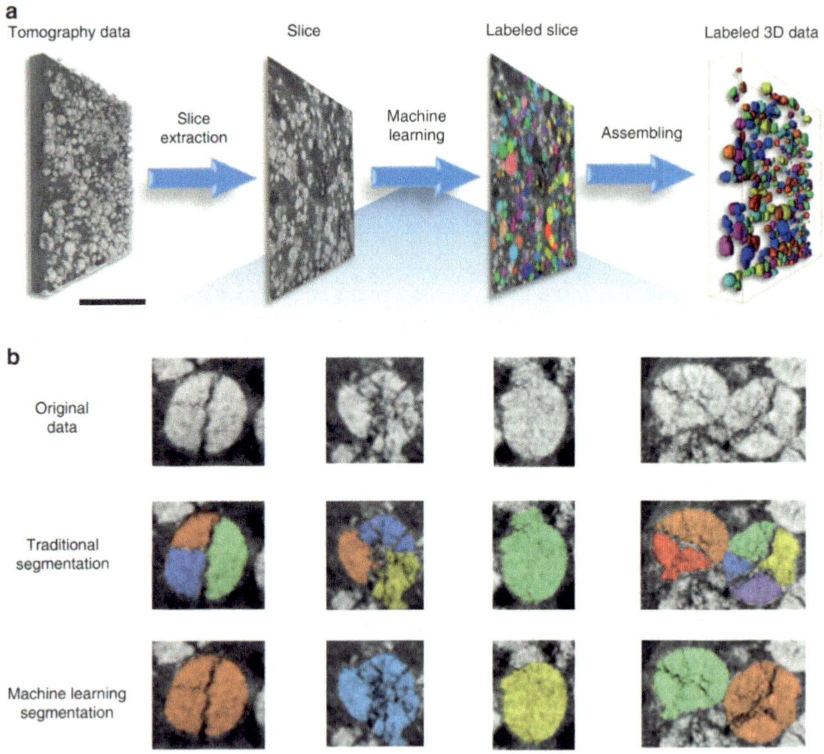

Fig. 9.9 **a** A workflow of machine learning-based segmentation. **b** Comparison of conventional segmentation results and the machine learning-assisted segmentation results for a few representative particles. Different colors denote different particle labels. Adapted from Ref. [50]. Copyright 2020, Nature

They have concluded deep learning methods outperform previously published methods by 10% on synthetic and real datasets. Timoshenko and co-authors [58] have reported a study of subnanometer substructures in nanoassemblies formed from clusters under a reactive atmosphere by combining small-angle X-ray scattering (SAXS), XANES spectroscopy, *ab initio* simulations, and machine learning-based on ANN techniques. Such a workflow is sketched in Fig. 9.10 (see Ref. [58]). In situ studies were carried out at different length scales in order to monitor the formation of such complex systems. They concluded that the combined study will have a significant effect on the understanding of particle sintering and assembly processes and of structure-property relationship in ultradispersed metal catalysts in reaction conditions. Their workflow can be used by a large number of researchers for studies of ultradispersed clusters in reaction conditions.

9.6 Summary and Outlook

In this book chapter, we summarized the latest progress in applying ML techniques to X-ray imaging and relevant applications for energy materials. Based on recent advances in GPU computing, deep learning has shown great promise in solving massive data-based tasks due to its ability of learning representations from complex data. Various deep learning tools have been successfully applied in speech recognition, visual object recognition, object detection, and many other areas such as materials discovery. As discussed in the above section, deep learning is particularly

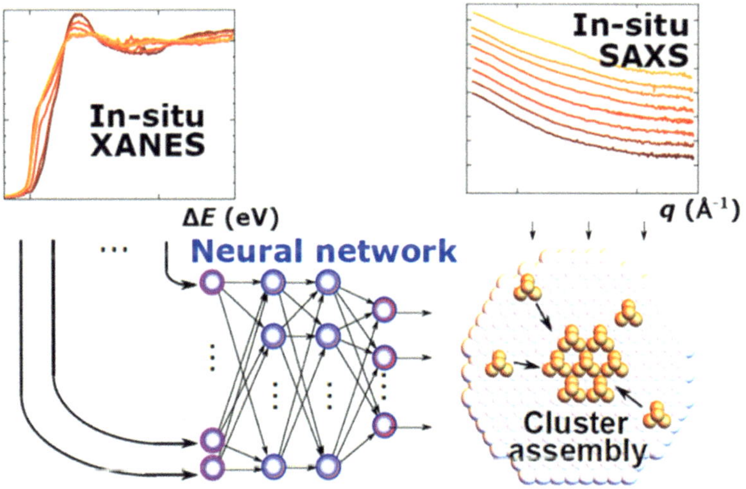

Fig. 9.10 A workflow of combined small-angle X-ray scattering (SAXS), X-ray absorption near-edge structure (XANES) spectroscopy, *ab initio* simulations, and machine learning (artificial neural network) techniques. Adapted from Ref. [59]. Copyright 2018, ACS

useful in the study and discovery of energy materials in recent years. However, unlike the application of X-ray imaging for medical sciences, combining deep learning and X-ray microscopy techniques is still in its infant in materials science. There are many opportunities for further breakthroughs in this emerging field and here we highlight some challenges. Firstly, the design of new algorithms and architectures should aim to improve the loss function that reflects the basic physical and chemical constraints. Currently, the applications of deep learning have mainly focused on structural information extraction (recognition and tracking of morphology, phase, defect, and so on). Knowing physical correlations between the structure and properties is necessary for meaningful analysis. In addition, it is suggested that multidisciplinary scientists (e.g., data scientists, theoretical physicists, microscopists) should work together to apply machine learning algorithms to solve materials questions. To achieve this goal, experts should learn languages of other fields, such as coding, modeling, characterization, architecture, to mention just a few. Last but not least, one of the biggest challenges in microscopic image processing is labeling data by human scientists, which is very time-consuming. So far, most training datasets are obtained from theoretical calculations. While computed data typically have certain discrepancies from realistic systems. Therefore, it is becoming important to build experimental imaging databases by different labs in the field. Our final remark is that machine learning techniques are indeed capable of accelerating materials discovery and their roles are clearly becoming more and more evident in all areas.

References

1. Berger M., Yang Q., Maier A.: X-ray Imaging. In: Maier A., Steidl S., Christlein V., Hornegger J. (eds) Medical Imaging Systems. Lecture Notes in Computer Science, pp. 119–145. Springer, Cham. (2018). https://doi.org/10.1007/978-3-319-96520-8_7
2. Maire, E., Withers, P.J.: Quantitative X-ray Tomography. Int. Mater. Rev. **59**(1), 1–43 (2014)
3. Baruchel, J., Bleuet, P., Bravin, A., Coan, P., Lima, E., Madsen, A., Ludwig, W., Pernot, P., Susini, J.: Advances in Synchrotron hard X-ray Based Imaging. C R Phys. **9**(5–6), 624–641 (2008)
4. Westneat, M.W., Socha, J.J., Lee, W.-K.: Advances in Biological Structure, Function, and Physiology Using Synchrotron X-ray Imaging. Ann. Rev. Physiol. **70**, 119–142 (2008)
5. Chen, R.C., Liu, P., Xiao, T.Q., Xu, L.X.: X-ray Imaging for Non-destructive Microstructure Analysis at SSRF. Adv. Mater. **26**(46), 7688–7691 (2014)
6. Mitchell, T.: Machine Learning. McGraw Hill, New York (1997)
7. Hey, T., Tansley, S., Tolle, K.: The Fourth Paradigm: Data-Intensive Scientific Discovery. Microsoft Research, Redmond, WA. USA (2009)
8. Koza, J.R., Bennett, F.H., Andre, D., Keane, M.A.: Automated Design of Both the Topology and Sizing of Analog Electrical Circuits Using Genetic Programming. Springer, Netherlands (1996)
9. Crevier, D.: AI: The Tumultuous Search for Artificial Intelligence. Basic Books, New York, NY, USA (1993)
10. Russell, S., Norvig, P.: Artificial Intelligence: A Modern Approach. Forth edition, Pearson (2021)

11. Butler, K.T., Davies, D.W., Cartwright, H., Isayev, O., Walsh, A.: Machine Learning for Molecular and Materials Science. Nature **559**(7715), 547–555 (2018)
12. Himanen, L., Geurts, A., Foster, A.S., Rinke, P.: Data-driven Materials Science: Status, Challenges, and Perspectives. Adv Sci **6**(21), 1900808 (2019)
13. Bishop, M.: Pattern Recognition and Machine Learning. Springer-Verlag, New York (2006)
14. Hoeser, T., Kuenzer, C.: Object Detection and Image Segmentation with Deep Learning on Earth Observation Data: A Review-part I: Evolution and Recent Trends. Remote Sensing. **12** (10), 43 (2020)
15. Schmidt, J., Marques, M.R.G., Botti, S., Marques, M.A.L.: Recent Advances and Applications of Machine Learning in Solid-state Materials Science. npj Comput. Mater. **5** (83), 36 (2019)
16. Chen, C., Zuo, Y.X., Ye, W.K., Li, X.G., Deng, Z., Ong, S.P.: A Critical Review of Machine Learning of Energy Materials. Adv. Energy Mater. **10**(8), 36 (2020)
17. Faber, F.A., Lindmaa, A., von Lilienfeld, O.A., Armiento, R.: Machine Learning Energies of 2 Million Elpasolite (ABC_2D_6) Crystals. Phys. Rev. Lett. **117**(13), 6 (2016)
18. Li, Z., Xu, Q., Sun, Q., Hou, Z., Yin, W.-J.: Thermodynamic Stability Landscape of Halide Double Perovskites via High-throughput Computing and Machine Learning. Adv. Func. Mater. **29**(9), 1807280 (2019)
19. Park, H., Mall, R., Alharbi, F.H., Sanvito, S., Tabet, N., Bensmail, H., El-Mellouhi, F.: Exploring New Approaches towards the Formability of Mixed-ion Perovskites by DFT and Machine Learning. Phys. Chem. Chem. Phys. **21**(3), 1078–1088 (2019)
20. Kim, S., Noh, J., Gu, G.H., Aspuru-Guzik, A., Jung, Y.: Generative Adversarial Networks for Crystal Structure Prediction. ACS Cent. Sci. **6**(8), 1412–1420 (2020)
21. Hohenberg, P., Kohn, W.: Inhomogeneous Electron Gas. Phys. Rev. **136**(3B), B864 (1964)
22. Gu, G.H., Noh, J., Kim, I., Jung, Y.: Machine Learning for Renewable Energy Materials. J. Mater. Chem. A. **7**(29), 17096–17117 (2019)
23. Sarma, D.S., Deng, D.L., Duan, L.M.: Machine Learning Meets Quantum Physics. Phys. Today **72**(3), 48–54 (2019)
24. Schütt, K.T., Chmiela, S., von. Lilienfeld, O.A., Tkatchenko, A., Tsuda, K., Müller, K.-R.: Machine Learning Meets Quantum Physics. Springer Nature, Switzerland (2020). https://doi.org/10.1007/978-3-030-40245-7
25. Raina, R., Madhavan, Ng, A.Y.: Large-scale Deep Unsupervised Learning Using Graphics Processors. In: Proceedings of the 26th Annual International Conference on Machine Learning. Montreal, Canada, 873–880 (2009)
26. Harju A., Siro T., Canova F.F., Hakala S., Rantalaiho T., Manninen P., Öster P. (eds): Computational Physics on Graphics Processing Units. In: Applied Parallel and Scientific Computing. PARA . Lecture Notes in Computer Science, Springer, Berlin, Heidelberg. 3–26. (2012). https://doi.org/10.1007/978-3-642-36803-5_1
27. Ziatdinov, M., Dyck, O., Maksov, A., Li, X.F., Sang, X.H., Xiao, K., Unocic, R.R., Vasudevan, R., Jesse, S., Kalinin, S.V.: Deep Learning of Atomically Resolved Scanning Transmission Electron Microscopy Images: Chemical Identification and Tracking Local Transformations. ACS Nano. **11**(12), 12742–12752 (2017)
28. Maksov, A., Dyck, O., Wang, K., Xiao, K., Geohegan, D.B., Sumpter, B.G., Vasudevan, R. K., Jesse, S., Kalinin, S.V., Ziatdinov, M.: Deep Learning Analysis of Defect and Phase Evolution During Electron Beam-induced Transformations in WS_2. npj Comput. Mater. **5**, 12 (2019)
29. Suzuki, Y., Hino, H., Hawai, T., Saito, K., Kotsugi, M., Ono, K.: Symmetry Prediction and Knowledge Discovery from X-ray Diffraction Patterns Using an Interpretable Machine Learning Approach. Sci. Rep. **10**(1), 21790–21790 (2020)
30. Liu, Y., Marcella, N., Timoshenko, J., Halder, A., Yang, B., Kolipaka, L., Pellin, M. J., Seifert, S., Vajda, S., Liu, P., Frenkel, A.I.: Mapping XANES Spectra on Structural Descriptors of Copper Oxide Clusters Using Supervised Machine Learning. J. Chem. Phys. **151**, 164201 (2019)
31. Alpaydin, E.: Introduction to Machine Learning. Fourth edition, MIT Press (2020)

32. Larmuseau, M., Sluydts, M., Theuwissen, K., Duprez, L., Dhaene, T., Cottenier, S.: Compact Representations of Microstructure Images Using Triplet Networks. npj Comput. Mater. **6**, 156 (2020)
33. Ge, M., Su, F., Zhao, Z., Su, D.: Deep Learning Analysis on Microscopic Imaging in Materials Science. Mater. Today Nano. **11**, 100087 (2020)
34. Rosenblatt, F.: The Perceptron: a Probabilistic Model for Information Storage and Organization in the Brain. Psychol. Rev. **65**(6), 386–408 (1958)
35. Nash, W., Drummond, T., Birbilis, N.: A Review of Deep Learning in the Study of Materials Degradation. npj Mater. Degrad. **2**, 37 (2018)
36. Pouyanfar, S., Sadiq, S., Yan, Y., Tian, H., Tao, Y., Reyes, M.P., Shyu, M.-L., Chen, S.-C., Iyengar, S.S.: A Survey on Deep Learning. ACM Comput. Surv. **51**(5), 1–36 (2019)
37. Asgari Taghanaki, S., Abhishek, K., Cohen, J.P., Cohen-Adad, J., Hamarneh, G.: Deep Semantic Segmentation of Natural and Medical Images: a Review. Artif Intell Rev **54**, 137–178 (2021)
38. Malik, J., Belongie, S., Leung, T., Shi, J.: Contour and Texture Analysis for Image Segmentation. Int. J. Comput. Vision **43**(1), 7–27 (2001)
39. Furat, O., Wang, M.Y., Neumann, M., Petrich, L., Weber, M., Krill, C.E., Schmidt, V.: Machine Learning Techniques for the Segmentation of Tomographic Image Data of Functional Materials. Front. Mater. **6**, 17 (2019)
40. Goodfellow, I., Bengio, Y., Courville, A.: Deep Learning. MIT Press (2016)
41. Guo, Y.M., Liu, Y., Georgiou, T., Lew, M.S.: A Review of Semantic Segmentation Using Deep Neural Networks. Int. J. Multimedia Inf. Retrieval **7**(2), 87–93 (2018)
42. Girshick, R., Donahue, J., Darrell, T., Malik, J.: Rich Feature Hierarchies for Accurate Object Detection and Semantic Segmentation. In: 2014 IEEE Conference on Computer Vision and Pattern Recognition, pp. 580–587 (2014)
43. Long, J., Shelhamer, E., Darrell, T.: Fully Convolutional Networks for Semantic Segmentation. In: 2015 IEEE Conference on Computer Vision and Pattern Recognition, pp. 3431–3440 (2015)
44. Girshick, R.: Fast R-CNN. In: 2015 IEEE International Conference on Computer Vision, pp. 1440–1448 (2015)
45. Ren, S.Q., He, K.M., Girshick, R., Sun, J.: Faster R-CNN: towards Real-time Object Detection with Region Proposal Networks. Adv. Neural Inf. Process. Syst. **28** (2015)
46. He, K.M., Gkioxari, G., Dollar P., Girshick, R.: Mask R-CNN. In: 2017 IEEE International Conference on Computer Vision, pp. 2980–2988 (2017)
47. Abdolhoseini, M., Kluge, M.G., Walker, F.R., Johnson, S.J.: Segmentation of Heavily Clustered Nuclei from Histopathological Images. Sci. Rep. **9**, 13850 (2019)
48. Tsai, H.F., Gajda, J., Sloan, T.F.W., Rares, A., Shen, A.Q.: Usiigaci: Instance-aware Cell Tracking in Stain-free Phase Contrast Microscopy Enabled by Machine Learning. SoftwareX **9**, 230–237 (2019)
49. Jung, H., Lodhi, B., Kang, J.: An Automatic Nuclei Segmentation Method Based on Deep Convolutional Neural Networks for Histopathology Images. BMC Biomed. Eng. **1**(1), 1–12 (2019)
50. Jiang, Z., Li, J., Yang, Y., Mu, L., Liu, Y.: Machine-learning-revealed Statistics of the Particle-carbon/binder Detachment in Lithium-ion Battery Cathodes. Nat. Commun. **11**(1), 2310 (2020)
51. Jain, A., Ong, S.P., Hautier, G., et al.: Commentary: the Materials Project: a Materials Genome Approach to Accelerating Materials Innovation. APL Mater. **1**, 011002 (2013)
52. The Novel Materials Discovery (NOMAD) Laboratory. https://nomad-coe.eu/
53. Curtarolo, S., Setyawan, W., Wang, S.D., Xue, J.K., Yang, K.S., Taylor, R.H., Nelson, L.J., Hart, G.L.W., Sanvito, S., Buongiorno-Nardelli, M., Mingo, N., Levy O.: AFLOWLIB.ORG: A Distributed Materials Properties Repository from High-throughput *ab initio* Calculations. Comput. Mater. Sci. **58**, 227–235 (2012)

54. Pizzi, G., Cepellotti, A., Sabatini, R., Marzari, N., Kozinsky, B.: AiiDA: Automated Interactive Infrastructure and Database for Computational Science. Comput. Mater. Sci. **111**, 218–230 (2016)
55. Ziletti, A., Kumar, D., Scheffler, M., Ghiringhelli, L.M.: Insightful Classification of Crystal Structures Using Deep Learning. Nat. Commun. **9**, 2775 (2018)
56. Park, W.B., Chung, J., Jung, J., Sohn, K., Singh, S. P., Pyo, M., Shin, N., Sohn, K.-S.: Classification of Crystal Structure Using a Convolutional Neural Network. Int. Union Crystallogr. **4**, 486–494 (2017)
57. Wang, B., Yager, K., Yu, D., Minh, H.: X-ray Scattering Image Classification Using Deep Learning. 2017 In: IEEE Winter Conference on Applications of Computer Vision, pp. 697–704 (2017)
58. Timoshenko, J., Halder A., Yang, B., Seifert, S., Pellin, M.J., Vajda, S., Frenkel, A.I.: Subnanometer Substructures in Nanoassemblies Formed from Clusters under a Reactive Atmosphere Revealed Using Machine Learning. J. Phys. Chem. C. **122**, 21686–21693 (2018)
59. Sindhu, V., Nivedha, S., Prakash, M.: An Empirical Science on Bioinformatics in Machine Learning. J. Mech. Continua Math. Sci. spl(7) 86–94 (2020)

Chapter 10
In-Situ/Operando Synchrotron X-ray Imaging Techniques for Energy-Related Applications

Lei Du, Nan Sun, Yajie Song, Hanwen An, and Jian Liu

10.1 Introduction

Sustainable energy-related applications (e.g., batteries) have been emerging as key technologies in remitting today's energy crisis. Scientists have devoted great efforts to developing advanced devices to control and promote the conversion between renewable energy and electronic energy and apply these devices to replace fossil energy in human activities. By virtue of this progress, the initial performance of energy-related applications has been significantly improved in the past decades. Meanwhile, the demands for these applications also sharply increase in terms of performance (i.e., high energy density, high power density) and durability/reliability (i.e., long lifetime). This triggers interest in getting more achievements and milestones in the near future.

Either performance or durability is closely relevant to the electrode materials in these devices. The properties of the surface, phase of electrodes, as well as the interaction between electrodes and electrolytes significantly determine the capacity and kinetics of electrochemical reactions and thus the energy density and power density in batteries. Therefore, electrode materials are essential for high-performance devices. Great efforts have been devoted to rational material design, which relies on the deep understanding of relationships between the electrode materials and electrochemical behaviors, i.e., the underlying reaction mechanisms. On the other hand, the electrode surfaces, electrode phases, and the interfaces of electrode and electrolyte always evolve during intercalation and deintercalation processes. This

L. Du · N. Sun · Y. Song · H. An
School of Chemistry and Chemical Engineering, Harbin Institute of Technology, Harbin 150001, China

J. Liu (✉)
School of Engineering, Faculty of Applied Science, The University of British Columbia, Kelowna, BC V1V 1V7, Canada
e-mail: jian.liu@ubc.ca

J. Wang (ed.), *Advanced X-ray Imaging of Electrochemical Energy Materials and Devices*, https://doi.org/10.1007/978-981-16-5328-5_10

evolution leads to performance changes and even failures of batteries over time and operation. The degradation phenomena trigger great interest in digging into the degradation mechanisms and how the electrode degradation fails the battery.

Various characterizations are helpful to understand the reaction mechanisms and battery failure mode. Advanced techniques, such as spectroscopy, microscopy, chromatogram, thermal-analysis, have been widely employed to monitor the initial states of the electrode and/or electrolyte and the interfaces, as well as the changes over time/operation. However, ex-situ characterizations might involve some uncontrollable factors caused by transportation to the site and remarkably different operating conditions compared to the real battery environments. The in-situ/operando techniques are thus pursued by the communities.

In other chapters of this book, the principles and fundamental knowledge for X-ray imaging have been well-reviewed. In this chapter, we mainly focus on the in-situ/operando applications of X-ray imaging techniques. We will start with the reasons why in-situ/operando X-ray imaging techniques are important for energy application study, followed by the design of testing cells to meet the requirements of imaging experiments, in terms of high resolution and real-time monitoring, as realistic as possible. The most recent progress in understanding the reaction mechanisms and performance failure will be critically reviewed. Finally, the perspectives on the development of in-situ/operando X-ray imaging will be provided.

10.2 The Importance of In-Situ/Operando Imaging Techniques

Most investigations in today's researches on battery reaction mechanisms and degradation mechanisms are based on ex-situ techniques, such as ex-situ spectroscopy, microscopy, chromatogram, thermal-analysis. The environments for the sample storehouse are usually required to be dry, high vacuum, or with a specific organic phase, which, however, are different from the real-world battery environments. The initial electrode materials are characterized, properties of which are linked with the battery performance; for the degradation mechanisms study, the materials after battery operation are disassembled from the battery and characterized. However, these processes lead to uncontrollable factors in the analysis, for example, electrode oxidation by the air when transporting the samples from chamber to chamber and laboratory to laboratory, electrolyte carbonation by carbon dioxide when collecting and transporting samples. Involving these uncontrollable factors, the characterization results, on one hand, cannot reflect all the properties of materials in battery environments; on the other hand, undesirably, these results might mislead the understanding of reaction mechanisms and degradation mechanisms.

In order to remit the uncontrollable factors and reflect the materials' properties under battery environments and operation, in-situ/operando characterizations are needed. We herein have to emphasize the definition of in-situ and operando. In-situ technique refers to the physical characterizations under the conditions similar/same with the real working conditions, while the operando reflects the physical characterizations in real running devices [1]. The in-situ/operando characterizations are still challenging—not all the techniques are suitable to be employed for in-situ and, especially, operando characterizations.

On the other hand, the characterizations should be readable and intelligible. Most spectroscopy, chromatogram, thermal-analysis provide indirect results, requiring professional knowledge to interpret the underlying information. However, microscopy is a kind of "what you see is what you get" method, as mentioned above, the operating conditions are far from the battery environments. Therefore, an in-situ/operando imaging characterization method is attractive and promising in studying reaction mechanisms and degradation mechanisms of batteries. Thanks to the synchrotron light sources with high-flux, high-brightness electromagnetic radiation, the in-situ/operando methods can be conducted and performed. The imaging strategy, particularly X-ray imaging, relying on contactless and scatheless photons and electrons, can convert the indirect spectroscopy information into the direct image information that people can see. Thus, in-situ/operando X-ray imaging has been attracting broad interest.

10.3 Synchrotron X-ray Imaging Techniques and In-Situ/Operando Cells

10.3.1 Transmission X-ray Microscopy

The imaging principle of Full-field Transmission X-ray Microscopy (TXM) is similar to that of traditional optical microscopes. An objective lens subjects to the imaging principle to magnify and image objects. The difference is that most TXM uses hard X-rays with shorter wavelengths (>2000 eV) as the light source. As shown in Fig. 10.1 [2], the synchrotron X-rays are focused on tens of microns through a focusing mirror in TXM. Then, the unfocused light is filtered through a beam interceptor and a small hole, and the filtered X-rays are applied to the sample, and the Fresnel zone plate is magnified on the CCD detector to form a projection image. TXM technology can ignore the absorption of hard X-rays by air and often does not require a vacuum environment, which reduces the complexity of the experiment. At the same time, it uses the full-field imaging mode, that is, the sample size is smaller than the spot size at the sample position so that it can complete the two-dimensional transmission image acquisition of the sample in one exposure. Therefore, its significant advantage is that the imaging speed is fast, which reduces the total experimental time. For the TXM technology, the spatial resolution is determined by the width of the Fresnel zone plate outer ring [3]. At present, the precise Fresnel zone plate can achieve a two-dimensional spatial resolution of 20–30 nm.

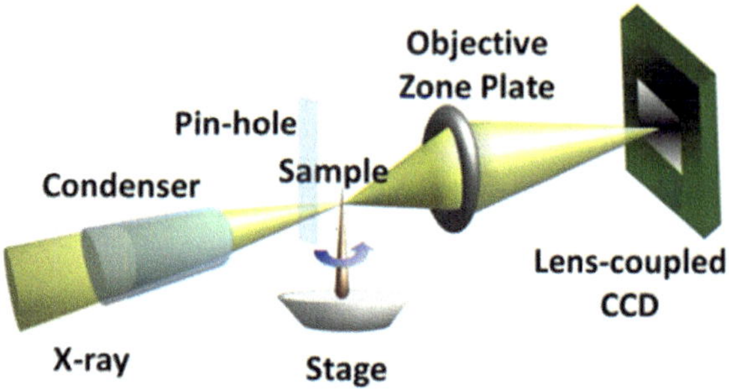

Fig. 10.1 Illustration of the TXM experimental setup [2]

Another unique feature of TXM is 3D imaging [4]. Generally, the three-dimensional image of TXM is obtained based on the superposition of a series of two-dimensional mapping images under different angles to obtain the internal structure information of the material. X-ray 3D tomography technology can realize the local morphology and chemical analysis by revealing the internal characteristics of the sample.

Considering the difference in sample thickness and chemical composition, the sample absorbs X-rays differently, resulting in TXM images often showing different contrasts, and the spatial distribution of chemical phases can be obtained through post-processing [5]. The principle is that the electrons in a specific shell of the atom to be measured can only be excited by a specific energy, thereby absorbing the X-rays of the corresponding energy. By continuously changing the X-ray energy before and after the absorption edge of the element to be measured, a series of absorption contrast images under different energies can be obtained. Based on the advantages of adjustable X-ray energy, combined with the development of the X-ray absorption spectrum spectroscopy imaging method, the element distribution, and valence distribution imaging of the elements of interest in the sample can be imaged. In this way, the correlation analysis between the function and the morphology of the sample is carried out on the mesoscopic scale to realize functional imaging.

10.3.2 Scanning Transmission X-ray Microscope

The scanning transmission X-ray microscope (STXM) uses a focused zone plate to focus X-rays to a spot with a size of tens of nanometers (far smaller than the sample size) and irradiates the sample at the focal point. The sample is moved or the spot is collected to scan the entire sample area. The detector records the rays passing through the sample to achieve the purpose of microscopic imaging. The schematic diagram of the imaging light path is shown in Fig. 10.2 [6]. After the X-ray is drawn from the accelerator, the monochromator selects a specific wavelength and then passes through a spatial filter to generate a secondary light source. Then the X-ray focusing element forms a concentrated spot. On the moving platform of dimensional scanning, a CCD detector is placed at the end of the system to collect the rays with sample information.

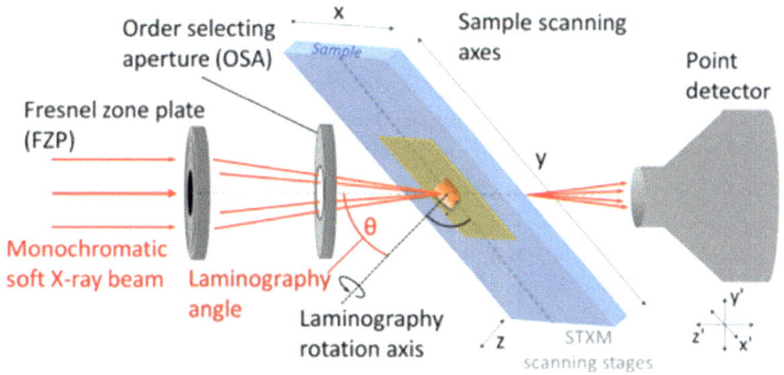

Fig. 10.2 Illustration of the STXM experimental setup [6]

The main differences between STXM and full-field TXM are as follows: First, the X-ray spot irradiated by the STXM system on the sample is much smaller than that of the full-field TXM. Usually, the full-field TXM system has a spot with size of between ten and tens of microns. The probe spot of the STXM system is below the order of micrometers, and it can even reach 10 nm at present, with higher resolution. Second, compared to TXM, STXM can study light elements (C, N, O, Li, etc.) The imaging resolution of STXM is higher. At the same time, the imaging area of STXM is more flexible and the X-ray dose is lower. Third, the STXM system does not have a zone plate as an objective lens behind the sample, and X-rays directly reach the detector after passing through the sample. Fourth, the STXM system detector records the sample information of each scan point, and finally stitches all records into a two-dimensional image, and the full-field TXM directly forms a two-dimensional image on the two-dimensional detector. Although STXM technology has many advantages, the sample generally needs to be placed in a vacuum chamber due to the severe absorption of soft X-rays by air [7]. Furthermore, since the penetration ability of soft X-rays to the sample is weak, this requires that the tested sample is very thin. Besides, STXM has exceptionally high requirements for synchronized radiation sources. This will bring great difficulties to the design and implementation of the experiment.

TXM and STXM are the most common synchronous radiographic microscope imaging systems. Many new synchrotron X-ray imaging technologies are developed based on these two. It can be said that TXM and STXM systems are synchronous radiographic imaging systems. The development and application of micro-imaging technology have laid a solid foundation.

10.3.3 X-ray Fluorescence Microscopy

For the detection of some battery materials modified by doping or surface modification and other trace elements, a highly sensitive analytical test method is required. X-ray fluorescence microscopy imaging technology (XFM) is effective testing technology in testing trace elements with high sensitivity. It can analyze the element concentration at the ppm level by detecting the excited secondary scattered electrons. When a substance is irradiated with X-rays whose energy exceeds the ionization energy of the electrons in the inner orbit of the atom, the inner electrons will be ionized, and vacancies will be generated in the inner orbit. At this time, the atom is in an unstable state, causing the electrons in the outer orbit of the atom to fall to the vacancy of the inner orbit. In this process, the potential energy of the electrons in different orbits will change in the form of secondary (fluorescent) X-rays. Launch out. Since the electron orbits of different atoms have specific energies, the fluorescent X-rays produced contain the intrinsic information of the material elements. The energy corresponds to the element type, and its intensity is proportional to the element content. Therefore, based on this, X-ray fluorescence can be used for elemental and chemical analysis of samples [8].

The experimental device of the X-ray fluorescence microscopy imaging system is shown in Fig. 10.3 [9]. Its main body is an STXM system. The difference is that the XFM system introduces a fluorescence detector near the sample grating. The basic idea is to obtain the morphology information of the sample through the ordinary detector of the STXM system, and use the fluorescence detector to receive the excited fluorescence signal of the illuminated sample area, to obtain the information of the sample morphology, element type, and chemical state at the same time. Therefore, in addition to element characterization, XFM can also be used for two-dimensional and three-dimensional imaging of target elements.

Nevertheless, caused by its low spatial resolution (generally micron and sub-micron) and time resolution (long exposure required), XFM severely limits its application in in-situ research on battery material reaction mechanisms [10].

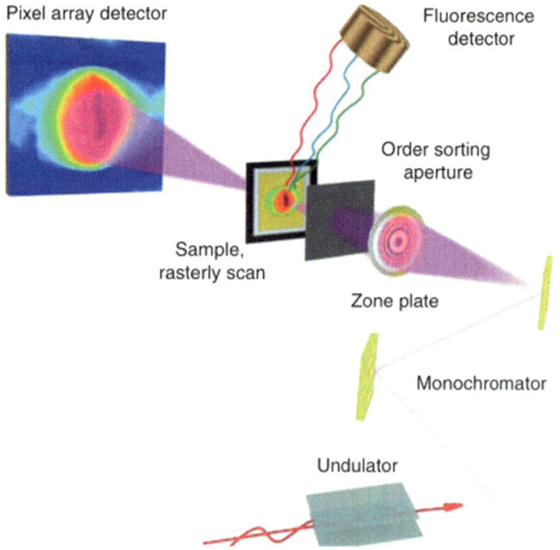

Fig. 10.3 Illustration of the XFM experimental setup [9]

However, with the development of the next generation of synchrotron radiation sources with low emission intensity, this technology will have great promise in the electrochemical reaction mechanism of battery materials.

10.3.4 Coherent X-ray Diffractive Imaging

Due to the introduction of lenses, the highest spatial resolution of the above-mentioned X-ray imaging systems is only about 10 nm, which is far from the limit of X-ray wavelength [11]. As shown in Fig. 10.4 [12], coherent X-ray diffractive imaging (CDI) is a "lensless" microscopy imaging method. Coherent X-rays are used to irradiate the sample, and a two-dimensional detector is directly used behind the sample to collect the far-field diffraction pattern. The rotation of the sample stage and the phase recovery algorithm can obtain the three-dimensional structure of the sample. Since there is no optical element inserted between the sample and the detector, CDI represents the X-ray imaging method with the highest photon efficiency.

CDI can measure the structure of amorphous and nanocrystals. By quantifying the incident and diffracted X-ray flux, the mass density of the material can be extracted to distinguish the different phases in the three-dimensional material. The biggest advantage of CDI imaging technology is its spatial resolution. The imaging resolution is only related to the wavelength and the diffraction angle of X-rays. At present, it has reached a spatial resolution of 10 nm. Besides, with the development of the next generation of coherent light sources (HHG and XFELs) to provide you with ultra-high brightness ultra-short X-ray pulses, CDI technology will achieve atomic-level theoretical spatial resolution and time resolution of <10 fs. On the other hand, extremely short X-ray pulses help CDI perform multi-site functional testing simultaneously on the nanometer scale, which will enable in-situ monitoring of the electrochemical reactions of battery materials during rapid charge and discharge to better understand the battery. The reaction mechanism of materials and the design and development of next-generation battery materials are guided [13].

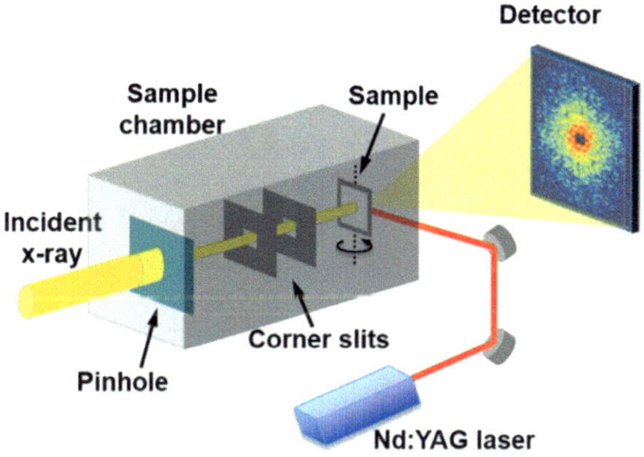

Fig. 10.4 Illustration of the CDI experimental setup [12]

CDI is a powerful tool for studying strain, dislocation evolution, electron density distribution, and lattice distortion under in-situ conditions. The strain diagram also shows how particle size, shape, and environment affect the strain distribution, which in turn affects electrochemical performance. Although CDI technology has such significant advantages in the research of battery material reaction mechanisms, the damage of synchrotron radiation light sources to battery materials, especially electrode materials in liquid electrolytes, will be a potential application of this technology in battery research. However, pre-cycle activation of the battery or the use of solid electrolytes may reduce the damage of the synchrotron radiation source to battery materials to a certain extent.

10.3.5 In-Situ/Operando Cells

In most cases, the design of the in-situ cell should suit the specific experimental technique. The design of the in-situ cell should follow the following principles: the cell is easy to disassemble and assemble in the synchronization device, the X-ray can reach the inspection area smoothly, and other components in the cell have no interference with the X-ray.

Several electrochemical cells have been used for in-situ/operando X-ray studies, including modified coin cells, Argonne's multipurpose in-situ X-ray (AMPIX) electrochemical cells, radially accessible tubular in-situ X-ray (RATIX) cells, as well as capillary type cells.

To facilitate the X-ray measurements, modified coin cells are widely used for in-situ experiments. As shown in Fig. 10.5a [14], modifications to the coin cell typically involve perforation of the X-ray attenuating, a conductive stainless-steel casing, and then sealing with Kapton (a thin polymer film) so that the X-rays can penetrate the cell. However, such a flexible window may cause uneven pressure distribution on the electrode, causing poor contact of the electrode close to the window, thereby affecting electrochemical performance. Cells for reliable operando measurements must mitigate these negative effects.

Recently, a multipurpose X-ray (AMPIX) cell has been developed by the US Argonne National Laboratory (ANL) with the capability to obtain high-quality X-ray data without compromising electrochemical performance. The main components of this cell include a cup-shaped body, top and bottom electrodes, an X-ray window, and a flat annular gasket as shown in Fig. 10.5b [15]. This cell can provide a more reliable electrochemical setup because of the uniform stack pressure applied by rigid X-ray windows and the formation of a high-fidelity hermetic seal.

In-situ study of the reaction process along the stacking direction will be helpful to improve understanding of the structure-activity relationship in electrochemical energy storage devices. However, the possible inhomogeneities in the electrode layer in traditional in-situ batteries might lead to difficulties in investigations. Recently, Chupas and colleagues in ANL designed a RATIX cell and successfully applied it to investigate the inhomogeneity of electrode materials along the

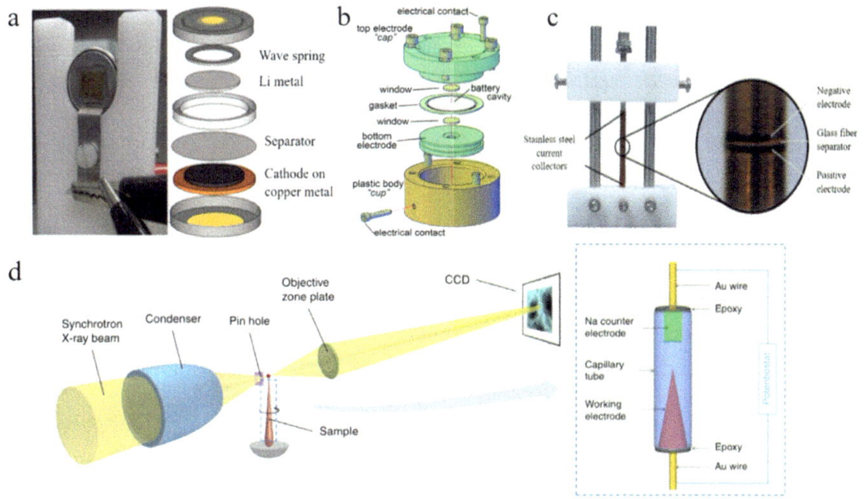

Fig. 10.5 Schematic illustration of the in-situ/operando cells for various synchrotron X-ray imaging techniques. **a** Modified coin cells [14]. **b** Argonne's multipurpose in-situ X-ray (AMPIX) electrochemical cells [15]. **c** Radially accessible tubular in-situ X-ray (RATIX) cells [16]. **d** Capillary type cells [17]

electrode stacking direction, as shown in Fig. 10.5c [16]. The RATIX cell has several key components: the electrode stack, sealing tube, two electrode pins, fixing screws, base, and top assembly. The cell is compatible with a variety of synchrotron-based scattering, absorption, and imaging methodologies.

Despite the advances in electrochemical cell design, most of the above electrochemical cells are designed for 2D measurements. Although some of them can achieve 3D imaging by rotating the plating cell, there will be some angular loss in this imaging method. Therefore, it is necessary to develop a working electrochemical cell that can perform X-ray imaging in the full-angle range. Recently, as shown in Fig. 10.5d [17], a novel electrochemical cell has been designed by the Brookhaven National Laboratory, where the column wall of the cell is a thin-walled quartz tube with low attenuation to X-rays. This structure allows full-angle X-ray imaging and can obtain high-quality 3D reconstruction images.

10.4 Applications Toward Reaction Mechanism Study

Scanning transmission X-ray imaging technology (STXM) is a powerful X-ray imaging technology with high spatial resolution and high chemical sensitivity, whose main application energy range is the soft X-ray range (200–2000 eV) [18]. This technology has been successfully applied in studying energy materials and can reveal the correlation between material properties and chemical structure.

For STXM technology, monochromatic X-rays are first focused to a small size (from a few hundred to tens of nanometers) through a rotating ellipsoid focusing lens, and then the focused along with the horizontal and vertical directions of the sample. The sample is scanned and the information is collected by recording changes in X-ray intensity before and after passing through the sample. For scanning imaging technology, compared to holographic imaging, the general imaging time is longer because STXM photos are combined and superimposed by scanning each pixel in turn. STXM has its unique advantages: on the one hand, the 10% effective rate of the zone plate brings less X-ray damage; on the other hand, it has relatively better sensitivity to lighter elements [19]. However, the detection thickness of the sample is relatively thin using STXM because the soft X-ray has to overcome the volatilization of liquid electrolytes during in-situ battery testing in a vacuum chamber. One of the strategies is to design an electrochemical reaction battery without liquid electrolytes, i.e., all-solid-state batteries. Ohmer et al. [20] studied micron-sized thin-film $LiFePO_4$ batteries by in-situ STXM method and observed the migration and diffusion process of the two-phase interface between single-crystal particles inside $LiFePO_4$. Combined with the Fe L_3 edge absorption spectrum, they observed a significant move of Fe absorption edge during lithium extraction from $LiFePO_4$, indicating the oxidation of Fe^{2+} to Fe^{3+}. The in-situ STXM chemical phase distribution map reveals the phase transition mechanism and the rapid movement of the two-phase interface along the (010) direction during the lithium extraction process for larger $LiFePO_4$ single-crystal particles, Therefore, the authors conclude that the growth of $LiFePO_4$ and $FePO_4$ phase is dominated by the stretching effect rather than diffusion control. This new discovery of the $LiFePO_4$ phase transition mechanism shows the importance of the in-situ STXM imaging method in the battery field (Fig. 10.6).

The development of an in-situ STXM method that can work in the liquid phase to study the battery reaction process is necessary since most of the electrochemical reaction process is carried out in the liquid phase (such as liquid electrolyte) and the gas phase (fuel cell electrocatalytic process). However, STXM needs to work in a vacuum environment, leading to a big challenge. de Smit et al. [21], for the first time, reported a method using in-situ STXM technology to observe the catalytic reaction process of catalyst on the nanoscale. The author designed a nanoreactor (micro-electrochemical reaction system) consisting of a reaction chamber connected to two flowing gas pipes. Two 1.2 μm thick SiN_x windows are used as windows to minimize the absorption of soft X-rays. An STXM vacuum chamber is used to separate them to allow gas to flow through, thus forming a nanoreactor. By using this gas-based reactor, the author successfully conducted an in-situ STXM experiment and analyzed the changes of the iron-based Fischer–Tropsch catalyst. They deeply analyzed the complex catalytic reaction of the process of local carbon group generation by combining with the captured conditions, when the temperature reached 350 °C under the environmental pressure. Therefore, further design of a nanoreactor, which can be used in more complex battery systems containing liquid electrolytes, will be necessary for the battery field study. Chueh et al. [22] reported an in-situ STXM study of $LiFePO_4$ materials via utilizing a multi-mode

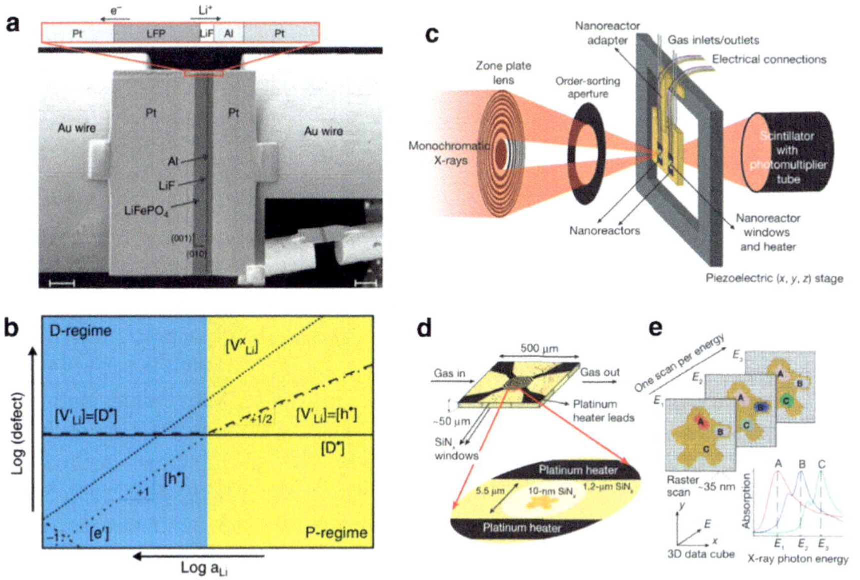

Fig. 10.6 **a** SEM image of the micrometer-sized battery; **b** diagram-illustration of the defect concentration as a function of lithium activity in LFP; **c** diagram of the in-situ STXM technique [21]; **d** close-up of the nanoreactor, showing the windows and the embedded heater spiral; **e** diagram of a typical STXM data acquisition method [20]

microfluidic electrochemical cell. The author developed a commercial micro-reactor (Hummingbird Scientific) by assembling two sandwich-structured Si flakes and a 75 nm SiN_x window. A single layer of $LiFePO_4$ particles is dispersed on the gold collector as the working electrode, and the lithium metal flakes are placed in the syringe cavity outside the STXM cavity as the counter electrode and reference electrode. In order to reduce the absorption of soft X-rays by other substances, the distance between the two Si wafers is only 1 μm to reduce the amount of electrolyte. The author observed dozens of individual $LiFePO_4$ particles, and the inhomogeneity of the electrochemical reaction on the $LiFePO_4$ crystal particles was clarified during the charging and discharging process. The study demonstrates the increased deintercalation rate of lithium on the $LiFePO_4$ particles as the charge and discharge current increase. As the degree of Li_xFePO_4 delithiation deepens, the heterogeneity in the spatial distribution of the chemical phase is caused by the uneven solid-liquid region formed at different magnifications. However, different reaction mechanisms and different heterogeneous reaction processes are presented during lithium insertion, which is caused by the heterogeneity of the chemical composition during the lithium insertion process. Consistent with the scalable phase transition process as described above, the ex-situ shelving experiment showed that a two-phase interface formed between the lithium-rich and the lithium-poor phases. This uneven chemical phase and composition distribution are related to the instability of the material during the cycle. Generally speaking, through the in-situ electrochemical reaction cells designed for different purposes introduced above, and further combined with XANES technology and in-situ STXM technology, it can be used to study the phase transition mechanism and other chemical information (such as the chemical phase distribution of different valence states, etc.), which is similar to the hard X-ray TXM technology. Because its working energy range belongs to the soft X-ray section, STXM has a unique advantage in detecting lighter elements such as carbon, sulfur, oxygen, etc., as well as the L-side chemical structure information of many transition metals as electrode materials [19, 23]. However, the poor sample detection depth of soft X-rays and the high vacuum working environment has brought huge challenges to the technical development of in-situ experiments, which requires more attention to sample preparation and battery structure design.

Synchrotron X-ray fluorescence imaging (XFM) is a technique that could scan a large sample area with a good spatial and temporal resolution to analyze the distribution of chemical elements [24]. Compared with other imaging technologies based on transmission mode, in the test of the electrochemical system, the influence of electrolytes on the transmission X-ray absorption can be eliminated. XFM technology has its unique advantages in the research of in-situ electrochemical reactions. In addition to detecting the types of chemical elements, this technology can also quantitatively analyze the concentration information of chemical elements. While the rays gradually scan the energy near the absorption edge of the element under study, the spatial information of the sample is scanned in scanning mode, XFM imaging technology can analyze chemical information, and the surrounding environment information of the studied metal atoms, especially the structure information of the transition metal atoms can be detected in the selected area, which also makes the technology can be better applied to the field of battery materials. Li et al. [18] developed an in-situ SXFM technology that can be used in liquid electrolyte systems, and the X-ray energy used in this technology is in the soft X-ray range. Through this online experiment, the author observed the migration process of lithium in $LiFePO_4$ particles (flaky and ellipsoidal) with different morphologies during charging and discharging. It is found that the reaction mechanism of this lithiation path is related to the morphology of the particles to some extent. For the ellipsoidal $LiFePO_4$ particles, a mosaic-shaped phase transition is formed in the particles during the process of deintercalating lithium, and the lithium-rich phase and the lithium-poor phase are scattered in the particles at intervals. However, for the flake-shaped $LiFePO_4$ particles, as the lithium extraction process continues, most of the flake-shaped particles exhibit a uniform phase distribution, indicating a higher electrode efficiency. This different lithium-ion deintercalation path and mechanism related to particle morphology may be caused by the respective surface reactivity of particles with different morphologies and the synthesis conditions of different samples. Yang et al. [25] used in-situ XFM imaging technology to directly track the evolution of the spatial distribution of sulfur and polysulfides during the reaction of lithium-sulfur batteries. During the charging and discharging process, the author directly showed the dissolution and deposition redistribution process of the sulfur electrode through the XFM image and combined the XANES technology to characterize the chemical information of sulfur and polysulfides. The in-situ XFM test results show that long-chain polysulfide compounds are formed at the beginning of the discharge process, and then these generated polysulfides chemically react with the electrolyte and dissolve into the electrolyte, which leads to a continuous loss of active electrode material and consequently a continuous decline in battery capacity. On the other hand, the polysulfide dissolved in the electrolyte may react with the lithium metal electrode under certain conditions, and then self-discharge occurs, resulting in the lower coulombic efficiency and poor cycle stability of the lithium-sulfur battery. With a larger observation area that can even be as large as centimeters, in-situ XFM imaging technology can deeply study the large-scale electrochemical reaction process of lithium-sulfur batteries, and even the reaction process of the entire electrode, so as to use this unique technology to better understand the reaction mechanism of lithium-sulfur batteries during the reaction process. In addition to being applied to lithium-sulfur battery systems, in-situ XFM imaging technology can also be extended to track the formation, growth, and dissolution process of zinc dendrites in zinc-air battery systems, providing a new way for in-depth study of the attenuation mechanism of zinc electrodes Strategies and methods.

Coherent light diffraction imaging (CDI) holds the advantage of high sensitivity and could characterize the three-dimensional image of the internal stress distribution of single-crystal particles on a three-dimensional scale [26]. Utilizing the third-generation synchrotron radiation source, coherent X-rays can be obtained by setting the width of the JJ slit in the X-ray incident direction to be smaller than its interface length. When a beam of coherent X-rays passes through a single-crystal particle, coherent diffraction spots are generated, which can be converted into real-space images by the phase recovery method [27]. Therefore, the local dislocation information on the single-crystal particles can be detected by this method. As a supplement to many of the above-mentioned imaging technologies, this new X-ray imaging technology can detect local defects and lattice distortions of battery materials during charging and discharging online [28, 29]. In 2015, Meng et al. [30] applied in-situ CDI technology to the field of battery materials for the first time. The author tracked the three-dimensional defect distribution of single-crystal $LiNi_{0.5}Mn_{1.5}O_4$ particles with disordered spinel structure in real time, revealing the uneven stress distribution of the material during electrochemical charging and discharging. It is well known that the presence of dislocations in crystalline materials will release the internal stress in the crystals to prevent the formation of possible cracks and the destruction of the microstructure [31]. The stress inhomogeneity of electrode materials under different charge and discharge voltages can be presented through the in-situ CDI imaging technology. This uneven stress distribution illustrates the phase-field model of the material in the process of lithium extraction. When the phase change occurs during charging and discharging, the boundary dislocations of the electrode material will transfer with the charge. The lithium-rich phase nucleates near the dislocation, but the growth process after nucleation is uneven. The dislocation field in the material indicates the elasticity of the material, and the Poisson coefficient in the high-voltage range is significantly different from the lower voltage range. These findings will have practical significance for the development of high-voltage cathode materials for high-capacity batteries with greater stress tolerance in the high-voltage range. Then, Meng et al. [32] also used CDI technology to characterize and analyze the correlation of the stripe morphology during the evolution of the three-dimensional stress distribution of single-crystal $LiNi_{0.5}Mn_{1.5}O_4$ nanoparticles during charge and discharge. When the size is greater than 50 nm, the material will form a striped phase during the charge and discharge phase transition. The author also analyzed the distribution of elastic energy. The results show that for a single particle, the charging process and the discharging process exhibit completely different elastic energy distributions, which is related to the difference in the three-dimensional stress field (Fig. 10.7).

10.5 Applications Toward Performance Failure Research

It is very important to trace the complex phase transition path in battery materials, and correlate the morphological change, chemical composition, and phase structure with electrochemical performance for exploring advanced battery systems with high energy density and safety. The emerging synchrotron radiation X-ray imaging technology with high spatial, temporal and chemical resolution provides a unique tool. Here, we introduce some typical examples of using field/operational synchrotron radiation X-ray imaging technology to understand performance failures.

Lithium nickel cobalt manganese oxide (NCM) materials have always been one of the most popular commercial cathode materials, and single-crystal materials have better performance than their crystalline counterparts. However, as the nickel content and the upper cut-off voltage increase, these single-crystal materials also exhibit severe structural instability and capacity fade. Ni-rich single-crystal $LiNi_{0.6}Co_{0.2}Mn_{0.2}O_2$(NCM622) was selected as a model electrode by Wang et al. [9]. There is the layered phase in the bulk of NCM and the rock-salt phase in the surface region. The failure of electrode material after 200 cycles is mainly due to the oxidation of nickel on the surface. In addition, the nickel rich rock-salt phase on the surface hinders the transport of lithium ions. Operands TXM were performed before and after 200 cycles to monitor composition and structure evolution. In the initial cycle, redox reaction occurs in all directions, and the front of phase transition propagates rapidly, as shown in the quantitative chemical diagram of operands in Fig. 10.8a. A nearly complete phase transition occurs in the final charge discharge stage, which indicates that the electrochemical reversibility is very strong. However, the high reversibility can not exist after 200 cycles, which is clearly revealed by the chemical phase mapping of the operating number TXM. In the 201th charging cycle, it is found that about 50% of the phase composition can not be restored to the original state, and a highly uneven phase distribution appears in the single-crystal NCM. In addition, in the 201th cycle, the phase transition at the same position is highly irreversible. The two-dimensional histogram highlights the inhomogeneity of the distribution of nickel oxidation states in single-crystal

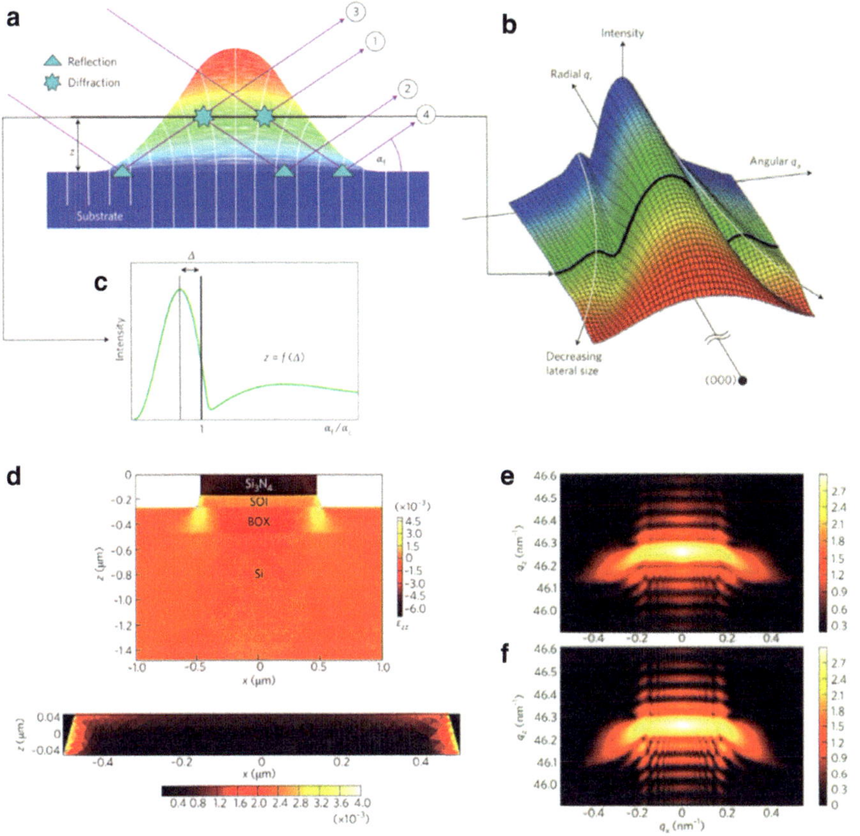

Fig. 10.7 **a** Cross-section of a quantum dot in which the lattice parameter increases from the bottom to the top; **b** diffraction pattern of the quantum dot, close to one of the substrate reflections. The deviation along the radial *qr* direction (black slice along the *qa* direction and color coding) identifies the contribution from a particular height *z*, in the structure. **c** Intensity as a function of the exit angle *αf* in units of the critical angle *αc*. **d** *xz* cross-section of the measured structure showing the FEA calculation of the *εzz* strain component shown on the indicated color scale. A magnified view of the silicon on insulator (SOI) portion is shown below on a finer color scale. The supporting oxide layer is denoted BOX. **e** Measured diffraction pattern surrounding the 004 reflection of the SOI structure as a function of its *qx* and *qz* components. **f** Corresponding diffraction pattern calculated for the structure [29]

particles. At the same time, the inhomogeneous phase transformation causes the inhomogeneous internal stress, which leads to a large number of cracks on the particles, and finally leads to the failure of the properties. This study elucidates the relationship between surface chemistry, phase transition, and performance maintenance, thus guiding the development of advanced single-crystal battery materials with improved electrochemical and safety characteristics.

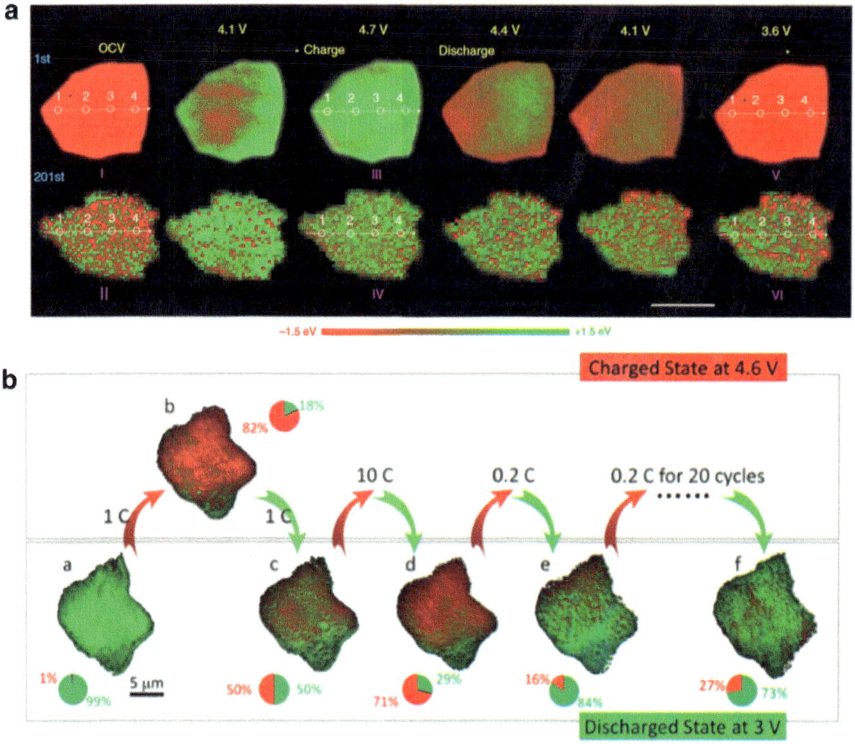

Fig. 10.8 **a** Spatial-resolved electrochemical evolution of Ni-rich single-crystal NCM622 after the 1st and 201st cycle. **b** In-situ monitoring of the chemical heterogeneity in a single particle of $LiCoO_2$ up to 20 cycles [33]

Layered material $LiCoO_2$ was chosen to study of degradation mechanisms of cathode electrode materials in LIB by Xu et al. In many microscopic studies, different particles are chosen at different SOCs and from different cycles due to the lack of capability to keep tracking the same particle over multiple charge-discharge cycles, causing some arbitrariness to the data and making the comparison more difficult. To tackle this challenge, Xu et al. designed an in-situ pouch cell for tracking and studying selected electrode particles over many cycles [33]. To unravel the degradation mechanisms of $LiCoO_2$ when operated over a voltage range of 3–4.6 V, in-situ transmission X-ray microscopy (TXM) is used to visualize the $LiCoO_2$ particle's response under various cycling conditions. As shown in Fig. 10.8b, a strong negative correlation was observed between the particle recovery and the cyclic C rate. After 20 cycles at 0.2 °C, the redistribution of inactive domains was observed at the sub particle level, indicating that the particles have the ability to readjust the repeated and stable reaction conditions. In addition, through the comprehensive analysis of 2D test and 3D data, it is found that the crack is closely related to the inactive region, which indicates that there is a close relationship between the morphological degradation and chemical inhomogeneity. This work also emphasizes the formation of metallic cobalt particles in cells due to the dissolution and precipitation of cobalt cations. These insights are of constructive significance for the design of new battery systems with higher energy density and longer cycle life.

Alloying anode materials have high theoretical capacity and conductivity and are considered as candidate materials for next-generation rechargeable batteries. However, the rapid decay of capacity in the cycle process has always been a major obstacle to its industrial application [34]. Through in-situ three-dimensional (3D) synchrotron radiation X-ray nanoimaging, Wang et al. Described the 3D quantitative analysis and nanoscale visualization of the microstructure evolution of tin electrodes during cycling [3]. We can directly observe the volume expansion in the first lithiation process, and the serious pulverization and fracture appear when the first lithium-ion extraction and volume contraction occurs. After the second diagenesis, the grains reach structural equilibrium, and there is no further obvious morphological change. This phenomenon indicates that the structural instability of Sn during the initial delithiation and subsequent lithiation is the main reason for the rapid capacity degradation and mechanical properties degradation (Fig. 10.9a, b). In addition, Wang et al. Also studied the situation of Sn as the anode material of sodium ion battery (NIB) [17]. After overcoming the technical/in-situ experimental difficulties in NIB system, the in-situ hard X-ray nano imaging technology was used to track the 3D morphology evolution of tin Anodes during (de)lithium and (de)-sodium processes. The different failure mechanisms of Sn anodes in NIB and LIB systems are demonstrated through comparative study. In LIBs, when a large number of particles are comminuted, the drastic change of their microstructure mainly occurs in the process of lithium removal. On the contrary, the insertion of sodium ions leads to the bulk expansion and fracture of nib (Fig. 10.9c, d). Interestingly, there is an unusual sodium removal equilibrium in multiple electrochemical cycles. Through the three-dimensional quantitative analysis after the first cycle, Sn particles show good microstructure stability, which is further supported by the same in-situ three-dimensional experiment. The reversibility of three-dimensional microstructure and the development of curvature characteristics explain the exciting structural equilibrium, which makes NIB have high electrochemical cycle stability. These findings may influence the development of alloy anode materials.

Due to the excellent theoretical capacity of silicon (4200 mAh g^{-1}, about 10 times the traditional graphite anode), the specific energy of silicon-based batteries can be significantly improved, so as to improve portability and prolong the service life after a full charge. The large volume change of 400% and the continuous development of SEI layer are widely considered to be the main reasons for the short life of silicon anode materials. Some studies have used atomic scale first principles or TEM methods to understand the changes of anode materials, but the periodic microstructure evolution of silicon, which is necessary to understand the failure mechanism of electrode, is still unclear. Dong et al. Used in-situ and operational synchronous X-ray imaging techniques to observe the evolution of internal

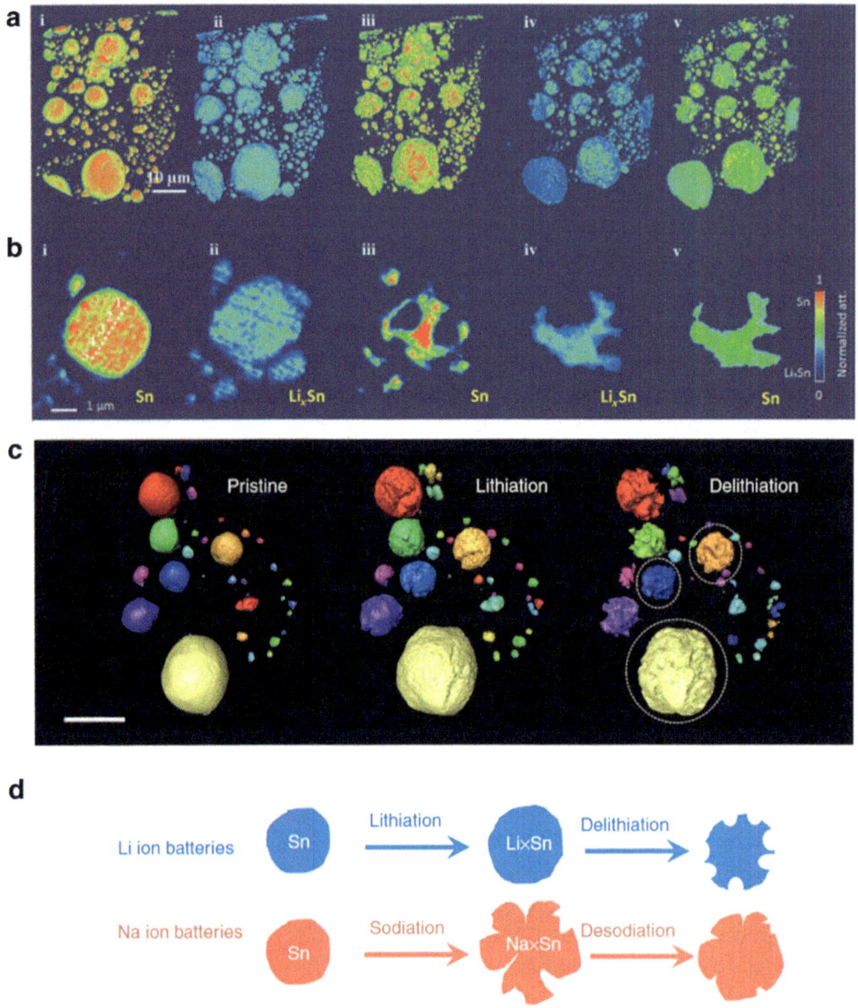

Fig. 10.9 **a** 3D morphological evolution of Sn particles during the first two lithiation–delithiation cycles. **b** Pseudo cross-sectional images of a single Sn particle during the first two cycles. **c** 3D visualization of Sn particles at the first lithiation-delithiation process. **d** Schematic illustration showing the difference of Sn microstructural change in NIB and LIB [17]

microstructure and quantify the volume change during battery operation [35]. The dynamic imaging clearly shows the microstructure evolution, volume expansion, and shrinkage of single silicon particle during the process of lithiation and delithiation. It is found that the expansion/contraction behavior is related to the initial particle size, but not to the depth of lithium removal. In actual battery operation, the changing conductive network and interface chemical environment (such as electrolyte migration and SEI propagation) may be the determinants of the volume change of silicon particles. The phenomenon of expansion and extension after diagenesis is observed and quantified. These results are helpful to understand the degradation and failure of silicon electrodes, and make the design and optimization of alloy anode with excellent performance more targeted.

For lithium-sulfur batteries, the shuttle effect of polysulfide is the main reason to hinder its commercialization. In order to design sulfur electrodes better, we need to have a more comprehensive understanding of the failure mechanism. However, many studies are limited to remote methods. Cui's team tracked the structural and morphological changes of individual particles in real time through the operator TXM [36]. As shown in Fig. 10.8, it can be observed from the operating number TXM micrograph that most of the morphological changes occur in images a–c, corresponding to the first discharge platform, on which elemental sulfur is reduced to higher-order lithium polysulfide. During this plateau, the particles expand slightly, but at the same time, some active substances are lost due to the dissolution of polysulfides, resulting in smaller particles with more porous appearance and increased background X-ray absorption. The results also show that the total particle size decreases and the porosity increases. These micrographs show that most of the soluble polysulfides are captured rather than lost to the electrolyte, which demonstrates part of the effectiveness of this form in reducing capacity decay. However, the escaping polysulfides solidify as insulating films, and this loss leads to the observed capacity decay. Therefore, the design of complex S/C Electrode morphology can completely capture the soluble polysulfides and improve the performance of lithium-ion batteries and battery life (Fig. 10.10).

In-situ/operational synchrotron radiation X-ray imaging technology has been further applied to the research of the most advanced solid-state batteries. The main advantage of solid-state batteries is the improved safety, which is due to the absence of typical flammable liquid electrolytes in lithium-ion batteries. However, the electrochemical performance of most solid-state batteries is still unsatisfactory, which is usually due to some interface problems. It is important to explain the behavior of electrode electrolyte interface, but it has been proved difficult. Sun et al. Carried out in-situ TXM measurements to track the morphological evolution of FeS_2 electrode particles in sulfide solid electrolyte and traditional liquid electrolyte systems [37]. At the initial stage of discharge, the conversion electrode particles show typical volume expansion and crack initiation and propagation in both solid and liquid systems. The difference is that in the solid-state battery, the subsequent phase propagation speed slows down and the phase transition almost ends at low discharge potential, while the continuous expansion of particles is observed in the traditional liquid electrolyte battery. The cross-section view of the discharged FeS_2 shows the core-shell structure, which further indicates the termination of the conversion reaction of the solid-state battery at low potential (Fig. 10.11). Therefore, the performance failure of FeS_2 in the solid-state battery system is due to the contact loss between FeS_2 and solid electrolytes in the initial electrochemical cycle caused by excessive volume expansion and non-uniform axial strain. Unlike liquid electrolytes, solid electrolytes can flow and penetrate the cracks or pores on the electrode. Once continuous ion/electron contact loss occurs, capacity attenuation is

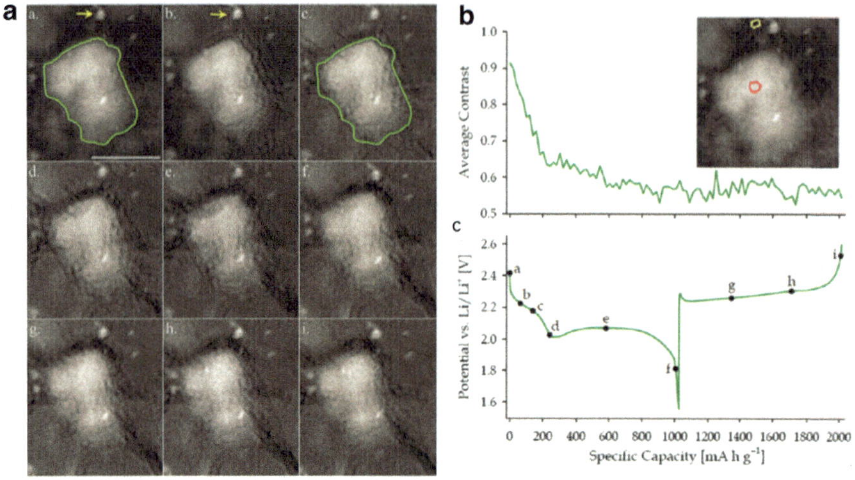

Fig. 10.10 **a** In operando TXM micrographs of a sulfur/super P composite particle during operation, where the letters correspond to points along the electrochemical cycle labeled a–i in Fig. 10.8c. **b** Change in average contrast versus specific capacity in TXM. **c** The cell potential [36]

inevitable, and solid electrolytes are not easy to fill the gaps. In-situ/operation synchrotron radiation X-ray imaging technology provides a new powerful tool for the phase transformation of solid-state batteries and reveals the solid-solid interface problems in solid-state battery electrodes.

For battery safety, thermal runaway is the biggest challenge. The performance failure of battery often comes from it, so it is very necessary to prevent and slow down the thermal runaway of battery. Shearing and his colleagues developed operado high-speed synchrotron radiation X-ray tomography to track the evolution of structural damage and thermal behavior during the initiation and propagation of thermal runaway in lithium-ion batteries [38] (Fig. 10.12). When the shell temperature exceeds 100 °C, the evolution of bubbles between the electrode layers is observed by tomography, which leads to the delamination of the electrode layer. The formation of cavitation reduces the interface area of electron and ion conduction between the electrode layers. The continuous gas generation leads to the exhaust of the battery, and the sudden release of pressure may lead to the collapse of the internal structure of the battery. It is envisaged that the use of an effective thermal management system and the optimal design of the battery structure will greatly improve its safety performance. At the same time, this study emphasizes that operator tomography as a tool to guide the development of lithium-ion battery safety applications.

The rapid development of synchrotron radiation X-ray imaging technology, especially the in-situ experiment of synchrotron radiation X-ray imaging, helps us to observe the dynamic evolution of electrode material morphology and structure intuitively, and deeply understand the performance failure of battery [39–41].

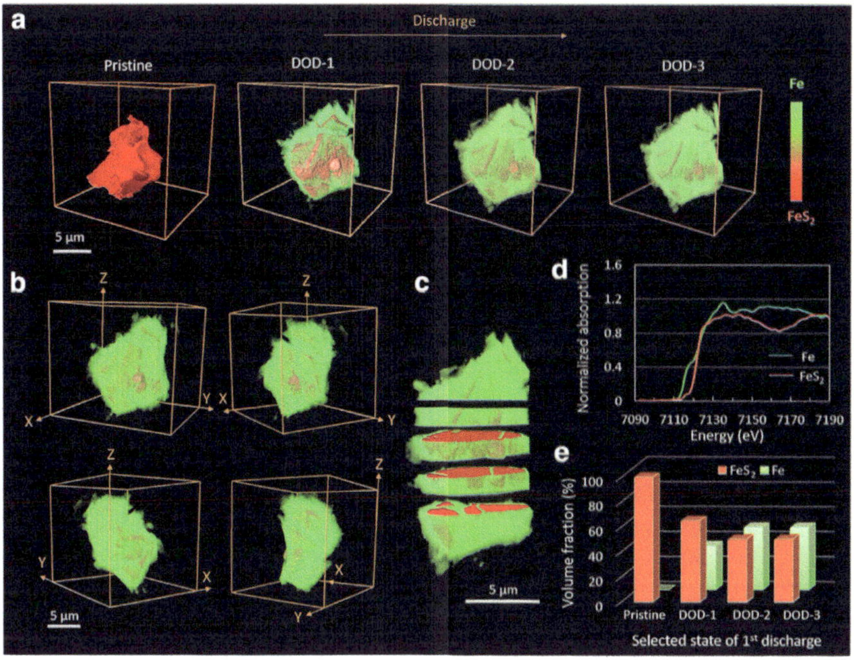

Fig. 10.11 **a** Phase distribution as a function of discharging states. **b** Discharged FeS_2 at DOD-3 from different orientations. **c** Cut-view of FeS_2 at lithiated state. **d** The XANES spectra of standard FeS_2 and Fe. **E** Phase volume fraction obtained from 3D quantitative analysis [37]

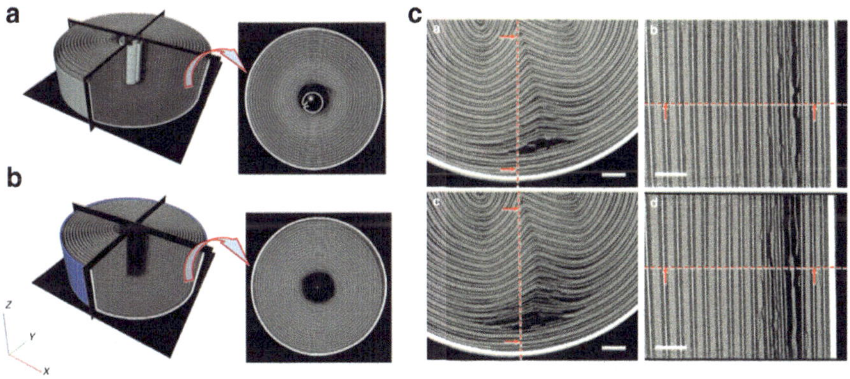

Fig. 10.12 **a** 3D reconstruction with orthoslices in the *XY*, *YZ* and *XZ* planes of the 2.6 Ah battery (Cell 1) with isolated XY slice. **b** 3D reconstruction with orthoslices in the *XY*, *YZ* and *XZ* planes of the 2.2 Ah battery (Cell 2) with isolated *XY* slice. **c** Grey-scale slices from 3D reconstructions during in-operando high-frequency tomography of Cell 2. Scale bar, 1 mm [38]

The elucidation of the failure mechanism provides theoretical guidance for the design of advanced materials, electrodes or batteries, and makes it possible to further realize the industrial application of high-performance batteries with high capacity and long cycle life.

10.6 Summary and Perspectives

The in-situ/operando X-ray imaging methods have been significantly developed in the past years, based on the fast development of synchrotron light sources. Today's in-situ/operando X-ray imaging can have a higher spatial resolution in nanometer and temporal resolution in sub-second, which have great advances in understanding both the reaction mechanisms and degradation mechanisms of energy materials in batteries. Due to the great progress in these fields, new knowledge and experiences in understanding existing materials, as well as designing and developing advanced battery materials, are achievable. To further develop this in-situ/operando X-ray imaging technique, exceptionally, in energy-related studies, some perspectives might be of help as follows.

(i) To design better in-situ/operando cells. The design of appropriate cells is critical for in-situ/operando X-ray imaging investigations, especially for 3D imaging. The cells should be similar/same with the battery pack, easily maintained, fabricated, and operated. Modified coin battery cells are workable, but the electrochemical behaviors of coin cells cannot completely be translated into the real-world battery, i.e., pouch battery, leading to a mismatch in the understanding of real-world battery in reaction and degradation mechanisms. For the real-world battery, it is still challenging to design desired cells for in-situ/operando imaging study.

(ii) To enhance the resolutions. Both spatial and temporal resolutions are critical for in-situ/operando imaging. The advanced fabrication methods for X-ray zone plate optics will be helpful in enhancing the spatial resolution, while the brighter light sources are powerful in improving the temporal resolution, i.e., probing fast electron transportation dynamics. Therefore, newly designed optics and light sources are essential in future research and developments.

(iii) To reduce the beam damage. As mentioned above, the higher-brightness light sources are helpful in the highly time-resolved probe, which, however, could damage the materials in batteries. For the organic Li-ion batteries, this problem is highlighted. This problem also brings another uncontrollable factor, i.e., the incent X-ray, on the performance and degradation of materials. To fully mitigate this beam damage is impossible, but some efforts such as using controlled beam exposure dose and stable electrolyte are promising in relieving this problem.

(iv) To develop maneuverable software. Besides the above-mentioned hardware, the coupled software for data collection, analysis, and report, which is easy to operate, is vital for the users. The technique serves research—a maneuverable software would help broaden the applications of the in-situ/operando X-ray imaging techniques.

(v) The most popular X-ray imaging methods have their unique pros and cons; the other synchrotron X-ray techniques such as X-ray diffraction, X-ray photoelectron spectroscopy are also influential in their fields. To compensate for the shortages of each X-ray imaging method and develop an advanced multiple characterization workflow to understand the energy-related materials deeply, the combination of different in-situ/operando X-ray imaging methods and other X-ray techniques is promising. Again, the rational design of test cells, how to enhance the resolution and reduce the beam damage are still challenging. The compatibility of the in-situ/operando cell for different methods is another urgent challenge.

The applications of in-situ/operando X-ray imaging methods in battery investigations have been becoming prospering. With the recent advances and hopeful progress in the near future, we believe that the in-situ/operando X-ray imaging will be a powerful strategy to understand electrochemical behaviors of battery materials and to provide solid foundations for rational battery material design. More breakthroughs will be led by these methods in the future. Beyond battery study (i.e., material and energy fields), the techniques can also be applied in other fields, including biological, medicinal, physical, and environmental fields.

References

1. Zhu, Y., Wang, J., Chu, H., Chu, Y.-C., Chen, H.M.: In situ/operando studies for designing next-generation electrocatalysts. ACS Energy Lett. **5**(4), 1281–1291 (2020)
2. Wang, J., Chen-Wiegart, Y.C., Wang, J.: In situ chemical mapping of a lithium-ion battery using full-field hard X-ray spectroscopic imaging. Chem Commun. (Camb) **49**(58), 6480–6482 (2013)
3. Wang, J., Chen-Wiegart, Y.C., Wang, J.: In situ three-dimensional synchrotron X-ray nanotomography of the (de)lithiation processes in tin anodes. Angew Chem Int. Ed Engl. **53** (17), 4460–4464 (2014)
4. Wang, J., Karen Chen-Wiegart, Y.C., Eng, C., Shen, Q., Wang, J.: Visualization of anisotropic-isotropic phase transformation dynamics in battery electrode particles. Nat. Commun. **7**, 12372 (2016)
5. Kuppan, S., Xu, Y., Liu, Y., Chen, G.: Phase transformation mechanism in lithium manganese nickel oxide revealed by single-crystal hard X-ray microscopy. Nat. Commun. **8**, 14309 (2017)
6. Witte, K., Spath, A., Finizio, S., Donnelly, C., Watts, B., Sarafimov, B., Odstrcil, M., Guizar-Sicairos, M., Holler, M., Fink, R.H., Raabe, J.: From 2D STXM to 3D imaging: soft X-ray laminography of thin specimens. Nano Lett. **20**(2), 1305–1314 (2020)
7. Spath, A.: Additive nano-lithography with focused soft X-rays: basics, challenges, and opportunities. Micromachines (Basel) **10**(12) (2019)
8. Petibois, C.: Imaging methods for elemental, chemical, molecular, and morphological analyses of single cells. Anal Bioanal Chem. **397**(6), 2051–2065 (2010)

9. Zhang, F., Lou, S., Li, S., Yu, Z., Liu, Q., Dai, A., Cao, C., Toney, M.F., Ge, M., Xiao, X., Lee, W.K., Yao, Y., Deng, J., Liu, T., Tang, Y., Yin, G., Lu, J., Su, D., Wang, J.: Surface regulation enables high stability of single-crystal lithium-ion cathodes at high voltage. Nat. Commun. **11**(1), 3050 (2020)
10. Barpanda, P., Oyama, G., Nishimura, S., Chung, S.C., Yamada, A.: A 3.8-V earth-abundant sodium battery electrode. Nat. Commun. **5**, 4358 (2014)
11. Jianwei Miao, T.I., Robinson, I.K., Murnane, M.M.: Beyond crystallography: diffractive imaging using coherent X-ray light sources. Science **348**(6234), 6 (2015)
12. Kim, Y., Kim, C., Ahn, K., Choi, J., Lee, S.Y., Kang, H.C., Noh, D.Y.: Laser-induced metastable mixed phase of AuNi nanoparticles: a coherent X-ray diffraction imaging study. J. Synchrotron Radiat. **27**(Pt 3), 725–729 (2020)
13. Yau, A., Cha, W., Kanan, M.W., Stephenson, G.B., Ulvestad, A.: Bragg coherent diffractive imaging of single-grain defect dynamics in polycrystalline films. Science **356**(356), 4 (2017)
14. Brant, W.R., Li, D., Gu, Q., Schmid, S.: Comparative analysis of ex-situ and operando X-ray diffraction experiments for lithium insertion materials. J. Power Sources **302**, 126–134 (2016)
15. Borkiewicz, O.J., Shyam, B., Wiaderek, K.M., Kurtz, C., Chupas, P.J., Chapman, K.W.: The AMPIX electrochemical cell: a versatile apparatus for in situ X-ray scattering and spectroscopic measurements. J. Appl. Crystallogr. **45**(6), 1261–1269 (2012)
16. Liu, H., Allan, P.K., Borkiewicz, O.J., Kurtz, C., Grey, C.P., Chapman, K.W., Chupas, P.J.: A radially accessible tubular in situ X-ray cell for spatially resolved operando scattering and spectroscopic studies of electrochemical energy storage devices. J. Appl. Crystallogr. **49**(5), 1665–1673 (2016)
17. Wang, J., Eng, C., Chen-Wiegart, Y.C., Wang, J.: Probing three-dimensional sodiation-desodiation equilibrium in sodium-ion batteries by in situ hard X-ray nanotomography. Nat. Commun. **6**, 7496 (2015)
18. Li, Y., Weker, J.N., Gent, W.E., Mueller, D.N., Lim, J., Cogswell, D.A., Tyliszczak, T., Chueh, W.C.: Dichotomy in the lithiation pathway of ellipsoidal and platelet $LiFePO_4$ particles revealed through nanoscale operando state-of-charge imaging. Adv. Funct. Mater. **25** (24), 3677–3687 (2015)
19. Bozzini, B., Kourousias, G., Gianoncelli, A.: In situ observation of dynamic electrodeposition processes by soft X-ray fluorescence microspectroscopy and keyhole coherent diffractive imaging. J. Phys. D Appl. Phys. **50**(12) (2017)
20. Ohmer, N., Fenk, B., Samuelis, D., Chen, C.C., Maier, J., Weigand, M., Goering, E., Schutz, G.: Phase evolution in single-crystalline $LiFePO_{(4)}$ followed by in situ scanning X-ray microscopy of a micrometre-sized battery. Nat. Commun. **6**, 6045 (2015)
21. de Smit, E., Swart, I., Creemer, J.F., Hoveling, G.H., Gilles, M.K., Tyliszczak, T., Kooyman, P.J., Zandbergen, H.W., Morin, C., Weckhuysen, B.M., de Groot, F.M.: Nanoscale chemical imaging of a working catalyst by scanning transmission X-ray microscopy. Nature **456**(7219), 222–225 (2008)
22. Lim, J., Li, Y., Alsem, D.H., So, H., Lee, S.C., Bai, P., Cogswell, D.A., Liu, X., Jin, N., Yu, Y.-S., Salmon, N.J., Shapiro, D.A., Bazant, M.Z., Tyliszczak, T., Chueh, W.C.: Origin and hysteresis of lithium compositional spatiodynamics within battery primary particles. Science **353**(6299), 6 (2016)
23. Olivares-Marin, M., Sorrentino, A., Lee, R.C., Pereiro, E., Wu, N.L., Tonti, D.: Spatial distributions of discharged products of lithium-oxygen batteries revealed by synchrotron X-ray transmission microscopy. Nano Lett. **15**(10), 6932–6938 (2015)
24. Boesenberg, U., Falk, M., Ryan, C.G., Kirkham, R., Menzel, M., Janek, J., Fröba, M., Falkenberg, G., Fittschen, U.E.A.: Correlation between chemical and morphological heterogeneities in LiNi0.5Mn1.5O4 spinel composite electrodes for lithium-ion batteries determined by micro-X-ray fluorescence analysis. Chem. Mater. **27**(7), 2525–2531 (2015)
25. Yu, X., Pan, H., Zhou, Y., Northrup, P., Xiao, J., Bak, S., Liu, M., Nam, K.-W., Qu, D., Liu, J., Wu, T., Yang, X.-Q.: Direct observation of the redistribution of sulfur and polysufi des in Li–S batteries during the first cycle by in situ X-ray fluorescence microscopy. Adv. Energy Mater. 1500072 (2015)

26. Ulvestad, A., Cho, H.M., Harder, R., Kim, J.W., Dietze, S.H., Fohtung, E., Meng, Y.S., Shpyrko, O.G.: Nanoscale strain mapping in battery nanostructures. Appl. Phys. Lett. **104**(7), (2014)
27. Ulvestad, A., Clark, J.N., Harder, R., Robinson, I.K., Shpyrko, O.G.: 3D imaging of twin domain defects in gold nanoparticles. Nano Lett. **15**(6), 4066–4070 (2015)
28. Newton, M.C., Leake, S.J., Harder, R., Robinson, I.K.: Three-dimensional imaging of strain in a single ZnO nanorod. Nat. Mater. **9**(2), 120–124 (2010)
29. Robinson, I., Harder, R.: Coherent X-ray diffraction imaging of strain at the nanoscale. Nat. Mater. **8**(4), 291–298 (2009)
30. Ulvestad, A., Singer, A., Clark, J.N., Cho, H.M., Kim, J.W., Harder, R., Maser, J., Meng, Y. S., Shpyrko, O.G.: Topological defect dynamics in operando battery nanoparticles. Science **348**(6241), 1344–1347 (2015)
31. Yau, A., Cha, W., Kanan, M.W., Stephenson, G.B., Ulvestad, A.: Bragg coherent diffractive imaging of single-grain defect dynamics in polycrystalline films. Science **356**(356), 739–742 (2017)
32. Ulvestad, A., Singer, A., Cho, H.M., Clark, J.N., Harder, R., Maser, J., Meng, Y.S., Shpyrko, O.G.: Single particle nanomechanics in operando batteries via lensless strain mapping. Nano Lett. **14**(9), 5123–5127 (2014)
33. Xu, Y., Hu, E., Zhang, K., Wang, X., Borzenets, V., Sun, Z., Pianetta, P., Yu, X., Liu, Y., Yang, X.-Q., Li, H.: In situ visualization of state-of-charge heterogeneity within a LiCoO2 particle that evolves upon cycling at different rates. ACS Energy Lett. **2**(5), 1240–1245 (2017)
34. Wang, G., Aubin, M., Mehta, A., Tian, H., Chang, J., Kushima, A., Sohn, Y., Yang, Y.: Stabilization of Sn anode through structural reconstruction of a Cu–Sn intermetallic coating layer. Adv. Mater. https://doi.org/10.1002/adma.202003684 (2020)
35. Dong, K., Markotter, H., Sun, F., Hilger, A., Kardjilov, N., Banhart, J., Manke, I.: In situ and operando tracking of microstructure and volume evolution of silicon electrodes by using synchrotron X-ray imaging. Chemsuschem **12**(1), 261–269 (2019)
36. Nelson, J., Misra, S., Yang, Y., Jackson, A., Liu, Y., Wang, H., Dai, H., Andrews, J.C., Cui, Y., Toney, M.F.: In operando X-ray diffraction and transmission X-ray microscopy of lithium sulfur batteries. J. Am. Chem. Soc. **134**(14), 6337–6343 (2012)
37. Sun, N., Liu, Q., Cao, Y., Lou, S., Ge, M., Xiao, X., Lee, W.K., Gao, Y., Yin, G., Wang, J., Sun, X.: Anisotropically electrochemical-mechanical evolution in solid-state batteries and interfacial tailored strategy. Angew Chem. Int. Ed Engl. **58**(51), 18647–18653 (2019)
38. Finegan, D.P., Scheel, M., Robinson, J.B., Tjaden, B., Hunt, I., Mason, T.J., Millichamp, J., Di Michiel, M., Offer, G.J., Hinds, G., Brett, D.J.L., Shearing, P.R.: In-operando high-speed tomography of lithium-ion batteries during thermal runaway. Nat. Commun. **6**, 6924 (2015)
39. Wang, L., Wang, J., Zuo, P.: Probing battery electrochemistry with in operando synchrotron X-ray imaging techniques. Small Methods **2**(8) (2018)
40. Cao, C., Toney, M.F., Sham, T.-K., Harder, R., Shearing, P.R., Xiao, X., Wang, J.: Emerging X-ray imaging technologies for energy materials. Mater. Today **34**, 132–147 (2020)
41. Yu, Z., Wang, J., Liu, Y.: High-dimensional and high-resolution X-ray tomography for energy materials science. MRS Bull. **45**(4), 283–289 (2020)

Printed by Printforce, the Netherlands